# a guide to amateur radio

EIGHTEENTH EDITION

**PAT HAWKER, G3VA**



RADIO SOCIETY OF GREAT BRITAIN

**Acknowledgements**
Thanks are due to the City and Guilds of London Institute for permission to reproduce the RAE syllabus in Chapter 8 and the sample RAE questions in Appendix 1. However, the Institute can accept no responsibility as to the accuracy of these or the answers given.

Acknowledgement is also made to the Home Office for permission to reproduce information from the leaflet *How to Become a Radio Amateur*.

Published by the Radio Society of Great Britain,
35 Doughty Street, London WC1N 2AE.

First published 1933
Eighteenth edition 1980
Reprinted 1981



ISBN 0 900612 50 9

Printed in Great Britain by Galliard (Printers) Ltd, Great Yarmouth.

# Preface

*A Guide to Amateur Radio* is intended to assist the newcomer to learn more about the hobby, and to help him or her to obtain a transmitting licence. The *Guide* also contains technical information and operating data of interest to all radio amateurs and listeners. The many editions of the *Guide* testify to its long-established role as an indispensable aid to all who want to learn about amateur radio—how the licences are obtained, how equipment is designed and built, how amateurs communicate.

The conditions for the UK amateur licences and the syllabus for the multi-choice questions form of Radio Amateurs' Examination are incorporated together with a set of typical questions. In the technical chapters the increased importance of vhf/uhf, ssb, nbfm and solid-state devices etc has resulted in many changes, though it is recognized that many amateurs continue to use thermionic devices, particularly for rf power amplification. Both newcomers and those seeking basic information on the very large range of equipment that has been produced for amateurs will find the enlarged chapter on factory-built equipment—popular receivers, transceivers and transmitters—particularly valuable. There is also now a chapter devoted to the principles of electronics.

At the important World Administrative Radio Conference held in Geneva during 1979, the role of amateur radio in providing an invaluable service of self-training ("technology transfer") to the peoples of developing and developed nations was recognized by the delegates from more than 140 countries. The new international Radio Regulations (beginning to take effect from 1 January 1982) make full provision for this unusual hobby with its many services to the community.

Amateur radio is a unique pastime. It offers to the newcomer an exciting and fascinating combination of a scientific hobby, a competitive sport and an entry into a world-wide fellowship which knows no boundaries of race, class or creed. It has its own customs and traditions, its own international language, and its "ham spirit". The future progress of amateur radio lies in the hands of those who shortly will be sending their first CQ calls and experiencing the thrill of hearing their brand-new callsigns repeated back to them from afar. Will you be one of that number?

*Pat Hawker, G3VA*

# RADIO CIRCUIT SYMBOLS

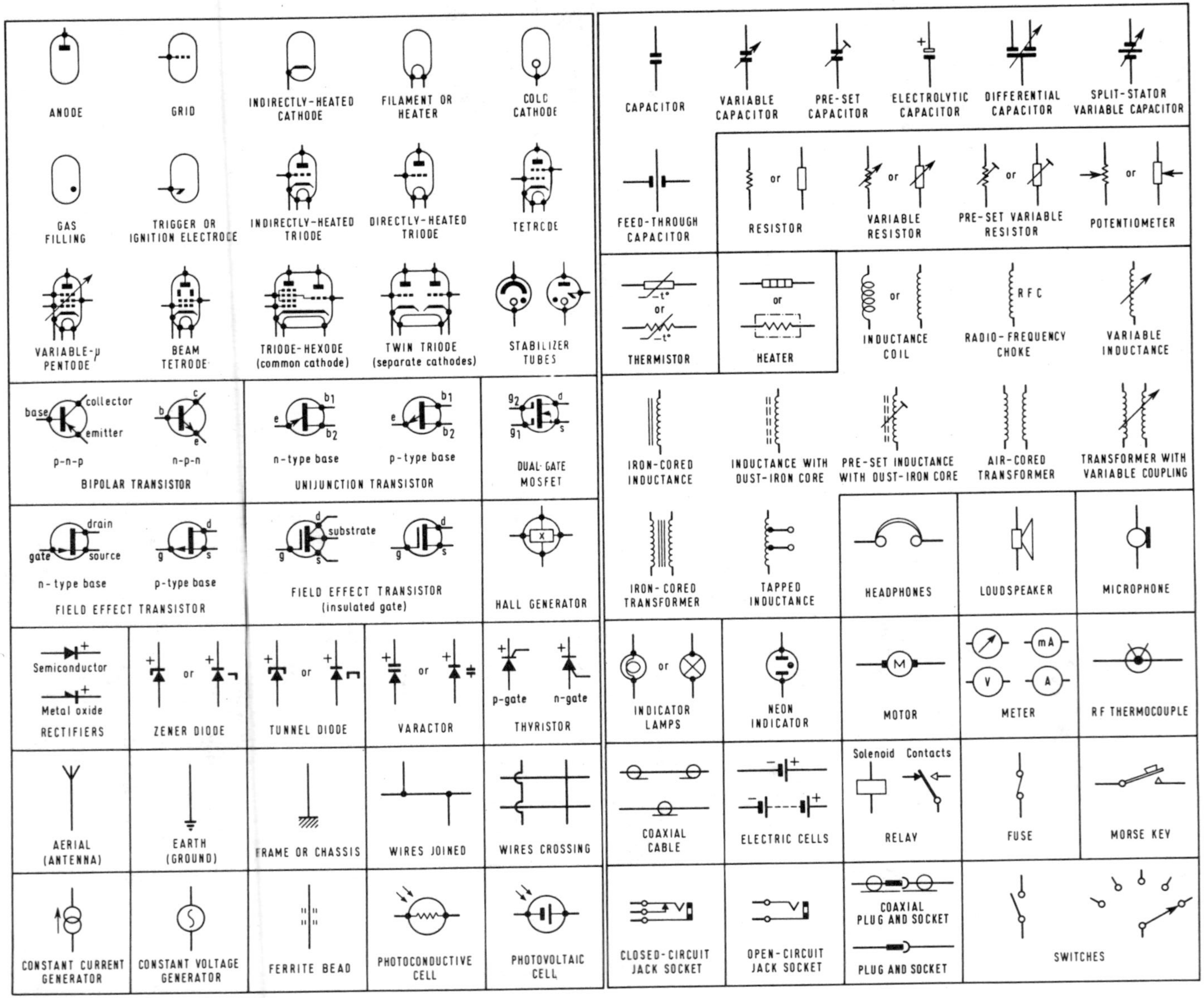

ANODE
GRID
INDIRECTLY-HEATED CATHODE
FILAMENT OR HEATER
COLD CATHODE

GAS FILLING
TRIGGER OR IGNITION ELECTRODE
INDIRECTLY-HEATED TRIODE
DIRECTLY-HEATED TRIODE
TETRODE

VARIABLE-μ PENTODE
BEAM TETRODE
TRIODE-HEXODE (common cathode)
TWIN TRIODE (separate cathodes)
STABILIZER TUBES

base, collector, emitter — p-n-p
b, c, e — n-p-n
BIPOLAR TRANSISTOR

e, b1, b2 — n-type base
e, b1, b2 — p-type base
UNIJUNCTION TRANSISTOR

g2, g1, d, s — DUAL GATE MOSFET

gate, drain, source — n-type base
g, d, s — p-type base
FIELD EFFECT TRANSISTOR

d, substrate, g, s — g, d, s
FIELD EFFECT TRANSISTOR (insulated gate)

HALL GENERATOR

Semiconductor
Metal oxide
RECTIFIERS

ZENER DIODE (+ or +)
TUNNEL DIODE (+ or +)
VARACTOR (+ or +)
THYRISTOR (p-gate, n-gate)

AERIAL (ANTENNA)
EARTH (GROUND)
FRAME OR CHASSIS
WIRES JOINED
WIRES CROSSING

CONSTANT CURRENT GENERATOR
CONSTANT VOLTAGE GENERATOR
FERRITE BEAD
PHOTOCONDUCTIVE CELL
PHOTOVOLTAIC CELL

CAPACITOR
VARIABLE CAPACITOR
PRE-SET CAPACITOR
ELECTROLYTIC CAPACITOR
DIFFERENTIAL CAPACITOR
SPLIT-STATOR VARIABLE CAPACITOR

FEED-THROUGH CAPACITOR
RESISTOR (or)
VARIABLE RESISTOR (or)
PRE-SET VARIABLE RESISTOR (or)
POTENTIOMETER (or)

THERMISTOR (−t° or −t°)
HEATER (or)
INDUCTANCE COIL (or)
RADIO-FREQUENCY CHOKE (R F C)
VARIABLE INDUCTANCE

IRON-CORED INDUCTANCE
INDUCTANCE WITH DUST-IRON CORE
PRE-SET INDUCTANCE WITH DUST-IRON CORE
AIR-CORED TRANSFORMER
TRANSFORMER WITH VARIABLE COUPLING

IRON-CORED TRANSFORMER
TAPPED INDUCTANCE
HEADPHONES
LOUDSPEAKER
MICROPHONE

INDICATOR LAMPS (or)
NEON INDICATOR
MOTOR (M)
METER (mA, V, A)
RF THERMOCOUPLE

COAXIAL CABLE
ELECTRIC CELLS
RELAY (Solenoid, Contacts)
FUSE
MORSE KEY

CLOSED-CIRCUIT JACK SOCKET
OPEN-CIRCUIT JACK SOCKET
COAXIAL PLUG AND SOCKET
PLUG AND SOCKET
SWITCHES

CHAPTER 1

# This is amateur radio

Almost a million people, of all ages and occupations, in almost every country in the world, are enthusiastic followers of a hobby that has many claims to being unique. These are the "radio amateurs" with their own transmitters in their homes, their amateur radio club rooms, often in their own cars, or light and compact enough to be taken almost anywhere—and are able to communicate with fellow enthusiasts over distances of hundreds and thousands of miles.

## The world at your finger-tips

Historically, this is a hobby that pioneered a whole new communication era when its followers first showed that the radio wavelengths of 200m and down, far from being useless as the early radio communication "experts" declared, could be used to propagate radio signals, even from low-power transmitters, right around the globe.

Later the amateurs were among the first to show that the very high frequencies (wavelengths below 10m), again written off by the experts as suitable only for "line of sight" communication, could—under certain conditions—be used to communicate over hundreds of miles.

Amateur radio—and the Radio Society of Great Britain (RSGB) which exists to serve it—began long before there were any short-wave or very-high-frequency (vhf) bands on radio receivers—indeed before there was any radio broadcasting as we now know it. But there were many young people who were interested in the new-fangled "wireless telegraphy" and began to experiment with all sorts of simple apparatus. They were soon shuttled into the unwanted wavelengths below 200m then considered useless—except perhaps for the enthusiasts to transmit in "their own backyards". To the surprise of almost everyone (except a few of the amateurs and one or two people such as Marconi who had suspected that these wavelengths were interesting) they were soon working other enthusiasts across the oceans.

This neat and efficient hf station at G3HTA is capable of world-wide communication by phone and cw. The two similar units in front of the operator are the separate transmitter and receiver

## Amateurs today

Today, although many official stations use the short-waves, certain frequencies are set aside by almost every country in the world for amateurs, so that they may continue their useful work. In the UK alone there are over 25,000 people who hold a special licence from the Home Office (Radio Regulatory Division) to operate an amateur station. There are even a number of enthusiasts who possess amateur stations fully equipped for transmitting their own television pictures. Throughout the world, there are more than 750,000 licensed radio amateurs. Apart from these, there are also many thousands of enthusiasts who, though not yet operating their own transmitters, share in the excitement and interest of amateur radio by listening to and building short-wave and vhf receivers, so gaining the knowledge and skill needed to obtain a licence.

## Expeditions, emergencies and space

Many well-known expeditions have used amateur radio stations to help keep them in touch with the world.

Then again, amateurs have frequently provided emergency communication services in towns stricken by natural disasters, and have aided in the rescue of survivors from crashed aircraft and ships in distress.

In recent years, in order to provide an efficient emergency service, the Radio Amateur Emergency Network has been formed by the RSGB and operates in conjunction with the British Red Cross Society, the St John Ambulance Brigade, County Emergency Planning Officers and the Police forces.

Amateurs have even designed, built and used their own communication satellites which have been put into orbit for them by the USA and USSR. These Oscar satellites, accepting and retransmitting signals from amateur stations, have enabled amateurs to effect long-distance communication on the vhf bands. The sixth Oscar satellite, launched in October 1972, remained in operation for more than four years. The eighth Oscar, launched in March 1978, carries two linear repeaters, one retransmitting 145MHz signals on 435MHz, the other retransmitting 145MHz signals on 29MHz. Amateurs also make contacts by "bouncing" signals off meteor trails and the moon.

## Amateur radio—questions and answers

The following sections give brief answers to most of the questions usually asked about amateur radio. More complete information can be found elsewhere in this book.

**What exactly is amateur radio?**

Amateur radio must surely be the only hobby to be formally defined by an international treaty drawn up by 150 nations. This

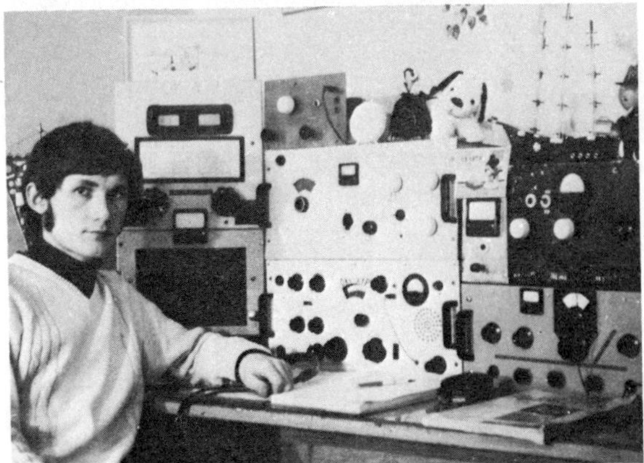

Frontiers are no barriers for amateur radio; this is the Czechoslovakian station OK1ATP. Radio amateurs are active in nearly all East European countries

was at the World Administrative Radio Conference held in Geneva during 1979, when amateur radio was redefined as "A radiocommunication service for the purpose of self-training, intercommunication and technical investigations carried out by amateurs, that is, by duly authorized persons interested in radio technique solely with a personal aim and without pecuniary interest."

Most amateurs, however, would probably define their hobby more simply as "the practice of two-way, short-wave radio communication not as a business or means of profit but as a spare time hobby pursued for the pleasure to be derived from an interest in radio technique, construction and operation and for the ensuing friendships with like-minded individuals throughout the world."

It should be noted that the term "amateur" is, strictly speaking, applied only to persons who hold official licences to operate transmitting stations, though there are of course many thousands of individuals who follow the hobby purely as listeners: such persons are known as "SWLs" (short-wave listeners) or—if members of the RSGB—as "BRS" (British Receiving Stations) or "ORS" (Overseas Receiving Stations).

Something approaching one-half of all amateurs are in fact persons whose work is in some way connected with radio, television or electronics—this does not of course prevent them from pursuing amateur radio as a hobby. It is perhaps an indication of the appeal of radio communication that so many who begin as amateurs later become "professionals", and vice versa.

Stations are often set up on remote islands to give others a chance to contact a new country. This is the site of the 1978 "dxpedition" to Clipperton Is in the Pacific Ocean

Equally very many amateurs have no professional ties with electronics but nevertheless often acquire considerable technical and operating skill from their hobby.

**How could I become a radio amateur?**

To operate an amateur radio station in the UK it is necessary to obtain a licence from the Home Office (Radio Regulatory Division). Applicants must be over 14 years of age, furnish proof of British nationality, and pass a technical examination. To obtain the full form of licence (Amateur Licence A, see below) it is also necessary to pass a morse code test. No morse test is required for the Amateur Licence B. The various examinations are described in detail in Chapter 8.

**Exactly what types of amateur licences are issued in the UK and elsewhere?**

In the UK two main types of amateur licence are issued: Amateur Licence A and Amateur Licence B. In addition special Amateur (Maritime) licences permit operation on sea-going vessels on 7, 14, 21, 28, 144MHz and 24GHz. Reciprocal licences are available for operation in the UK to holders of amateur licences issued in some other countries.

Amateur Licence A covers all available bands and all specified modes of operation, including morse telegraphy, telephony, radio teleprinting, slow-scan and high-definition television, facsimile and data from fixed, portable, pedestrian and mobile locations. Amateur Licence B does not authorize the use of frequencies below 144MHz or the use of morse telegraphy; otherwise its conditions are broadly the same as those for Amateur Licence A.

Other countries have different forms of licences, and with different technical and morse requirements, although in most countries these do not differ greatly from those of the UK. Some countries, including the USA, also issue special "Novice" licences which provide certain facilities after passing only a very simple morse test. There is also, in some countries (but not the UK), a radio service known as "Citizen's Band" (CB) covering certain limited types of two-way radio communication, mostly on about 27MHz.

**Are the examinations difficult to pass?**

Not really—provided that you are genuinely interested. Of course, they require a certain amount of preparation and willingness to learn; but with care, and regular spare-time practice and training, there is no reason why anyone should not feel confident of obtaining a licence. Many do so within a year of starting from scratch. Some amateurs obtained their licences when they were 14–16 years of age; others have not become interested until late in life—and have found the hobby an excellent means of enlivening their retirement.

**Is it necessary to go to classes to prepare for the examinations?**

No, though the evening courses provided by a number of local education authorities and by radio clubs form an excellent preparation for those who have little technical background. There are also a number of correspondence courses. But many amateurs are entirely self-taught, particularly on the technical side. It is certainly more difficult to teach oneself the morse code—though many have done so—and assistance from a licensed amateur, an ex-commercial or a Service operator is most helpful. Then again, two would-be amateurs often learn together. For those without such aids, morse code disc records and tapes

Home construction is again flourishing. Here G8CMB is constructing a 50W portable 144MHz transmitter

are available, and the RSGB regularly publishes a list of amateur stations which send special slow morse practice transmissions.

**How much does an amateur transmitting licence cost?**
The annual UK licence fee is currently (1980) £6.40 for either the Amateur Licence A or the Amateur Licence B.

The A and B licences permit operation as required from a fixed address, from a temporary or alternative address, as a pedestrian, or in any vehicle or vessel (but not on the sea or within any estuary, dock or harbour). A station may not be established or used in an aircraft or a public transport vehicle.

A charge of several pounds is made for the City and Guilds technical examination and for the Post Office morse test.

**What information can be gathered from an amateur callsign?**
The callsigns are allocated by the licence issuing authority which in the UK is the Home Office (Radio Regulatory Division). Every callsign includes at the beginning an *international prefix* of the country concerned. For example a station in England will start with G, in Scotland GM, in France F, etc (a detailed list is given on p112).

In some countries (but not in the UK) the callsign also includes a "district" identification showing the particular region of the country in which the station is located.

In countries such as the UK where most callsigns are issued in alphabetical order it is often possible to gain a rough idea of when the licence was issued (see accompanying table for some selected dates).

When the station is operated away from its usual location, or for certain other reasons, a *suffix* is added at the end (see table). Sometimes when an amateur is operating in a country or district other than his or her usual one the suffix takes the form of a country or regional identification. For example the callsign WB2AAA/VE2 would denote that a station normally in New York or New Jersey is actually operating in the Province of Quebec, Canada. An exception is a foreign amateur operating in the UK who uses a special G5 call.

**How much does an amateur station cost?**
A small but complete and reasonably efficient amateur station—transmitter and receiver—can be built for about £50, or even less if you already have some radio components available. On the other hand, an amateur may spend say, £1,000 or more on

### BRITISH AMATEUR CALLSIGNS

| Prefix | Country | Prefix | Country |
|---|---|---|---|
| G | England | GM | Scotland |
| GD | Isle of Man | GU | Guernsey |
| GI | Northern Ireland | GW | Wales |
| GJ | Jersey | | |

GB Exhibition and Special Stations

**Approximate Date of Issue**

| Two letters | |
|---|---|
| G2AA, etc | 1920–39 |
| G3AA | 1937–8 |
| G4AA | 1938–9 |
| G5AA | 1921–39 |
| G6AA | 1921–39 |
| G8AA | 1936–7 |

| Three letters | |
|---|---|
| G2AAA, etc | pre-war "artificial antenna" licences issued as full radiating licences |
| G3AAA | 1946 |
| G3BAA | 1946–7 |
| G3CAA | 1947 |
| G3DAA | 1947–8 |
| G3EAA | 1948 |
| G3FAA | 1948–9 |
| G3GAA | 1949–50 |
| G3HAA | 1950–1 |
| G3IAA | 1951–2 |
| G3JAA | 1952–4 |
| G3KAA | 1954–6 |
| G3LAA | 1956–7 |
| G3MAA | 1957–8 |
| G3NAA | 1958–60 |
| G3OAA | 1960–1 |
| G3PAA | 1961–2 |
| G3QAA | not issued |
| G3RAA | 1962–3 |
| G3SAA | 1963–4 |
| G3TAA | 1964–5 |
| G3UAA | 1965–6 |
| G3VAA | 1966–7 |
| G3WAA | 1967 |

| Three letters (cont.) | |
|---|---|
| G3XAA | 1967–8 |
| G3YAA | 1968–9 |
| G3ZAA | 1969–71 |
| G4AAA | 1971–2 |
| G4BAA | 1972–3 |
| G4CAA | 1973–4 |
| G4DAA | 1974–5 |
| G4EAA | 1975–6 |
| G4FAA | 1976–7 |
| G4GAA | 1977 |
| G4HAA | 1978–9 |
| G4IAA | 1979 |
| G5AAA, G5MAA, etc | "Reciprocal" licences issued to foreign amateurs |

| Class B licences (vhf/uhf only) | |
|---|---|
| G8AAA | 1964–7 |
| G8BAA | 1967–8 |
| G8CAA | 1968–9 |
| G8DAA | 1969–70 |
| G8EAA | 1970–1 |
| G8FAA | 1971–2 |
| G8GAA | 1972–3 |
| G8HAA | 1973 |
| G8IAA | 1973–4 |
| G8JAA | 1974–5 |
| G8KAA | 1975 |
| G8LAA | 1975–6 |
| G8MAA | 1976–7 |
| G8NAA | 1977 |
| G8OAA | 1977–8 |
| G8PAA | 1978 |
| G8SAA | 1979 |
| G8TAA | 1979 |

**Suffixes**

| | |
|---|---|
| /A operation from temporary address | /P Portable operation |
| /M mobile operation | /MM maritime mobile |
| | /MA maritime alternative (at anchor or berthed) |

**Note:** A few British licences are special cases issued "out of sequence"—for example a close relative may take over a callsign or a club may sometimes obtain appropriate letters with the permission of the former holder of a discontinued licence.

complex equipment. In practice, most amateurs begin in a very modest way and add to their equipment as they go along. Fortunately, results depend as much upon the skill and knowledge of the operator as upon the possession of a well-filled wallet.

**Are most amateur stations purchased as commercially-manufactured units or are they home constructed?**
At one time almost all amateur equipment was home-built, but today most of the receivers, transmitters and the increasingly popular transceivers (combined transmitter-receivers) are factory made. This does not mean that home construction is out of date; many of the keenest amateurs still take great pride in building much of their own equipment. Even when the main

These illustrations underline two features of vhf/uhf operation on such bands as 70, 144 and 432MHz—home construction and portable operation. This all-transistor 144MHz transceiver was built by Bill Scarr, G2WS, and the collapsible four-element antenna is carried in a fishing-rod bag. Portable operation for Class A and Class B amateurs is encouraged by annual field-days and contests

units are factory built, there is much interest in building the auxiliary units and test equipment that form an important part of the amateur station. Factory or home-built "transverters" allow operation on vhf/uhf in conjunction with hf transceivers.

On vhf home-built converters are often used ahead of factory-built receivers. Most microwave transmitters are home-built.

For those who wish to construct their own equipment to designs evolved and tested by others, there are a number of kits available for a wide range of amateur equipment.

**Does an amateur station take up a lot of room?**

Photographs of amateur stations may give a misleading impression because they often show a room crammed full of apparatus. An operating room—or "shack" as it is universally called—is fine for those who have the space to spare, but there are plenty of amateurs who are confined to a small hut in the garden or an unused attic. Others again find it possible to fit all their equipment into a cupboard, a desk or a bureau so that it is acceptable in the average living room.

Amateur stations differ enormously. Some amateurs gradually build up stations comprising several different receivers, an hf single-sideband transmitter or transceiver plus a higher-power "linear amplifier", a separate low-power "top-band" (1·8MHz) transmitter which is still sometimes amplitude-modulated, fixed and/or portable vhf transmitters, vhf receivers or a vhf converter which can be used in conjunction with an hf receiver, test instruments, antenna tuning units and the like. On the other hand, a station may comprise just a compact transceiver or receiver-transmitter combination. Each approach can provide plenty of interest—and the chance to take a keen interest in the technical and operational sides of the hobby.

Trends and operating patterns in amateur radio gradually change: for example not so many years ago almost all phone operation was by means of conventional double-sideband amplitude modulation; today single sideband (ssb) has largely replaced a.m. on the long-distance hf bands, while narrow-band frequency-modulation (nbfm) and ssb have become popular on vhf. The use of nbfm has been encouraged by the appearance in the UK of "talk-through" vhf repeater stations, operated by groups and clubs, which enable an amateur with mobile or modest portable equipment to take advantage of the good site and high antennas usually associated with a repeater station. Then again, there is increasing use of radio-teleprinter systems (rtty) and the very different techniques associated with microwave operation.

**What can amateurs talk about—and what topics must they avoid?**

The UK licence regulations state that messages must be "about matters of a personal nature in which the Licensee, or the person with whom he is in communication, has been directly concerned" and do not allow the use of the station for "business, advertisement, or propaganda purposes or for the sending of news or messages of or on behalf of, or for the benefit or information of, any social, political, religious or commercial organisation, or anyone other than the Licensee or the person with whom he is in communication." An exception is made for official "emergency" communication.

In practice, this leaves plenty of scope for the amateur to exchange opinions on technical matters, talk about his equipment and compare notes on its performance with his opposite number. It leaves him full liberty to endeavour to get in touch with other amateurs all over the world and obtain reports on the strength and quality of his transmissions.

**Not all stations consist of a room packed with equipment. This is the compact set-up at G4DQF**

Similarly he may, if he so wishes, exchange information on what he has been doing recently—technically or personally—make arrangements to meet his friends, or even, should he so desire, just "ragchew" on almost any subject that may be of mutual interest to the two persons concerned.

**Is most amateur communication by speech?**

While speech is today the most commonly used system, there is also very considerable use of morse (cw) by holders of A licences. Morse is particularly effective for communication over long distances with low power by stations having simple wire antennas; it also makes it easy to communicate with foreign amateurs who may not speak English; many contests are held on cw as a means of improving operating speeds etc. On vhf and uhf, the vast majority of transmissions are by speech, although morse is extremely useful for long distances, for auroral-type propagation and for moonbounce etc.

On both hf and vhf bands amateurs also make considerable use of radio-teleprinting (rtty) where the messages are typed out on a keyboard and then printed out automatically at the receiving end. There are also quite a lot of amateurs using slow-scan television (sstv) techniques on both hf and vhf to exchange still images, and some are equipped for fast-scan, high-definition television for the 430–440MHz (70cm) band. Facsimile (picture-telegraphy) and, above 144MHz, various forms of data transmission are also permitted.

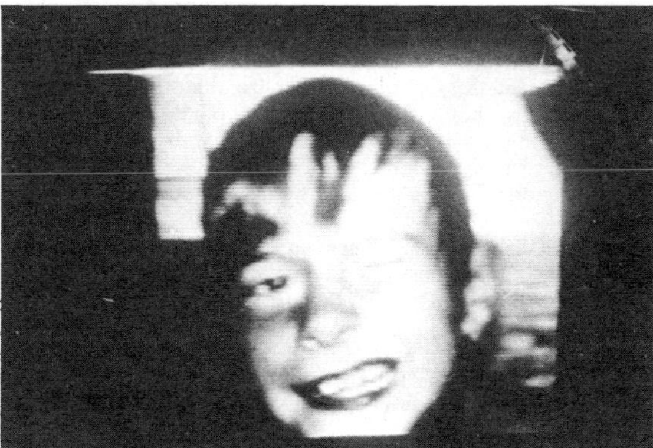

**A typical 128-line slow-scan television picture**

**AMATEUR BANDS (UNITED KINGDOM)**

| Frequency limits | Popular designations | Approx. wavelength |
|---|---|---|
| 1,800–2,000kHz | 1·8MHz, "top band" or "160" | 166·7–150m |
| 3,500–3,800kHz | 3·5MHz or "80" | 85·7–78·95m |
| 7,000–7,100kHz | 7MHz or "40" | 42·86–42·25m |
| 14,000–14,350kHz | 14MHz or "20" | 21·43–20·9m |
| 21,000–21,450kHz | 21MHz or "15" | 14·29–14·0m |
| 28,000–29,700kHz | 28MHz or "10" | 10·71–10·1m |
| 70·025–70·700MHz | 70MHz or "4" | 4·27–4·24m |
| 144–146MHz | 144MHz or "2" | 2·08–2·05m |
| 430–440MHz | 432MHz or "70cm" | 69·77–68·18cm |

In addition there are six microwave bands above 1,000MHz (1GHz); for details see Chapter 8. In some ITU Regions and countries the frequencies available to amateurs differ slightly from the table given above (see Chapter 11). All frequencies were subject to review at the 1979 World Administrative Radio Conference and eventually extra bands will become available (see p19).

**Are amateur stations confined to certain fixed frequencies (wavelengths)?**

By international agreement, the amateur service—like every other radio service—is allotted certain fixed bands of frequencies, and amateurs may not operate outside the limits of these bands. The amateur, however, may at any time choose which band he wishes to use, and is permitted (subject to certain voluntary band-planning schemes) to select his own frequency and to change it whenever he so desires. This is an important point on some of the more crowded bands since it allows an amateur to search for and then use a clear frequency (or "channel") and, should heavy interference be experienced later, to move immediately to another position in the band. It also enables what is known as "single-channel operation" (with both stations on approximately the same frequency) to be used.

**What is the maximum range of an amateur transmitter?**

There is no simple answer to this question. On some bands, in good radio conditions, the only limit is the size of the globe. Many quite low-powered stations are in regular contact with stations all over the world. Some British amateurs have been in contact with more than 300 different countries! But, on other frequencies, distance can still be of great consequence: an amateur who contacts a North American station on top band (1·8MHz) has accomplished a more difficult task than, say, in working an Australian on "twenty" (14MHz), while on uhf he may feel equally pleased if he is able to make contact with a station only a few miles away. Each band has its own limitations and special problems, and thus the amateur has always some new field to conquer.

**Do amateur stations use transistors or valves?**

Either or both. In recent years the introduction of transistors and other semiconductors has opened up entirely new fields for the development of equipment. So far, this has had most influence upon mobile and portable gear since the cost of high-power transmitting transistors has remained higher than their valve equivalents—though more and more amateurs are finding that the building, for instance, of fully transistorized hf and vhf transmitters and receivers offers much scope—and this is spreading gradually throughout all new equipment. All-transistor transmitters and transceivers with powers of from about 2 to 100W are being increasingly used for mobile, portable and fixed-station operation.

**Is amateur radio nowadays just a matter of operating factory-built equipment or is it a hobby for those with an interest in technical development and home construction of equipment?**
Some enthusiasts are interested primarily in *operating* equipment and in making friendly two-way contacts with other amateurs, "chasing dx" or taking part in the many operating contests; but they find that technical knowledge and expertise is needed to get the best out of their equipment and antennas and to avoid causing interference. Others are still basically more interested in technical development, equipment construction and the scientific investigation of unusual modes of radio propagation: but these find that operating their equipment is often the best way of judging the merit of or furthering their technical work. Either way, amateur radio reflects a combination of operating and technical aspects of radio communication. Even when using factory-made equipment there is always scope for improving your technical knowledge and keeping it up-to-date, as well as polishing your operating skills.

Operating a home-constructed vhf mobile station, similar in size to an ordinary car radio

**Would you recommend a newcomer to forget all about morse and try for a Class B licence (above 144MHz)?**
The B licences offer a most useful facility for those who are concerned primarily with developing the technical side of vhf, uhf and microwave equipment, and who are not unduly interested in hf world-wide amateur contacts. But even on uhf there are many applications—such as moon-bounce, meteor scatter, etc—where morse is invaluable. Our advice would therefore be always to work towards the full A licence—but this can be backed up with the confident knowledge that if morse has to be put off for any reason, it will still be possible, once the technical examination has been passed, to obtain a B licence with its specialized but very interesting possibilities. Quite long distances are regularly covered on vhf and uhf, and most "mobile" operation is on vhf.

The inclusion of the 144MHz band in this licence has opened the way for the B licence-holder to join in amateur operation on the most popular of all vhf bands. This has led to a rapid increase in Class B licences which now constitute about one-third of British licences. Many of these find vhf operation entirely satisfying.

**Can an amateur operate in another country?**
In an increasing number of cases, amateurs can obtain special permits or temporary licences without taking any further examinations to allow them to operate in another country, for instance while on holiday.

Hand-portable vhf transceivers are now very popular, often used with relay stations (repeaters) to give greater range

**Is it possible to transmit over long distances from a car or from compact transistorized portable equipment?**
Most of the mobile transmissions from cars are on frequencies more suitable for making contacts over relatively modest distances of a few miles. But by operating on the hf bands or by using the "talk-through repeaters" (p15) longer distances are possible.

**What are the QSL cards often mentioned by amateurs?**
These are usually individually designed printed postcards which can be completed to provide details of contact made by radio. These cards are often exchanged—usually via the various QSL bureaux—so that operators can have a permanent record of particular contacts, for interest or so that they may apply for achievement certificates. By sending QSL cards in bulk via the RSGB QSL Bureau—one of the most comprehensive services of its type in the world—the cost is very much less than sending cards direct to the amateurs concerned. Some typical QSL cards can be seen in the photograph on p5.

**Are newcomers to amateur radio welcome?**
Yes, most certainly. Although some of the amateur bands become very crowded at popular operating times, such as weekends, there is still room for those who are willing to take care that their transmissions do not occupy more ether-space than is technically necessary. Amateurs realize that their hobby can flourish only by the continual recruitment of newcomers, and are always willing to extend the "helping hand" of advice and practical assistance to all who show a genuine interest in radio communication.

**What is the first step that a newcomer should take?**
Fortunately this is an easy question to answer. For almost 70 years the Radio Society of Great Britain (founded in 1913 as the London Wireless Club) has been recognized, officially, as representing the radio amateur not only in the UK, but—as a member society of the International Amateur Radio Union—throughout the world. By joining the Radio Society of Great Britain, the newcomer will have the experience, guidance and backing of over 20,000 other members to help him or her before and after obtaining an amateur radio licence.

CHAPTER 2

# Getting started

There is no better way of starting amateur radio than to spend some time with a short-wave receiver, listening to the transmissions on the amateur bands. For this purpose even a simple and relatively inefficient receiver will suffice to begin with; though as your interest is aroused and skill acquired you will soon want to listen to the distant stations using ssb. Then a good receiver and an efficient antenna will be in demand.

Right from the start your listening should not be entirely haphazard but should, to some extent, be planned to help you to understand more about the peculiar and remarkable characteristics of hf and vhf propagation. Such knowledge will be a tremendous help later on when you begin to operate an amateur transmitting station. In this country we are all used to, and take for granted, the ever-changing weather conditions—but it will probably come as a surprise to many to find that there is just as much variation in the radio conditions on hf and vhf. Some days you will hear a particular amateur band full of long-distance stations at good strength—the next day the same band may be devoid of all but local signals. The range of a band may suddenly lengthen or fade out altogether as night falls. At first there may seem to be little rhyme or reason about it all, and it may seem as though forecasting radio conditions must all be just a matter of luck. But, if you keep careful records, you will soon notice that, just as there are daily and seasonal variations in weather, so equally are there regular recurrences of likely radio conditions—although in this case there is one even longer cycle that plays an important part; this is the fairly regular increase and decrease of sunspot activity—a normal cycle of maximum activity through minimum to a second maximum taking approximately 11 years.

So if the first requirement of the newcomer to amateur radio is a receiver and an antenna, the second is a well-kept log book of stations heard and notes on their strength. You can never keep in your head the thousands of callsigns, the times when you receive stations in the various parts of the world and their relative strengths from day to day. It will also be good practice for later on, as an accurately-kept log book is required by the authorities for all amateur transmitting stations. A typical extract from a receiving log is given in Fig 1 and indicates the type of information worth recording. You can either rule up an exercise book, or buy a radio log already ruled.

As your knowledge of the various bands increases you will find definite patterns emerging, along the following lines, which will give some indication of the type of stations you can normally expect to hear on each amateur band in the UK.

**1·8MHz (160m):** This band is shared by amateurs with coastal shipping and coast stations as well as other commercial stations. You can hear mainly semi-local (up to about 50–75 miles) telephony and cw (morse) amateur stations during daylight, with the range lengthening at night to cover the British Isles and many parts of Europe. In the winter some long-distance stations, including American, can be heard in the early hours. This band is not so susceptible to radio disturbances as some others but the atmospheric noise increases considerably during summer months when amateur activity is mainly during the week-ends or late evenings or by mobile stations operating from cars.

**3·5MHz (80m):** This band is also shared with commercial stations. Many Western European telephony and cw stations can usually be heard, particularly after nightfall. The occasional long-distance (dx) station can be heard at night and early morning, particularly in the winter.

**7MHz (40m):** This band varies markedly at various periods of the 11-year sunspot cycle. Towards the peaks (1958, 1969, 1979–80)

| Date | Time (GMT) | Band (MHz) | Station heard | Calling | Mode of emission | Signals R | Signals S | Signals T | Notes | QSL Sent | QSL R'cd |
|---|---|---|---|---|---|---|---|---|---|---|---|
| 24·10·19[illegible] | 1530 | 21 | WB4GDP | CQ | cw | 5 | 7 | 9 | Melbourne, Florida, Fading | | |
| | 1537 | 21 | N4OL | HA6NB | cw | 5 | 8 | 9 | Bob, Stone Mountain, Georgia | | |
| | 1548 | 21 | HBØNL | GM5BGD | cw | 5 | 9 | 9 | Franz, Nr. Vaduz, Liechtenstein | ✓ | |
| | 1732 | 7 | UA6PAM | CQ | ssb | 5 | 7 | | Wit in Grozny, 200W | | |
| | 1744 | 7 | OE6ESG | SM7BYU | ssb | 5 | 8 | | Sepp in Weisskirchen, 75W | | |
| | 2058 | 7 | VE1SU | GM3JZK | cw | 5 | 6 | 9 | Ed in Loggieville, N.B. 200W dipole | ✓ | |
| | 2105 | 3·5 | OZ2NU | OK1AUN | cw | 5 | 7 | 9 | Borge in Aalborg | | |
| | 2110 | 144 | G3HBN | CQ | nbfm | 5 | 9 | | Jimmy, London SW7 | | |
| | 2120 | 144 | F2MA | G3TMC | ssb | 4 | 4 | | Andy, near Paris | | |

Fig 1. An extract from an amateur receiving log. For transmitting, the following extra details would be required: (1) A column to show time of ending each contact; (2) a set of three columns to record the incoming RST report received on your own signals

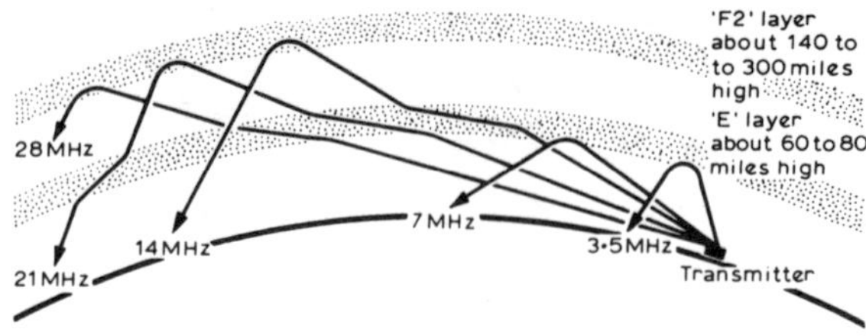


Fig 2. Typical paths of amateur signals during daylight. The 14, 21 and 28MHz signals may make further "hops" and, under good conditions, may travel thousands of miles. Direct "ground" signals from the transmitter will grow very weak within a few miles, and from there onwards up to the point where the signals bounce back from the ionized layers there exists a "skip zone" where only very weak signals are likely to be heard

UK stations can be heard at good strength during most of the day, giving way to mainland Continental signals towards dusk. With less sunspots UK stations tend to be heard less frequently and most of the signals will be from the rest of the Continent, although many long-distance stations will come through in the evening and early mornings. The main drawback to this band is the many very powerful broadcasting stations which transmit in it, often in breach of international regulations.

**14MHz (20m):** Of all the amateur bands, this one is the most consistently suitable for long-distance transmissions. There will be few days when at least some dx signals cannot be heard on both ssb and cw. In addition many European stations can be heard at very good strength during daylight. In sunspot-minimum years, the band tends to fade out completely a few hours after nightfall; in other years it may stay open all round the clock except in mid-winter when it usually closes late at night. Dawn and dusk are the traditional times for hearing the best dx (over 5,000 miles).

**21MHz (15m):** This is an extremely good daylight dx band except during sunspot-minimum years when activity drops sharply. It tends to be best in the spring and late autumn when in the afternoons and evenings the band will be full of loud American signals, with fairly consistent signals from Japan and Oceania in the mornings. Signals tend to fade out a few hours after darkness falls. The band is subject to severe disturbances and conditions therefore vary considerably from day to day. It is often open over north/south paths (eg UK to South Africa or South America) when closed for east/west paths.

**28MHz (10m):** This is rather like 21MHz. In sunspot-maximum years it may be very good indeed with loud signals even from low-power dx stations; in the sunspot-minimum periods few signals may be heard for days on end. Even in the sunspot-maximum years results tend to fall off in the summer months; after darkness the band usually fades right out. More consistent over north/south than east/west paths.

**70 and 144MHz (4m and 2m).** These are vhf bands in which signals are not reflected from the ionosphere in the same way as on the hf bands. Local and semi-local signals up to about 100 miles can be heard consistently and stations from several hundreds of miles away—including mainland European—can be received when conditions are good.

**432MHz (70cm):** A uhf band which attracts the experimentally minded and those interested in amateur television. Most contacts are between local stations, but distances of many hundreds of miles have been covered.

**Above 1,215MHz (23cm):** These bands are for the technically expert; but for those who can build suitable equipment they offer limitless opportunity for original experimental work.

Sunspot maxima occurred around 1947, 1958, 1969, 1980. We are now seeing a period of decreasing sunspot activity; this will be regretted by the users of 14, 21 and 28MHz—but will bring a parallel benefit on the lower-frequency bands. On vhf certain weather conditions may result in temperature inversions and consequent "tropospheric propagation"—one of the conditions allowing vhf and uhf contacts to be made over hundreds of miles: see later.

## Listener reports

The diligent listener will soon begin to wonder if reports on the transmissions which he (or she) hears would be of assistance to the amateurs concerned. Then also he will probably feel that it would be interesting to receive confirmation—in the form of QSL cards (Fig 3)—of having heard certain stations; by obtaining such verifications he will be able to qualify for the special awards offered by the RSGB to listeners. These include the BCRRA (British Commonwealth Radio Reception Award) requiring cards from 50 call areas in the British Commonwealth, and the "DX Listeners' Century Award", for those who submit proof of the reception of stations located in at least 100 different countries. "Four Metres and Down" listener awards mark outstanding reception achievements on 70, 144 and 432MHz.

The newcomer, however, should always use his discretion when sending reports of reception, or he will find the percentage of answers rather low. Some years ago, owing to many listeners with little basic knowledge of amateur radio sending reports of no real interest, the practice fell into some disfavour with amateurs, and the very real value of detailed or rare reports was sometimes forgotten. Since then most listeners have shown a greater awareness of the amateur's requirements but even today stations using high-power telephony on 14, 21 and 28MHz, particularly those located in countries where there is little amateur activity, receive many listener reports every month from regions with which they are in regular two-way communication. Sooner or later such stations usually give up any attempt to answer all such reports. On the other hand, a newly licensed amateur or one who seldom works over long distances may be really pleased to know how far his signals are reaching or to have detailed observations covering a period of time. Even amateurs who have been operating for many years have been known to prize highly, for example, a listener report from Australia on a 7MHz transmitter used for local working.

With the object of providing a rough guide to whether or not a

**BRS 53001**

**35 DOUGHTY STREET, LONDON, ENGLAND**

To Radio .................... your .................... MHz ssb/a.m./cw
signals received here at .............. GMT on .............. 198...
You were calling/working ............ and you were RST ............
QRM .................. QRN .................. Conditions ..................
Other countries audible at time were ..................................
Receiver .............................. Antenna ..............................
Remarks ..........................................................................
.......................................................................................
I hope this report is useful. Do you need further reports?
Please QSL direct or via RSGB. 73, Philip Hamm

Fig 3. Layout of a typical short-wave listener's QSL card

reception report is likely to be considered of real interest, the following list of recommendations has been drawn up, though of course there is no need to stick too closely to the suggestions if you have reason to believe that a detailed report may be useful or if you really need a QSL card from the country concerned.

**1·8MHz:** Stations heard during daylight more than 150 miles distant, or more than 750 miles away during the night.

**3·5 and 7MHz:** Stations outside Europe not observed to be in regular communication with the UK.

**14MHz:** Newly licensed or low-power stations more than 4,000 miles distant not observed to be in regular communication with Europe.

**21 and 28MHz:** Overseas stations not regularly in contact with Europe, particularly when heard at unusual times of the day or night or during seasons when their region is usually inaudible.

**70 and 144MHz:** Any station newly active on the band and more than about 150 miles away. Reports on such signals are usually most welcome.

**Above 432MHz:** Any station heard, except locals.

In general, reports covering a period are of greater value than those giving a single instance of reception, and a careful comparison with other signals heard from the same region are usually of interest to the amateur.

Listen particularly for the weaker stations who often call CQ without getting replies—they will be glad to learn their signals were reaching you and the chance of their sending QSL cards will be good.

Information supplied should always include:

Date; time (in 24h gmt system); frequency; signal strength in RST code (see p111); fading, atmospherics and quality of modulation; comparison of signals with other stations audible at the time; details of receiver and antenna; any special remarks.

If you wish to receive cards direct from the amateur it will usually be necessary to enclose reply postage in the form of the rather expensive International Reply Coupons which are available at all main post offices. However if you are in no particular hurry it is possible, when you are a member of the RSGB, to send and receive cards for certain countries through the Society's QSL Bureau, and this will reduce considerably the cost. The addresses of British amateur stations can be obtained from the *RSGB Amateur Radio Call Book*.

The really keen listener will find addresses of amateurs throughout the world in the *Radio Amateur Call Book Magazine* (issued in two parts, one of which covers the USA only). It is available in the UK from the RSGB.

## Decibels

One cannot go very far in radio communication without encountering the term *decibel*, used to indicate the difference between two different power or signal levels: often to indicate amplification (*gain*) or its converse *attenuation*. However it should be understood that this is a *ratio* based on the *logarithmic* scale; this often confuses the newcomer who is more used to thinking in terms of linear units, such as volts or amperes, or inches, miles or metres.

One of the fundamental laws of hearing and sight and many other forms of physiological stimulus is that the effect produced by an increase of the stimulus does not follow a simple arithmetic or linear scale. We have to keep doubling the stimulus to produce a sensation increasing in unit steps (Weber's Law). To illustrate this: if we stick two pins in our arm we would find it hurts twice as much as one pin; but it takes four pins to hurt three times as much; and eight pins to provide four times as much sensation of pain as a single pin!

A system which measures in unit steps in this way is the logarithmic scale. The log of 1 is 0; the log of 2 is 0·3010; that of 4 is 0·6021; that of 8 is 0·9031.

So the *bel* was chosen to represent a power ratio of one logarithmic unit, noting that 1 is log 0, 10 is log 1, 100 is log 2 and 1,000 is log 3. Thus a change of power level of 1 bel represents a power ratio of 10 times. But for many applications a bel is rather a large unit, so we use instead the *decibel* (dB) which is simply *one-tenth of a bel*. Thus a power gain of 10 times is 1 bel or 10dB: a power gain of 100 times is 2 bel or 20dB; a power gain of 1,000 times is 3 bel or 30dB.

Now if we only double the power, we can use the fact that the log of 2 is 0·3010, so this would be very nearly a gain of 0·3 bel or 3dB. Similarly if we were to halve the power this would be 3dB attenuation (or we can think of this as −3dB gain). Those readers who use log tables will also recognize that we can in effect multiply our gains by adding the logarithmic units. Thus a power ratio of 20 times (2 × 10) is the same as 3 + 10dB or 13dB, and so on. This is very convenient when thinking of say a three-stage amplifier, each stage providing 5dB power gain. The total amplification will be 15dB or a power ratio of 31·6 times. In practice we might however think in terms of voltage rather than power ratios, with rather different results (it is permissible to use decibel notation for voltage ratios provided that the impedance remains unchanged and we remember that if we double the voltage across a resistance we also double the current and hence cause the power to increase by four times).

It is worth noting that Weber's Law indicates that a 3dB change of power level at the input of a receiver is roughly the smallest change that our ears will readily detect: this tells us that if, for instance, we double the power of a transmitter it results only in a just perceptible increase of the audio output of the other man's receiver (or in practice because receivers use automatic gain control a just perceptible improvement in signal-to-noise ratio). This is much less an improvement than many newcomers hopefully expect from quite modest increases in transmitter power.

For example, if we have a transmitter with 100W output, we should realize how insignificant would be the effect of increasing the power to say 120 or even 150W. To give a really worthwhile improvement more like 4 or 5dB increase of power is needed, that is from 2·5 to 3·2 times the power; in this case representing powers of 250–320W. It is often far more sensible to try to obtain this sort of gain by improving the antenna, rather than increasing the transmitter power. However too much should not be made of this, since when your signals are competing with others every decibel counts (but again it may be easier to gain 1dB by reducing feeder losses than increasing the power of a 100W transmitter to 130W).

*Examples*

| Power ratio | Voltage ratio | dB |
|---|---|---|
| 10,000:1 | 100:1 | +40 |
| 100:1 | 10:1 | +20 |
| 4:1 | 2:1 | +6 |
| 1:100 | 1:10 | −20 |

## HF conditions

The listener soon comes to appreciate that hf propagation is subject to many changing factors that influence the state of the ionospheric layers and vary their ability to reflect or to absorb (attenuate) hf signals. The normal pattern of changing conditions can also be upset by *ionospheric storms* or *fadeouts*, often induced by the passage of a large sunspot.

During an ionospheric storm the higher frequencies (14MHz and particularly 21 and 28MHz) tend to fade out but conditions may often continue to be quite good on the lower-frequency amateur hf bands: 7, 3·5 and 1·8MHz. Such ionospheric storms may last for several days and their effects tend to be most noticeable at night.

The other form of ionospheric disturbance—the fadeout or *sudden ionospheric disturbance* (sid)—is more spectacular but seldom lasts for long. In an extreme form it can result in the virtual disappearance or greatly reduced strength of all sky-wave signals over a wide band, although signals may continue to come through on the *higher* frequencies. It occurs quite suddenly and seldom lasts longer than an hour or two; it takes place during daytime and affects transmissions passing through the daylight zone.

In the absence of a disturbed ionosphere, the same pattern of conditions tends to re-occur daily, gradually changing with the seasons and with the general level of sunspot activity.

Since ionospheric storms are caused by the passage across the sun of an intense sunspot, it is quite common for a storm to re-occur after 27 days. It is also interesting to note that the first effect of an ionospheric storm is to *raise* the maximum usable frequency (muf) and to produce a day or two of abnormally *good* conditions on the 14, 21 and 28MHz bands, which suddenly stops as the lower layers begin to become highly ionized and attenuate the signals, producing in turn a spell of poor conditions.

Although in general terms sunspot activity follows an 11-year cycle, the period and maximum activity cannot in fact be predicted with any degree of certainty, and there have been periods of history when such cycles may have disappeared for years on end. Although, throughout the 19th and so far the 20th centuries, sunspot cycles have been fairly regular in nature they have varied significantly in maximum intensity. For example the maxima of 1958 and 1979 were the highest since regular records began more than 200 years ago, and it seems unlikely that we shall see again such high sustained MUFs of the F2 layer for many years. A curious disturbance to the normal decline of sunspot activity occurred in 1972 and the subsequent minimum period lasted an unusually long time with low average MUFs over the years 1974–6. Fortunately this does not mean that high MUFs (resulting in good "openings" on 21 and 28MHz) do not occur occasionally, even in these low-sunspot years, but it does mean that there are long periods (particularly during the hours of darkness) when 14, 21 and 28MHz are virtually "closed" to long-distance signals. It should also be noted that during the summer months the 21 and 28MHz bands provide many signals at distances up to about 1,000 miles by sporadic E reflection.

For long-distance signals, which depend on the F layers, good paths will occur much more often in a north-south direction than for east-west paths. This is because the layers tend to be more highly ionized over the tropical and equatorial regions than in northerly latitudes. For example on 21 and 28MHz a listener in the UK will be able to hear stations in southern Africa or South America much more often than in, say, Canada or the Middle West of the USA where the signals have to traverse paths close to the North Pole, where the layers are less densely ionized.

A particularly interesting "grey line" path occurs between places where dawn and dusk, dawn and dawn, or dusk and dusk coincide and such conditions may, for example, provide a good path from the UK to New Zealand and Australia on 1·8 or 3·5MHz. Such paths may "open" for about 30min, so that dawn and dusk are times of considerable interest.

In brief: in daylight, on all-daylight long-distance paths towards the south, look on the highest frequencies; at night, for relatively short distances, look on the very lowest frequencies.

## VHF conditions

The propagation of vhf and uhf signals is often considered by professional communication and broadcasting engineers as virtually "line-of-sight". While this is a convenient assumption for receiving sound and television broadcasting, or for police mobile radio networks—which have to be sure of good communication under all propagation conditions, it is far from true for amateur operation. Indeed much of the fascination of vhf/uhf to the serious experimenter lies in his ability to exploit "abnormal" modes of propagation that make possible contacts between low-power stations over hundreds and even sometimes thousands of miles.

These special modes include:

### Ionospheric reflection

During periods of high sunspot activity it is not unknown for signals up to above 50MHz to be reflected back from the F layers of the ionosphere. Of particular interest is *transequatorial* propagation, usually between places in a north-to-south line, one on each side of the equator. Long-distance transequatorial contacts are possible up to beyond 144MHz, and are most likely about mid-day and the evenings around the spring and summer equinoxes. Unfortunately the UK is a little far north for transequatorial propagation at the highest frequencies, but it is this type of propagation (originally discovered as a result of amateur operation) that can, for example, result in the reception in Cyprus of Rhodesian television signals.

### Sporadic E

Intense ionization of the E layer occurs quite often in the UK (and is almost a daily occurrence in some countries nearer the equator) and can result in contacts over distances of up to 1,000 miles in the 28, 50MHz (not available in the UK) and occasionally 70 and 144MHz bands.

### Tropospheric ducting

The majority of extended-distance contacts at 70, 144MHz and above result from the presence of water vapour in the lower atmosphere at heights up to about 6,000ft above sea level. The presence of warm moist air at these heights can significantly extend the range of vhf stations so that this form of working is directly linked with specific meteorological conditions, particularly fine warm anticyclonic settled weather. Tropospheric propagation extends right up to uhf and microwaves. A good "tropo opening", as it is often called, can easily extend the range of contacts by about four times that of normal, so that several hundreds of miles are often covered on say 144MHz.

### Auroral reflection

An interesting mode of propagation becomes possible when there is marked auroral activity in the polar regions. Ionized sheets occur in the atmosphere, usually well to the north of the UK, and signals up to about 150MHz can be "bounced" off these. This means that stations working by auroral reflection transmit towards and receive signals from the north, no matter in what direction the station they are working is located. Such reflected signals take on a curious rapid fluctuation which makes such signals very distinctive and tends to make phone operation very difficult, because of the buzz-like tone.

### Meteor scatter

When a meteor strikes the upper atmosphere it produces a temporary trail of ionization which for a matter of a few seconds will partially reflect vhf signals, permitting short "bursts" of contact with stations up to about 1,200 miles away. For amateur contacts of this type it is usually necessary to know exactly what frequency the station is going to use—and contacts are usually limited to just an exchange of reports.

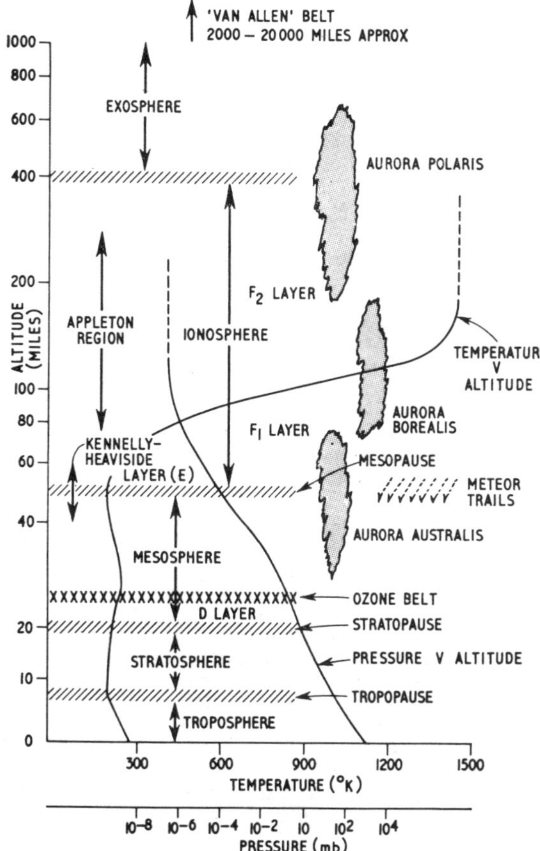


Fig 4. Earth's atmosphere and structure—a basic diagram for all interested in radio propagation. From the book *VHF radio wave propagation in the troposphere* by W. G. Burrows

**Moonbounce**

An interesting form of operation consists of using the moon as a giant passive reflector, bouncing back signals beamed on to its surface. While a number of amateurs have shown that it is quite possible to achieve long-distance contacts by this means on the vhf and uhf bands, it does require the use of high-gain antennas and very sensitive and stable receivers. However as a reward it is possible to work Australia, California or New Zealand at these frequencies, although one must remember such results cannot be easily achieved and call for elaborate high-gain directional antennas.

For the newcomer with a vhf-only Class B licence for 144MHz and above the most useful of these special modes of propagation is tropospheric ducting, allowing him to work several hundreds of miles using amplitude or frequency modulated speech without any special equipment other than a good antenna, low-noise receiver and transmitter of moderate power.

Typical of the type of weather conditions that bring about the "inversions" that result in an increase in vhf range are: a cold calm night following a warm day; the edge of a high pressure area; ducting over sea paths in summer; paths nearly parallel to a cold front shortly after a cool change without rain. In general, a low barometric pressure is associated with a "dead" band; if the barometer is high and steady, conditions may not be much above normal, but look for long-distance stations when the barometer is rising or falling as an area of high pressure passes over your station; even if your own barometer is not affected because the high pressure is passing to the north or south of you, the band may be "open". Generally tropospheric openings occur in the late autumn, the spring, and (particularly in coastal districts) in the summer. But even in mid-winter, good "tropo" openings can occur.

## Microwaves

The six uhf/shf microwave bands—1,215MHz (23cm), 2,300MHz (13cm), 3,400MHz (9cm), 5,650MHz (6cm), 10GHz (3cm) and 24GHz (12mm)—are basically for those with a real interest in the rather special techniques required for successful operation at these frequencies. There is at present very little equipment manufactured specifically for amateur use, so that these bands attract those who really enjoy building their own equipment or adapting that made for other applications. This does not mean that all the equipment is complex, indeed some of the wideband equipment is relatively simple, but it does often involve precision mechanical work.

In the UK considerable efforts have been made to exploit the 10GHz band with transmitter powers of only a few tens of milliwatts, using klystron or Gunn-diode oscillators but with the effective radiated power along the required direction raised by 100–1,000 times (20–30dB) by the use of compact parabolic dish antennas.

Although microwave range is often restricted to optical paths, very much longer distances are possible using super-refraction ducts or (with higher powers) by scattering from the troposphere. During 1976, UK amateurs made contacts exceeding 500km over the sea path from Cornwall to Scotland on 10GHz, and contacts across the North Sea to Sweden, Denmark and Holland have been made on the 1·3, 2·3 and 3·4GHz bands respectively.

## Using "all-band" receivers

For many years, some domestic and portable broadcast receivers have included one or more short-wave bands. These receivers, sometimes called "all-band sets", vary in hf performance from the very bad (difficult to tune accurately, drift, poor "image" performance, lack of sensitivity etc) to the very good (these are rare but they do exist). Some of the better receivers are virtually the equivalent of general-purpose communication receivers and can provide very adequate reception of weak signals. However unless fitted with a *beat frequency oscillator* (bfo) they are not suitable for the reception of either ssb or cw signals.

This can be overcome in two ways. A bfo operating at the intermediate frequency of the receiver (usually between 455–475kHz) can be added; or alternatively an oscillator providing output at the signal frequency of the incoming signal can be used. This latter form of carrier re-insertion has several advantages and can readily be provided by a small external oscillator with good bandspread tuning, without any modification to, or even any connections to, the broadcast receiver. For example a small stable oscillator based on a field-effect transistor can be used, powered by a small 9V battery. It need only be placed within a few feet of the receiver; often the harmonics provide sufficient signal to eliminate the need for any bandswitching. The main receiver is tuned to maximum "monkey chatter" of an ssb signal, and then resolved into speech by carefully tuning the small external oscillator. With this form of signal-frequency carrier re-insertion, the stability of the broadcast receiver's local oscillator is much less critical than when a conventional bfo at the intermediate frequency is used, since moderate drift will not affect reception.

Using this technique, some of the better all-band portable and domestic receivers can give results good enough for quite serious reception, once one gets used to the two-knob tuning technique.

The same technique can also be used with those older communication receivers where oscillator drift or insufficient bfo injection would otherwise make ssb reception difficult. Such receivers should be used with their bfo switched off.

It is easier to build an external bfo of good stability and an excellent "tuning rate" than to improve the hf oscillator and tuning arrangements on an old receiver. You may be surprised at the results that are possible on receivers never intended for ssb, although the bandwidth of these receivers will seldom be ideal for ssb reception in crowded bands.

## Receiving antennas

Listeners to sound-radio broadcasting stations seldom worry overmuch about the efficiency of their antennas (unless they are keen short-wave listeners) and usually depend on internal ferrite-rod antennas for medium and long wavebands, or short telescopic-rod ones for hf and vhf. However, when searching for low-power amateur signals, the antenna becomes extremely important and a good one can result in a tremendous improvement in the signal-to-noise ratio, so that many signals which would be inaudible or unintelligible when using a telescopic rod come through "loud and clear" on a good outdoor antenna. However a word of warning is also necessary; many transistor-type general-purpose receivers are very easily overloaded by the strong broadcast signals that come from an efficient antenna, even when these are many kilohertz away from the required station. Similarly, many receivers intended for use with short antennas such as telescopic rods have the antenna very tightly coupled to the first stage of the receiver and the direct connection of a long wire antenna may detune the first stage, so rendering a set even more vulnerable to "second-channel" image responses (see p 24) unless care is taken to ensure that the wire antenna is only loosely coupled to the receiver.

It is worth noting, even when an antenna is intended only for reception, that antennas are governed by the *law of reciprocity* which means that they possess the same fundamental characteristics and directivity for reception as for transmission; this means that any of the information provided in the journals on transmitting antennas can be utilized for the construction of efficient receiving antennas. One exception is that "ohmic losses" (resulting from the conductivity of the wire and the high currents in some portions of a transmitting antenna) are less important for reception. This means that antennas having low *radiation resistance* can be used successfully for reception, whereas they would tend to be inefficient for transmission unless extremely heavy conductors and excellent insulation were used. This is one reason why the loop or frame antenna can be very effective for mf reception, but needs considerable care when used for transmission.

The tuned loop is in fact a most useful receiving antenna for 1·8MHz since it can be turned to provide a deep rejection "null" on strong signals from a fairly local transmitter, while receiving on the same frequency much weaker signals from a more distant station in a different direction.

For serious reception on hf it is desirable to use a good outdoor antenna which will usually prove less susceptible to local mains-borne electrical interference and be less affected by the shielding effects of buildings and trees etc than would a wire strung within a building (although a loft antenna can be a good second choice if an outdoor antenna is not possible).

For a single hf band, it is relatively simple to design an efficient receiving antenna, such as a resonant half-wave dipole with low-impedance feeder. This type of antenna may be less useful on other frequencies unless some form of antenna matching or tuning unit (atu) is connected between the receiver and the antenna.

For general purposes a random length of wire, mounted as high and as clear as possible of the buildings and carefully insulated from the supporting wires will usually give good results on all bands unless the receiver is intended for use only with a low-impedance feeder (50 to 75Ω) in which case an atu will be required and will improve results on some frequencies.

Typical outdoor antennas (based on designs recommended by the BBC External Services) are shown in Fig 5. These comprise about 10 to 30m of wire (or vertical rod about 5m long plus lead-in wire).

The centre-fed dipole with balanced twin or open-wire feeder (Figs 5(c) and(d)) can be particularly effective when used with an atu since it will then function well over a wide band of frequencies. It should be noted that a half-wave ($\lambda/2$) dipole with low-impedance feeder is effective (without atu) not only at the resonant frequency but at *odd* multiples of this frequency; for example a 7MHz dipole will also function well at 21MHz

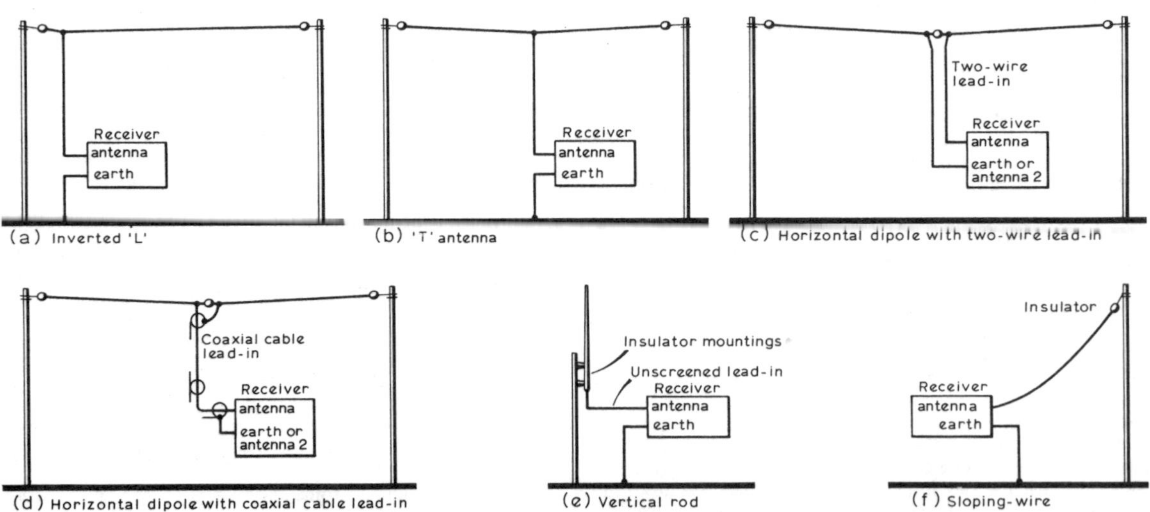


**Fig 5. Typical outdoor antennas based on designs recommended by the BBC External Services**

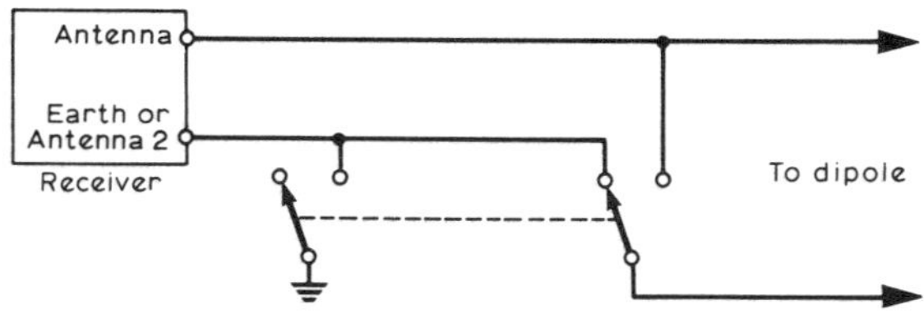


Fig 6. Switching arrangement in order to convert a dipole antenna to a T-type antenna

although the directivity of the antenna will change from that of a $\lambda/2$ dipole to that of a $3\lambda/2$ dipole.

The broadly directive properties of the $\lambda/2$ resonant dipole are very useful when receiving stations located in directions roughly broadside on to the direction in which the antenna is erected; however this may be a disadvantage when attempting to receive long-distance signals coming in from the end-on directions or on frequencies well away from resonance. A useful modification, suggested by the BBC, is to incorporate a double-pole, double-throw switch near the receiver so that the dipole antenna can optionally be converted into a T-type antenna operated against earth: see Fig 6.

Some portable receivers intended for use with a telescopic rod antenna do not have any socket provided for the connection of an external antenna or earth. However a few turns of insulated wire can be wrapped around a paper sleeve which is made free to slide on the rod: see Fig 7. The height of the rod and the position of the paper tube can then be varied for optimum results, remembering that too tight a coupling of a long wire antenna may degrade rather than improve performance. For reception on the higher frequencies signals may sometimes be improved by connecting point A to earth.

For 1·8/3·5MHz reception on portable receivers incorporating "shipping band" facilities but dependent upon an internal ferrite-rod antenna, improved performance is often possible by inductively coupling the rod antenna to a tuned Marconi-type antenna (see Chapter 5) simply by standing the antenna loading coil near the receiver, in the same plane as the ferrite-rod windings.

For vhf reception on 144MHz an efficient antenna is most desirable, particularly for reception of more distant stations by means of tropospheric propagation; local repeater stations however can usually be received with very simple rod or wire antennas. For serious vhf operation the highly directional multi-element Yagi type of antenna array (as generally used for tv reception), preferably with means of rotating the array, is widely used by amateurs. However it should not be forgotten that there are many other effective forms of antenna that can be used at vhf and that for general reception it may be desirable to have a lower gain but less directive antenna, if this cannot be rotated.

A vertical rod antenna can be very effective on long-distance signals but may also prove rather too efficient in picking up local electrical interference, so that it may prove better to use a mainly horizontal wire in districts where there is such interference.

Radio transmissions are in the form of *polarized* waves and receiving antennas similarly respond best to a particular type of polarization. For instance an upright rod antenna responds best to vertically polarized waves; an antenna parallel to the ground responds to horizontally polarized waves, and exhibits *polarization coupling losses* to signals with the opposite form of polarization. At vhf it is thus important for the listener to know and use an antenna polarized similarly to the transmitting antenna. On hf, however, where the signals have been bounced back from the ionosphere, the polarization of the incoming signals becomes mixed up and virtually random, and constantly changing so that there is little benefit to be obtained by matching the plane of the receiving antenna to that of the transmitting

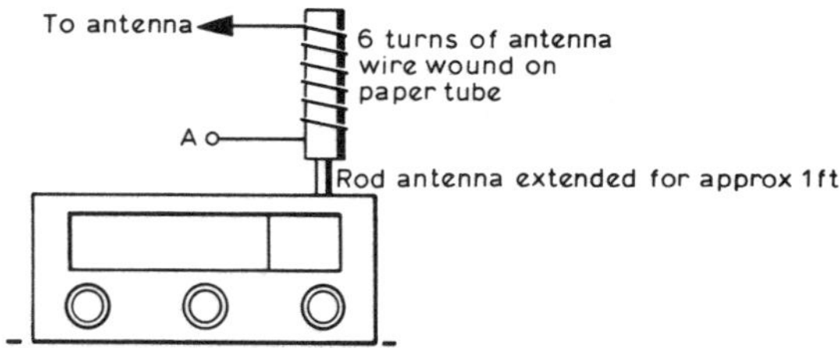


Fig 7. Using an external antenna when no socket is provided. Reception of frequencies above 15MHz is usually enhanced by connecting point A to earth

antenna. This does not mean that polarization coupling losses do not occur on hf but in this case they are temporary and indeed are one of the main reasons for the deep fading of hf signals (it is possible to reduce such fading by means of polarization diversity techniques but this is usually rather too costly for the amateur to implement).

For space and some other systems the problem of changing polarization can be largely overcome by the use of *circular polarization*, in which the plane of polarization is continuously rotated.

## PREPARING FOR TRANSMISSION

After a spell of regular listening, the itch to be able to transmit back to the stations heard soon becomes irresistible, and the attention of the newcomer turns to the problems of obtaining a licence. If he has not already done so to help his listening, he begins to study the elements of receiving and transmitting theory and to learn morse so as to be able to pass the Post Office test. At the same time he will continue to keep in touch with current amateur activities by means of his receiver and, if possible, by joining a local group or club, which may have special classes for those studying for the examinations.

It is quite likely that by the time the newcomer begins to think about transmitting equipment, certain preferences and interests will already have become apparent. There are some amateurs who find time to have a crack at almost every type of amateur activity: local as well as long-distance contacts; several types of telephony transmissions and cw (morse); constructional work and the detailed study of propagation; rtty and slow-scan tv; mobile and fixed-station operation; contests and club networks. But for the average amateur there is just not time—at least in any given period—to indulge in all the many facets of amateur radio, and he will tend to specialize in those aspects which appeal to him most. These particular interests will determine the type of equipment which he will build and operate. However it is as well when planning your station to provide for a later shift of interest and to make your equipment as flexible as possible.

### VHF operation

The popularity of the Class B licence (and the comparable Technician licences in other countries) has meant that very many newcomers now *begin* by operating on 144MHz. This is the reverse of former practice when even those operators primarily interested in vhf would usually start off with a spell on 1·8MHz or the hf bands.

There are a number of significant differences between hf and vhf operation. In the first place a high proportion of all vhf activity is based on fairly low-power nbfm or a.m. phone, but using appreciable antenna gain to provide quite high effective radiated powers. It is much easier to provide high antenna gain on vhf than hf because of the much shorter antenna elements. Most vhf operators use high-gain Yagi antennas, similar to those used for tv reception, which may be equivalent to using a transmitter 10–20 times as powerful with a dipole antenna.

Then again, although the vast majority of amateurs on hf use some form of vfo to change frequency readily and to *net* with the other amateur, until recently most operation on the 70 and 144MHz bands has used fixed-frequency crystal-controlled transmitters, with the frequency chosen to comply with the European band-planning system, which helps to separate the strong local signals from the weak longer-distance signals. Recently, however, there has been a tendency towards variable frequency operation at vhf using either a highly stable vfo multiplied up in frequency through a succession of doublers or of the heterodyne (mixer) type, or alternatively by using techniques of "pulling" the frequency of a crystal oscillator (known then as a *vxo*) or by using a *frequency synthesizer*.

For reception, it has been common practice to use an hf communication receiver in conjunction with a crystal-controlled converter; such an arrangement provides continuous tuning on the vhf band by tuning the hf receiver. However more and more amateurs use receivers specially designed for vhf often with 10·7MHz intermediate frequency. Increasingly, and particularly for mobile use, these special vhf receivers may be of fixed-channel type, where to change receiver tuning may require the insertion or switching in of a different crystal. Fixed-channel operation of this type is normal practice in business and police mobile networks, and quite a lot of equipment originally designed for such services is now used by amateurs, sometimes modified to provide variable tuning on the receiver.

Most vhf converters are based on semiconductors—suitable bipolar and field-effect transistors now being readily available. In areas where there are strong local vhf signals (local amateurs, mobile base stations, broadcasting stations etc) the problems of cross-modulation may be very severe, and for this reason some amateurs still use valves in the signal path of the converters; the miniature Nuvistor valves can give excellent results.

Because of the much lower atmospheric noise on vhf bands, the *noise factor* of the convertor or receiver is much more important than on hf. Amateurs often go to considerable trouble to ensure that they achieve in practice the lowest noise factor that their front-end devices are capable of (it should be appreciated that the ultimate sensitivity of a receiver is almost always determined by the performance of the first stage). To achieve optimum performance may require careful *neutralization* of the internal capacitances of the valve or transistor, very careful setting up of the operating point of the device by adjustment of the biasing network, and careful impedance matching of the antenna input to the first stage. Whereas the noise factor of an hf receiver virtually never needs to be less than about 8 or even 10dB, on 144MHz such a figure would be regarded as extremely "noisy"; every effort is usually made to achieve a noise factor of around 3 to 5dB. As with hf receivers a good dynamic range, good selectivity and freedom from spurious responses are very important: unfortunately a very low noise factor seldom goes with good dynamic range. The use of mosfet devices tends to provide a better dynamic range than bipolar transistors; dynamic range can also be extended by the use of balanced stages or, for example, by use of Schottky diodes (hot-carrier diodes) in diode ring modulators. In fact for the construction-minded amateur there is still a great deal of interesting development work to be done in improving the performance of vhf receivers and converters.

Detailed information on circuit techniques used at vhf and uhf will be found in the RSGB publication *VHF/UHF Manual*.

## FM or a.m.?

Until the seventies, the vast majority of phone communication on vhf was based on double-sideband amplitude modulation using high-level modulation of the power amplifier stage. Yet within a few years the dominant modes have become ssb (quite often through the use of *transverters* in conjunction with hf transceivers) and narrow-band fm. As on hf, the ssb mode has established itself as a highly effective long-distance mode. Its use has been made possible by the improved frequency stability of vhf equipment.

For local and medium-distance contacts, with fixed, mobile and hand-portable equipment, nbfm is now extremely popular. The reasons why this should be so are rather more complex, since inherently *narrow-band* fm has rather *less* communications effectiveness than conventional a.m.; further it is also highly dependent upon the use of a good fm discriminator in the receiver. It is important to distinguish between fm systems with low modulation index, as in nbfm, and wideband *broadcast-type* fm where significant advantages over a.m. can be achieved, at considerable cost in bandwidth.

Among the practical advantages of fm are:

(1) The constant power output of an fm transmitter is more suited to the power-limiting characteristics of transistor power amplifiers than the peaky nature of a.m. systems.

(2) The constant power output of an fm transmitter has been shown to cause less interference to nearby tv and vhf/fm receivers and audio equipment. Typically broadcast receivers can tolerate a 10–20dB *higher* level of interfering signal on nbfm than on a.m., provided always that the interference is not being caused by an in-band harmonic or spurious signal.

(3) Although fm signals may be heavily distorted at low levels (below the *threshold* of the discriminator), on the other hand the so-called *capture effect* of an fm signal means that a reasonably strong signal is much more immune to interference from other (weaker) signals than an a.m. system. The capture effect, however, is more pronounced where the modulation index (ie the deviation) of the signal is high. On the other hand the capture effect is a *disadvantage* when trying to receive a weak signal in the presence of a stronger signal.

(4) There is no requirement for a high-power audio amplifier since a transmitter can be frequency-modulated at low-level, yet subsequent stages need not be operated in a low-efficiency "linear" mode.

(5) The signal/noise ratio of the higher audio frequencies can effectively be increased by *pre-emphasis*, and subsequently this form of deliberate "distortion" eliminated by means of *de-emphasis* in the receiver's af stages.

(6) Since an fm signal can be passed through non-linear rf amplifiers, it is much easier to provide "repeaters" for fm transmissions than for a.m. or ssb signals. All the vhf and uhf repeaters so far commissioned in the UK are intended for use with nbfm transmissions.

## Receiving nbfm

Frequency-modulated signals are received by a process of conversion into a.m. form. NBFM can be received on most a.m. envelope detectors by slightly off-tuning the carrier and using the slope of the i.f. response. However this system is relatively inefficient and better results can be obtained using other techniques, including fm discriminators such as the Foster-Seeley, the Weiss and the ratio detector. While such circuits can be readily implemented at an i.f. of 10·7MHz for wideband broadcast fm it is however usually necessary to convert nbfm signals to 455–475kHz in order to obtain good sensitivity, unless a *crystal* is used to provide a 10·7MHz discriminator sensitive to nbfm.

An alternative form of fm detection is the use of a *phase-locked loop*. Originally this form of detection was too complex for general use but the availability of integrated-circuit devices makes this approach possible for amateur receivers.

For the amateur using a standard hf communication receiver with a vhf converter, it may be better to receive phase-modulated nbfm signals as though they were ssb signals. In fact nbfm signals can be thought of as ssb with the carrier re-inserted 90° out of phase. By re-inserting a local stronger signal a form of *exalted-carrier* reception is possible. However if the deviation of the signals is excessive, the audio signals will be heavily distorted if the receiver has an effective ssb filter.

## Mobile operation

Over 5,000 British amateurs—or roughly 20 per cent of the total—operate compact amateur stations in their cars. In the UK much of this mobile operation takes place on the vhf bands (70MHz and particularly 144MHz) although some use is also made of the 1·8MHz band ("top band"). The hf bands from 3·5 to 28MHz are also used by some mobile operators, although to a lesser degree in the UK than elsewhere even though a number of the popular compact hf transceivers were designed with mobile operation in mind.

Transmitter power on 1·8MHz is limited by the UK licence to 10W (26·7W p.e.p. ssb) and a similar order of power is often used on vhf. Operation is almost always on phone, mostly using amplitude (a.m.) or narrow-band frequency modulation (nbfm) although the use of ssb is increasing both on 1·8MHz and vhf.

Ranges of 10 to 20 miles are usual; much longer distances can be achieved on hf bands, or in favourable circumstances. Consistent area coverage at up to 50 miles radius is possible using the increasing number of fm repeaters, mostly on 144MHz and 432MHz, that are now in operation in many countries, including the UK. Talk-through repeater stations are normally installed by clubs or groups at favourable high sites and are "accessed" (made operational) by the mobile station sending a short audio tone blip (1,750Hz) for half a second. This turns the repeater transmitter on and causes it to re-radiate the incoming signal. Because of the high antennas the repeaters provide longer reliable range than is usually possible from a car.

Most repeaters are intended only for nbfm operation although a few "linear" repeaters will also handle a.m. and ssb. In the UK vhf repeaters retransmit the signals on a channel spaced by 600kHz from the incoming signal; the uhf spacing is 1·6MHz.

Because a repeater is often used by different mobile units it imposes a need for good operating discipline, and transmissions should be kept very short: some repeaters automatically cut out if a transmission continues beyond a specified period.

Transmissions can be made through a repeater from one mobile to another, from a mobile to a fixed station, or, for example, from a mobile to an amateur using a small hand-held portable transceiver.

Whereas most vhf fixed stations use antennas with horizontal polarization, vertical polarization is often preferred for mobile working since this allows the use of short whip antennas (19in for 144MHz) with the roof of the car forming a good ground plane.

Mobile vhf operation tends to be *channelized*, that is to say the equipment is intended to receive and transmit only on certain fixed (crystal-controlled) frequencies, with moderately broad selectivity in the receiver used to accommodate minor frequency differences. Contact is often established using a national or regional calling frequency (see p115) and then switching to working frequencies to allow others to use the calling frequency. This procedure does not apply on 1·8MHz or on hf where mobile stations operate in a similar manner to fixed amateur stations.

Equipment for mobile operation conforms in many ways to low power fixed-station practice; though a mobile station is often designed for use on a single band, and the receiver may consist simply of a converter in conjunction with a standard car radio receiver. The limited space in small family cars calls for considerable ingenuity if the best use is to be made of what is available, but the paramount consideration should always be that of safety: if the station is to be operated by the driver, the layout and controls must not in any way distract his attention from his prime task of driving the car. See RSGB recommendations on this page.

**RSGB**
**MOBILE SAFETY RECOMMENDATIONS**

1. All equipment should be so constructed and installed that in the event of accident or sudden braking it cannot injure the occupants of the car.
2. Mobile antennas should be soundly constructed, taking into account flexing at speed and possible danger to other vehicles or pedestrians. The maximum height must not exceed 14ft above ground.
3. Wiring should not constitute a hazard, either electrical or mechanical, to driver or passengers.
4. All equipment should be adequately fused and a battery isolation switch is desirable.
5. The transmit/receive switch should be within easy access of the operator and one changeover switch should perform all functions.
6. The microphone should be attached to the vehicle so that it does not impair the vision or movement of the driver.
7. A driver/operator should not use a hand microphone or double headphone.
8. All major adjustments, eg band change by a driver/operator, should be carried out whilst the vehicle is stationary.
9. Essential equipment controls should be adequately illuminated during the hours of darkness.
10. Logging must not be attempted by the driver whilst the vehicle is in motion.
11. All equipment must be switched off when (i) fuelling, (ii) in close proximity to petrol tanks and (iii) near quarries where charges are detonated electrically.
12. A suitable fire extinguisher should be carried and be readily accessible.

The power source for low power operation is almost invariably the normal car battery, often providing a nominal 12V (usually in practice about 13·5V). While this voltage is sufficient to operate fully transistorized equipment directly some mobile operators still use valves for power amplification in their transmitters, and this means that a transistorized *dc–dc inverter* will often be used (occasionally older forms, including *vibrators* or *rotary converters*, may still be found). The transistorized dc–dc inverter consists basically of a power oscillator with two or more power transistors oscillating rapidly between their "on" (low resistance) and "off" (high resistance) states, and so converting the dc supply into a form of ac at roughly 200–1,000Hz so that it can be stepped up by transformer action, and then rectified and smoothed.

Since many cars are now fitted with alternators having

relatively high outputs, and since fully-transistorized amateur equipment often imposes relatively light loads on the battery, there is much less risk than formerly that an amateur mobile station will impose excessive drain on the car battery. Some installations for hf are of high power.

A major problem with all mobile installations is that of suppressing to a high degree all the electrical interference which stems from the ignition system and other car electrics.

Another problem, particularly for 1·8MHz operation, is that of achieving maximum radiation from a short whip antenna, which can seldom exceed 7 or 8ft in length; usual practice is to inductively load the antenna with a weatherproof coil inserted either in the centre or at the base of the whip. On vhf rather more efficient antennas can be realized: one popular type is the omnidirectional "halo" which consists essentially of a $\lambda/2$ dipole bent round in the form of an open circle about 1ft in diameter. The "halo" is mounted some 2ft above the car roof and has horizontal polarization. The $\lambda/4$ ground-plane ("whip") antenna is a popular vertically polarized antenna.

## The station layout

From the beginning it should be appreciated that an amateur station is more than just a number of separate items of equipment —receiver, transmitter, frequency meter and the like—gathered together haphazardly in the same room. There should always be forethought and planning, bearing in mind the ultimate *operation* of the equipment as a single unit. An efficient transmitter is useless without good receiver performance; a stable variable frequency oscillator unit is wasted unless the output frequency can be readily adjusted to any required channel from the operating position; slowness in changing from transmission to reception and vice versa can spoil endless contacts.

A station should be planned not only to keep abreast of technical developments, but also with the objectives of comfort, rapid change-over, and general ease of operation. Many hours will be spent searching for and copying weak signals; yet, too often, the newcomer gives little thought to his own comfort. For example, the change-over from transmission to reception may require a sudden flurry of effort and concentration.

The following are some of the factors which govern the "operability" of an amateur station:

The operating table should be as large as space permits, and never allowed to become cluttered up with stray pieces of equipment or spare parts. There should be plenty of room for the log and scribbling pads directly in front of the operator.

The receiver should be placed comfortably back on the lefthand side of the operating table so that the forearm may be rested while tuning. The height of the most frequently used controls is important and can often be adjusted advantageously by mounting the receiver on rubber blocks, which also reduce the effects of vibration. Four to six inches is normally a good height for the main tuning and gain controls.

Where variable-frequency control is employed, the oscillator unit should be within easy reach of the operator, preferably in the back right-hand corner of the table. Switching arrangements should permit the operation of the oscillator unit by itself (ie with the exciter and amplifier stages inoperative) in order to provide a weak signal on the receiver to allow adjustment to any desired frequency.

Change-over from transmission to reception should be by means of not more than one switch: for telegraphy it will be better still if this is accomplished solely by the action of the morse key; for telephony the ideal is a voice-operated change-over device (vox) or a single push-to-talk switch on the microphone. Where, as is usually the case, it is desired to switch a number of separate circuits, this should be done by means of a multi-contact relay operated by a single change-over switch.

Where the main transmitter is not constantly under view from the operating position, it is advisable to provide some indication on the operating table to show whether the transmitter is functioning correctly. This can be either a morse or telephony monitor providing a low-level indication of the transmitted signals in the operator's headphones, or alternatively a meter or bulb indicating the current being delivered to the antenna feeder.

Finally, look round for the most comfortable cushioned chair (preferably one with arms), provide yourself with a handy bookshelf and space for pencils, reference lists, a good atlas—and do not forget to leave plenty of room for your legs and feet!

## Test equipment

Little can be done in radio construction and operation without some items of test equipment, although fortunately an amazing amount can be achieved with very little.

The following is a list of test equipment and instruments most often used by amateurs—although it is not suggested that newcomers would need to have all these units.

1. *Universal testmeter* (ie combined voltmeter-ohmmeter-milliammeter) preferably with a sensitivity of about 10,000 or 20,000Ω/V on dc ranges. Higher-sensitivity electronic testmeters are useful if a good deal of constructional work is undertaken.

2. *Grid dip oscillator* (gdo) or gate dip oscillator. These units, often home-constructed, are extremely useful for checking the resonances of tuned circuits, antennas etc.

3. *Crystal calibrators.* These provide a series of accurate marker signals at selected frequencies—for example at intervals of 1MHz, 100kHz or 10kHz. They may consist of just an oscillator producing harmonics, but more versatile units can now be made using decade divider integrated circuits (eg SN7490). The lower cost of digital integrated circuits also now makes it possible for amateurs to build or buy *frequency counters* which provide a direct display of the frequency of any signal fed to the meter.

4. *SWR meter.* These meters provide an indication of the standing-wave ratio present on a coaxial cable fed with power from a transmitter at one end and connected to the antenna at the other end. Knowledge of the approximate swr will help to show whether the impedance matching is correct (ideally when the swr is unity or 1:1). A common form of swr meter is a simple "reflectometer" which detects the forward and reflected power in the cable. Many modern transmitters are not suitable for operation into a mismatched transmission line, and this makes it important to have some means of detecting a high swr.

5. *RF noise bridge.* For some investigations of antenna impedances, antenna matching, receiver-input matching and the like, a useful tool is a simple rf noise bridge comprising a wideband noise generator and simple bridge using the receiver as the "detector". Whereas the gdo allows investigation of parallel-resonant circuits, the rf noise bridge will deal with series-resonant circuits.

6. *Capacitance meter.* The home constructor finds it very useful to be able to determine approximately the value of small unknown capacitances, since many modern capacitors may not be clearly marked with their values. Simple but effective units can be based on measuring the rf current flowing through the unknown capacitor, deriving the rf signal from a crystal oscillator with the meter calibrated from capacitors of known value.

Many other simple test units, including for example "go/no-go" testers for transistors, crystals and so on, or simple noise generators for rapidly checking the functioning of the various stages in a receiver, are all described frequently in the amateur journals.

Similarly for adjustment of ssb transmitters it is useful to have

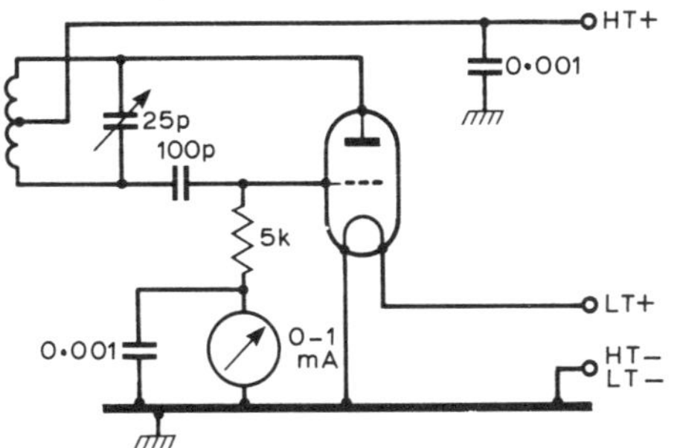


Fig 8. One of the most useful pieces of test gear is a grid-dip oscillator. In this circuit, almost any small triode or pentode connected as a triode may be used. Approximate coil dimensions: 3·5MHz, 60t 28swg; 7MHz, 34t 24swg; 14MHz, 14t 24swg; 21MHz, 10t 24swg. All coils are centre-tapped and closewound on 1½in dia formers. Valve GDOs tend to have better performance than those using bipolar transistors, but field-effect devices (see p26) are extremely good and operate from a single 9V supply. They require a more sensitive meter (200μA) and higher-value gate resistor (about 47kΩ)

two-tone test oscillators or oscillators which can be adjusted to provide variable on/off ratios. For amateurs designing and developing advanced equipment, a good oscilloscope and wobbulator is desirable; in its more refined form—*spectrum analyser*—detailed information can be obtained on spurious emissions etc.

But it must be stressed that many amateurs manage quite well with an extremely limited amount of test equipment, making use of their receiver, neon bulbs as rf voltage indicators, torch bulbs as rf current indicators etc.

## Semiconductors

A factor which nowadays must be taken into account by newcomers planning future equipment is the gradual trend towards transistors and other semiconductor or solid-state devices, including for example silicon power diodes.

Completely solid-state stations, except for fairly low power portable or mobile operation, are still quite rare, although now technically feasible for fixed-station operation at medium powers. The main problem is that hf power transistors still tend to be considerably more expensive and provide lower gains than valves of equivalent power rating. Recent years have seen a steady reduction in the prices of these special semiconductors. Again, most high-power transistors call for power supplies of 28–40V at fairly heavy currents, so that there is only limited operational and size advantage in solid-state transmitters, the major bulk represented by power supplies being little affected. It is also easy to damage or destroy the expensive hf transistors unless care is taken—normally, for example, such transistors should never be operated without a correctly matched load unless adequate protection circuits have been incorporated.

On the other hand for audio work and for receivers and low-power stages in exciters, the compact size, simple circuitry and low price at which many transistors are available all make these attractive for current designs. We can therefore expect to see more "hybrid" equipments—that is units employing a combination of transistors and ICs—with valves being retained mainly for rf power amplification.

Then again, there are some applications in which semiconductors offer new facilities and so are particularly attractive. One example is the use of semiconductor diode switches to overcome the problems which have always attended the use of mechanical switches for switching low-power circuits. Semiconductor diodes also possess the important characteristic that their capacitance can be changed by altering the reverse voltage applied across them, providing a form of variable capacitor which can be "tuned" by altering a dc voltage—particularly useful for some forms of bandspread tuning or for remote operation of a vfo, etc. Semiconductors are also, of course, extremely useful for compact test instruments needed in amateur stations, including grid dip oscillators, field-strength meters and monitors.

Although at present many amateur stations use valves for normal fixed station operation, this is sure to change gradually with increasing use of semiconductors during the next few years. Already most of the compact transceivers use transistors in most stages.

A transmitting application for which transistors have proved their worth is for stable variable frequency oscillators. Since very little heat is generated in a low-power transistor, a transistor vfo, if mounted well away from other sources of heat, will show very little frequency drift.

Certain precautions must always be taken in any equipment using semiconductors: transistors can be damaged or destroyed by applying supply voltages the wrong way round, by excessive transient voltages, or by subjecting the junctions to excessive heat or mechanical shocks. The internal resistances of transistors are low, and even quite low voltages applied across the wrong electrodes can cause excessive currents which can burn out or damage the junction. The high-power transistors are among the most susceptible to this hazard because of their very low internal resistances. Some of the ways in which these conditions can accidentally occur include: (i) leakage voltages to earth from the bit of an electric soldering iron (the answer is to earth the casing of the iron) or from the output leads of mains-operated test gear such as signal generators (always include isolating capacitors in the leads between such test gear and the transistor equipment); (ii) voltages from internal batteries in low-resistance continuity testers or low-resistance ohmmeters (internal resistance of instruments should exceed 10,000Ω or an external resistor of this value should be connected in series), or from the sudden discharge of a large-value electrolytic capacitor; (iii) excessive currents arising from too great an input from a signal generator or nearby transmitter.

There is also some hazard arising from a sudden disturbance in operating conditions: do not disconnect a transistor until all potentials have been removed; do not run a transistor with its collector circuit open; do not short-circuit base to collector while a transistor is operating. With power transistors, both af and rf types, it is important not to operate them without a properly matched load, since otherwise excessive high voltages may appear on the collector. Power transistors normally have to be mounted in *heat sinks;* these are pieces of metal which help to dissipate internal heat to the surrounding air.

The use of *field-effect* devices and microelectronic *integrated circuits* is discussed on p 28.

## Bandwidth

An important communication principle is that all transmissions containing information do not occupy just a single frequency but a band of frequencies (denoted by its *bandwidth*). This bandwidth depends on the rate of information transmitted and the modulation system being used. For example, conventional double-sideband a.m. having maximum audio frequencies of 3,500Hz would require a bandwidth of at least 7,000Hz. In other words if the carrier frequency were 14,200Hz, the signal would occupy 14,196·5 to 14,203·5kHz. On the other hand, an ssb signal transmitting audio frequencies between 300 to 3,500Hz would have a bandwidth of 3,200Hz (assuming complete suppression of all unessential frequencies). Even cw, which might appear to be just the switching on and off of a single carrier frequency, does in fact require a minimum bandwidth which depends on the speed of the keying (a bandwidth of about 20 to 50Hz in typical cases).

Optimum signal/noise will be achieved if the bandwidth of the receiver "matches" that of the transmission. However in practice receiver bandwidth can be reduced to the theoretical minimum only if the stability of the receiver is extremely good and the tuning arrangements enable the receiver to be set precisely to the correct frequency. For example, a cw bandwidth of about 300Hz is usually regarded as about the minimum practicable at the present state of the art for normal operation. On vhf receivers, it becomes even more difficult to achieve the high order of stability which would allow exact matching to the transmission bandwidth, and for this reason the i.f. bandwidth of a vhf receiver is often made deliberately broad to accommodate some oscillator drift.

Nevertheless the concept of matching the receiver selectivity to the mode of transmission has become increasingly important, and this is why it is convenient for the selectivity to be determined primarily in a single "block filter" rather than achieved gradually by a succession of i.f. transformers; this then allows the receiver bandwidth to be changed by providing more than one filter—though as indicated in the section on ssb filters this tends to make receivers expensive.

## SSB transceivers plus "linears"

Recent years have seen a very marked growth in the popularity of conveniently packaged ssb/cw transceivers. At first such equipments often covered only one band, or a few segments of a few bands, but many modern designs now offer virtually complete coverage of all hf bands, and some even include the 1·8MHz band. Some transceivers are all-semiconductor apart from a driver (often 12BY7A) and power amplifier (usually 6146B or colour tv "sweep" valves). Dual-gate field-effect devices are generally used in the front-ends of all-semiconductor receivers, though even these may provide less dynamic range than a good valve receiver. For ssb generation the filter technique is generally used, though a wide range of filter frequencies is still found. Some equipments use mechanical filters at 455kHz, but more models are now using crystal filters at about 3,180, 5,200 or 9,000kHz since these reduce the number of frequency conversions needed to provide output on all bands up to 30MHz.

Output of many transceivers is of the order of 150 to 250W peak envelope power (p.e.p.) While this is more than adequate for many purposes, the higher-power stations often use the transceiver to drive a separate higher-power linear amplifier (generally known just as a "linear").

A fundamental object of a transceiver is that this provides common tuning of the receiver and transmitter, so that transmission is always on the same "net" frequency as the incoming signals. In practice it is often an advantage to be able to tune the receiver slightly away from the transmitter frequency and vice versa. Provision for this "incremental" tuning is made in many hf transceivers; alternatively some amateurs prefer to use a separate vfo unit so that the two frequencies can be separated when required.

## Good equipment

What are the factors that decide whether amateur equipment is good or bad? Obviously an answer to such a question could include many different factors—for example the reliability (as determined by the time between any failures), the ease of servicing when the inevitable fault does occur, the mechanical precision of the tuning arrangements, the good placing of the controls etc. Many other points will be brought out elsewhere in this book. But there are some factors which may be less obvious to a newcomer, yet can greatly affect the performance of the equipment.

For example, when considering a transmitter, it is important to think in terms of the degree to which (unfortunately) it is likely to radiate some signals on frequencies other than those necessary; this includes the suppression of the carrier and the unwanted sideband on any ssb transmitter; its short and long term stability; any spurious emissions including key clicks on cw.

At one time, spurious emissions from transmitters were generally those on harmonics of the signal, or caused by unwanted parasitic oscillation. Typically with a transmitter on 14,100kHz one strove to reduce the second harmonic on 28,200kHz, and even more so the third harmonic on 42,300kHz (since in the UK this was liable to cause interference to television reception). While these harmonics are still very important, the wide use of mixer stages in modern transmitters may mean that it can produce signals on many different frequencies unless care is taken to ensure that these unwanted mixer *products* are adequately suppressed.

When two frequencies $f1$ and $f2$ are mixed together, then the output contains signals at $f1, f2, f1+f2, f1-f2, 2f1, 2f2, 2f1+f2, 2f1-f2, f1+2f2, f1-2f2$ and so on, in fact at any frequency of the form $mf1 \pm nf2$, where $m$ and $n$ are any whole numbers. This means that after any mixer stage there will be many unwanted frequencies present in the output, so that it is essential to have sufficient tuned circuits or other types of filters to reduce to the maximum possible extent the likelihood of any of these passing through the remaining stages of the transmitter and reaching the antenna.

Unwanted frequencies will also be introduced if a linear amplifier in any ssb transmitter is operated in a non-linear fashion, by over-driving or by incorrect biasing. This will tend to increase the harmonic and spurious outputs—but will also cause *splatter*, the radiation of an excessively broad signal due to *flat-topping* (so called because the amplifier output does not increase linearly with the input waveform but flattens the peaks).

Many of the same problems of non-linearity are reflected in receiver design; here again the presence of mixer(s) and local oscillator(s) can, unless the design is a very good one, result in many forms of *spurii*. In other words, the listener may hear the wanted signal on more than one frequency, or more usually he hears broadcasting and commercial stations as though they were operating in the amateur bands (sometimes unfortunately they are—but that is another question). Or he may tune in *birdies* (unmodulated signals that come from the oscillators in his own receiver). Or he may find that really strong unwanted signals affect the weak signal he is listening to, even if these signals are well outside the passband of his receiver. Many of these effects are due to *cross-modulation* and *blocking* or *intermodulation*, all of which arise because some stages in the receiver are not linear or do not have sufficient *dynamic range* (in other words they become non-linear when confronted with a very strong signal).

There are very many factors which govern the linearity and dynamic range of the various stages, and a great deal of effort is being expended by both amateur and professional designers to reduce the problems that arise from present limitations. What we are trying to indicate here is that there are finer points of design—particularly in ssb equipment and receivers using semiconductors—which may not be immediately apparent from a casual study of advertisements. The firms catering for the amateur market generally have a pretty good reputation for giving "value for money" but they are faced with the very difficult problem that amateurs seek standards of performance which may well be beyond what is possible if the selling price is to remain within their reach. It may often be better for an amateur to be satisfied with equipment based on valves and fairly simple variable frequency oscillators than to strive after all-semiconductor units,

## WORLD ADMINISTRATION RADIO CONFERENCE 1979

At the World Administrative Radio Conference of the International Telecommunication Union, held at Geneva from September to December 1979, many revisions and amendments were agreed to the international Radio Regulations and the International Table of Frequency Allocations. These regulations have the force of an International Treaty and provide the framework under which all amateur operation is permitted by the national administrations (in the UK by the Home Office Radio Regulatory Department) of the different countries. Important changes were made to the Amateur Service although amateurs will not be able to take advantage of these until they are officially confirmed by the licensing authority; in some cases this may not be possible until some time after the new Regulations come into force on 1 January 1982, since provision has been made for the "fixed" stations currently using frequency allocations to continue to do so for a number of years.

(1) It will become possible for administrations to issue amateur licences without requiring a morse test for operation on frequencies above 30MHz, ie this may make it possible for the UK Class B licence to cover 70MHz operation, and in Region 2 may apply to 50MHz and above.

(2) A new system for the classification of emissions will be introduced which differs from that currently used in amateur licences (see p105). In the new system the first symbol denotes the type of modulation of the main carrier; the second symbol the nature of signal(s) modulating the main carrier; and the third symbol the type of information to be transmitted (ie telephony, manual telegraphy, automatic telegraphy etc). For example, amateur ssb telephony, currently designated A3j, will become J3E; fm telephony, currently F3, becomes F3E; while a.m., currently A3, becomes A3E.

(3) *New hf bands.* An important result of WARC79 is that three new hf bands will gradually become available for the amateur service. These are:

10·100–10·150MHz (about 29·7–29·55m)
18·068–18·168MHz (about 16·04–16·51m)
24·890–24·990MHz (about 12·05–12·00m)

In addition part of the 1·8MHz band (1,810–1,850kHz, about 166–162m) has been restored to the International Table as an exclusively amateur band and should gradually become available to amateurs in all countries, while it is hoped that UK amateurs will continue to have access to the wider band of 1,800–2,000kHz shared with other users. Another concession is that considerably more of the amateur bands (including several at hf) will become available for amateur satellites.

Although the new bands are very narrow and will require self-discipline by all amateurs, nevertheless they provide most useful bands intermediate between the existing ones. It should mean, for example, that long-distance operation will be possible during virtually 24h over a much wider portion of the sunspot cycles, enabling amateurs to choose frequencies near the muf. For example, the new 10MHz band can be expected to be open for dx after dark, and for long-path transmission to Australia/New Zealand around dawn and dusk with great regularity. 18MHz should prove a useful "daylight dx" band for contacts with North and South America in sunspot minimum years when 21MHz is liable to "open" only occasionally. Similarly 24·9MHz should be often "open" in high sunspot years at times when the muf fails to reach 28MHz.

The new bands will thus present a rewarding challenge to amateurs to acquire the fullest knowledge of propagation conditions that will be needed to exploit fully these new frequencies. They also of course present a challenge to manufacturers and amateurs in the design, modification or adaptation of designs for receivers, transmitters, transceivers, linear amplifiers, antennas and antenna tuning units so that amateur stations can work effectively on the new bands. Since these are not in direct harmonic relationship with current hf bands, it may prove difficult to modify directly old-style transmitters, but equipment based on heterodyne mixing rather than frequency multiplication may require only the fitting of additional crystals (though this may mean less-than-optimum LC ratios etc). It should also prove possible to design hf transverters covering the new bands when used with an existing hf transceiver.

Perhaps one of the most challenging requirements will prove to be the evolution of antennas, particularly beam antennas, covering both existing and the new bands with reasonable efficiency. It may prove necessary to break away from current multi-banding techniques such as "traps" (W3DZZ etc); however forms of "loop", "delta-loop" and "quad" arrays, fed with open-wire feeders, may prove suitable. Wideband arrays such as log-periodic, vee, rhombic etc should be effective, and simple "long-wire" antennas can be matched to transmitters over the entire hf range by using suitable antenna tuning units: this also applies to centre-fed dipoles fed with open-wire feeders.

(4) *New microwave bands.* For British amateurs there seem likely to be few changes affecting the existing vhf or uhf bands, but a number of entirely new amateur allocations have been made at the extremely high frequencies above 24GHz (currently the highest frequency available to amateurs in the UK). These include bands near 47, 75·5, 120, 142 and 241GHz. These will require the development of new techniques if effective use is to be made of them for communication over other than line-of-sight paths at frequencies approaching those of infra-red radiation, and where such factors as attenuation from fog and rain become very important. However there are "windows" and ducting modes that may provide surprising results even with very simple equipment (though this will need to be made with precision).

or frequency synthesizers and the like. This advice applies not only to factory-made equipment but also to home-construction where the amateur may not have the test and measuring equipment that would allow him to check how the equipment is *really* working in respect of dynamic range and spurii. This is not intended to discourage anyone from building advanced equipment—but is a re-statement of the age-old principle of learning to walk before trying to run; and when buying factory-made equipment it may be wise to seek out the views of somebody who knows the equipment well and whose judgement you respect. Some further guidance on this subject is given in Chapter 6.

CHAPTER 3

# Communication receivers

The first requirement of every newcomer to amateur radio is a good receiver covering the amateur bands—and the skill to make the best use of it. Almost any "all-band" broadcast receiver or simple one- or two-transistor "straight" receiver will bring in amateur signals on some bands, but will usually prove unsatisfactory for serious listening in one or more respects. The use of ssb, the crowded bands, the narrow spaces they occupy in the full short-wave spectrum, and the extremely wide variation between strong and weak signals (which may easily differ in voltage by 10,000 times and can occasionally be of the order of 500,000 times) all impose very stringent demands upon the receiver. Over the years high performance receivers which are not unduly expensive and which have become known as "communication receivers" have been evolved to meet these special needs. Although there are now many users of such sets other than amateurs—including the Services and commercial organizations—it is to the credit of the radio amateurs that this type of receiver was originally produced in large numbers.

Today, it has been said that receiver design and construction by the amateurs themselves is a lost art; and it is true that only a small minority use receivers entirely constructed at home. But whether you build or buy your receiver, it is certain that you will not obtain the best results unless you have a sound understanding of the main features of a communication receiver, how it differs from a standard broadcast set, and the purpose of the various refinements and controls.

## "All-band" superheterodyne receivers

For broadcast entertainment usually only a few stations of relatively high signal strength are wanted and the user is satisfied if he can obtain these few stations at a steady volume, with good quality reproduction, and as free as possible from electrical interference or interference from other stations. The majority of domestic receivers use an almost standard six-transistor circuit which will meet these demands reasonably well: see Fig 2. The first transistor is a combined mixer and local oscillator which changes the incoming signals to the intermediate frequency of about 470kHz and is coupled to the next transistor by means of the first intermediate transformer containing two inductively coupled resonant circuits. This second transistor amplifies the i.f. signals and passes them on, through a second i.f. transformer, to another amplifier stage and finally to a diode detector. Either the same diode or a second diode is used to provide a dc voltage which varies with the strength of the signal and this voltage is used to control automatically the gain of the i.f. amplifier transistors so as to keep the output from the receiver fairly constant despite any fluctuations in the strength of the incoming signals. The final three transistors are used as an audio amplifier providing sufficient power to operate a loudspeaker.

Six-transistor broadcast receivers of the type outlined above sometimes incorporate a short-wave band or bands and many newcomers to amateur radio receive their first amateur transmissions on such receivers; indeed many local and long-distance stations can often be received successfully. It will soon be found however that many transmissions are spoilt by other stations on adjacent frequencies, or lost altogether in attempts to tune them in more accurately or because the receiver drifts in frequency between transmissions. When the stations are weak and the volume is turned up there is either insufficient amplification available or transistor noise drowns the stations. It will also soon be noticed that most receivers will tune to the louder commercial and broadcast stations at more than one point on the dial, and this will make the bands sound even more

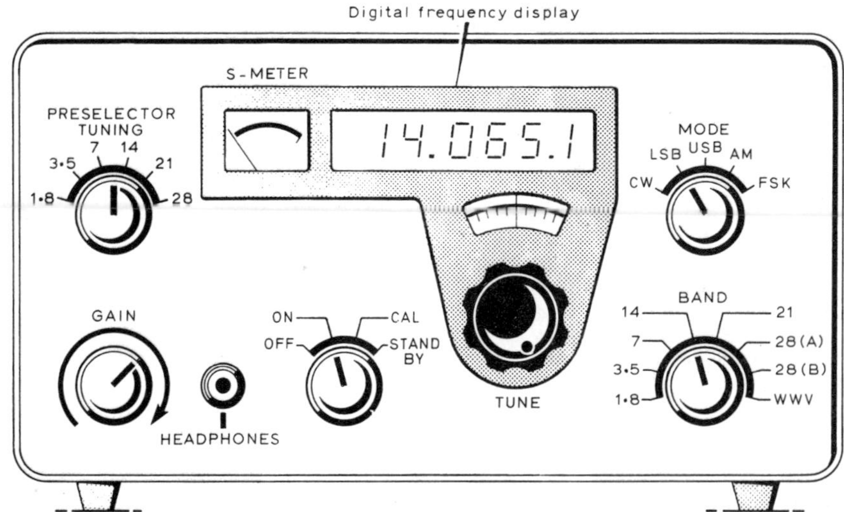


Fig 1. Panel layout of an amateur communication receiver, showing the useful controls

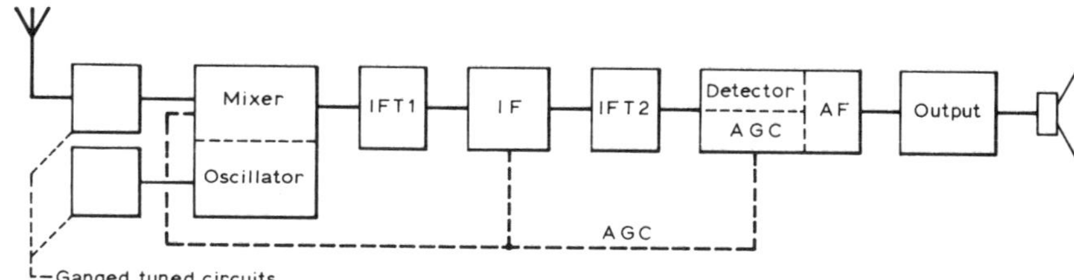


Fig 2. Representative design of a conventional broadcast-type superhet receiver using valves or transistors

crowded with stations than they really are, particularly on the higher-frequency bands. Amateur telegraphy stations can usually be heard only as a series of "thumps" which are extremely difficult to copy even if you can read morse. There will also be many signals which will be completely unintelligible; these emanate from stations using suppressed-carrier (ssb) systems. Thus if we were to attempt to use a receiver of this category for serious amateur operating we should be able to hear and work only the louder stations and would miss altogether many of the weaker long-distance signals. It would be rather like entering for a TT race on a motor scooter.

It will not be long before we begin to sum up the needs of a receiver for communication purposes. Our list will be more or less as follows: (1) very good *selectivity* to separate the stations; (2) a high-degree of *sensitivity* to permit reception of weak stations; (3) *ease of tuning*, with the ability to return quickly to a given frequency; (4) *stability*, so that the set does not drift off a station or lose it suddenly due to shock or vibration; (5) absence of *spurious* signals, so that even the loudest stations are heard at only one point on the dial and with no tunable whistles generated within the receiver; (6) an internal oscillator (called a *beat frequency oscillator*) for listening to morse signals and ssb telephony stations.

## Selectivity

The ability to separate stations on closely adjacent frequencies is governed by the selectivity of a receiver. A high degree of selectivity is extremely difficult to achieve in straight receivers and in the superheterodyne receiver this characteristic will be determined very largely in the intermediate frequency section. It has been stated earlier that the standard broadcast set usually has two intermediate frequency amplifying stages with an i.f. transformer, comprising two tuned circuits, in both the input and output circuits of each stage. This provides altogether six resonant circuits peaked to accept signals at a given frequency, usually between 455 and 470kHz (about 640m). Now six circuits at this frequency *can* provide a fair degree of selectivity even in crowded amateur-bands, although the most powerful stations will usually tend to spread out over a number of kilohertz. But the type of i.f. transformers fitted in broadcast receivers are not generally the most efficient for communication purposes. For good quality broadcast reception we need to receive without undue loss frequencies up to about 6,000Hz (6kHz) on each side of the nominal station frequency; for amateur telephony this figure could be reduced by about half, that is to some 3,000Hz either side, and for telegraphy we need a bandpass of only 100–200Hz. Broadcast i.f. transformers are often deliberately "over-coupled" or peaked to different frequencies in order to produce a broader response and are not of high-$Q$ construction ($Q$ is a term denoting the "goodness" of a coil and it is this which determines the selectivity of a given tuned circuit).

Thus as we tune our broadcast receiver across a powerful signal, we shall hear the station without distortion over a fairly narrow band and then for some kilohertz on either side we shall continue to hear the transmission with increasing sideband distortion until the set is between say 15 and 25kHz away from the carrier frequency. Thus a really powerful signal may block out weak stations over a channel up to 50kHz wide. There is also the possibility that a very loud signal may affect the receiver over an even wider band, due to a form of interference called *cross-modulation*.

There are a number of ways in which the selectivity can be improved.

(1) The coupling in the i.f. transformer can be decreased by placing the windings farther apart.

(2) The i.f. transformer windings can be of higher $Q$, or some other form of tuned circuit can be used having a much higher $Q$ (in practice this may be either a quartz crystal filter, a mechanical filter, or a device known as a $Q$-multiplier which depends on the fact that a circuit near oscillation has a very high $Q$).

(3) The total number of tuned circuits can be increased; this is most easily done by adding i.f. amplifiers, with additional i.f. transformers to provide the coupling between stages.

(4) The i.f. can be lowered; this will increase selectivity because the response of a tuned circuit to a signal off resonance depends upon the percentage difference between these two frequencies. For example a signal at 490kHz fed to a 470kHz i.f. amplifier represents a percentage difference of $20/470 \times 100$, approximately 4 per cent. If we lower the i.f. to 50kHz and feed into the amplifier a signal the same number of kilohertz off tune (ie 20kHz), the percentage difference is 40 per cent, and the rejection will be correspondingly greater. We shall see later,

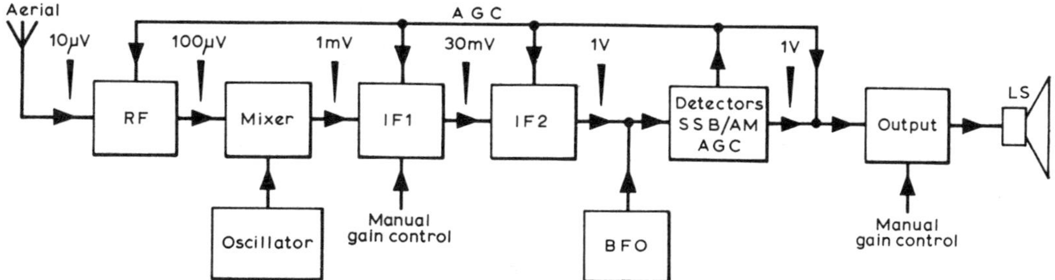


Fig 3. Block diagram of a basic superhet receiver suitable for general purpose short-wave reception and amateur communication

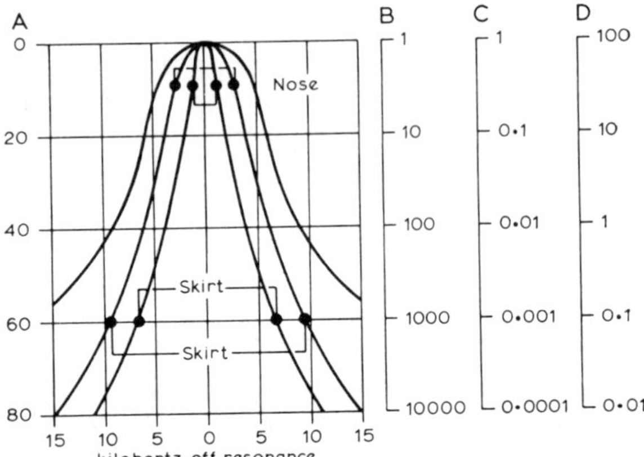


Fig 4. One of the most important characteristics of a good communication receiver is adequate selectivity. This graph shows three different selectivity curves varying from the "just adequate" to the type of results obtainable in a good modern receiver. A, B, C and D indicate four different scales in which such curves are often given: A is a scale based on the attenuation in decibels from maximum response; B is relative inputs for a constant output; C is output voltage compared with that at maximum response; D is the response expressed as a percentage

however, that merely lowering the i.f. brings in other problems that offsets this advantage.

In practice a modern communication receiver will often use a combination of all these four methods of achieving high selectivity: specially designed under-coupled i.f. transformers of high-$Q$ construction, with a very high-$Q$ crystal filter or $Q$-multiplier for optional use; two or even three stages as i.f. amplifiers and as many as 12 circuits resonant at the i.f.; such a receiver may achieve its main selectivity at a very low i.f. (40–100kHz). Several different degrees of selectivity can sometimes be selected by a switch.

To compare the selectivity of different receivers, or receivers having more than one position of selectivity, we can use selectivity curves such as those shown in Fig 4. There are two ways in which we must consider these curves: the first is the *nose* figure which represents the bandwidth in kilohertz over which a signal will be heard with relatively little loss of strength; the other figure—and many amateurs will consider this the more important—is the bandwidth over which a really powerful signal is still audible (often taken as a reduction of 1,000 times (60dB) on the peak strength) and this is often termed the *skirt* performance. These two sets of figures are related by what is known as the *shape factor* of the receiver. Since, even with a very good receiver, the skirt bandwidth will tend to be from 2·5 to 5·5 times as great as the nose bandwidth, it is most important always to distinguish between these two sets of figures. In practice, any receiver which has a skirt bandwidth of less than about 10–15kHz can be considered a very good one, though the nose figures for even medium grade receivers will look much more impressive!

## Sensitivity

Weak signals clearly need to be amplified very much more than strong ones in order to make them easily readable. Now it is quite easy to amplify any signal by means of a transistor or valve, so at first glance it might seem that all we need to do in order to receive weaker and weaker signals would be to add more and more stages of amplification. Unfortunately it is not so simple as this—as we can quickly observe if we turn up all the gain (ie amplification) on any receiver having more than about six transistors or four valves. What happens is that the noise from the set rises and masks any very weak signals that may be present. No matter how much further we increase the gain we shall simply hear more and more noise and the signal itself will remain inaudible. How weak a signal can be heard on a receiver then is not governed by how much the set will amplify but by how small a signal can be heard above the general noise level. This characteristic of a receiver is described by stating the minimum signal voltage required to produce a given output for a certain ratio of signal above the noise level at a specified selectivity, and is known as the *signal-to-noise ratio* (or more strictly as the *signal-plus-noise to noise ratio*).

To see how the signal-to-noise ratio, or in other words the sensitivity, of a receiver can be improved it is necessary to discover where this background noise comes from, since if this can be reduced then clearly weaker signals will be heard. Without going too deeply into the subject, it is possible to distinguish between a number of different sources of noise.

(1) There will be some noise from atmospherics, radiation from space and other natural phenomena. There is not much that any of us can do about this type of noise, which would in fact limit the usable sensitivity of an ideal noise-free receiver.

(2) There will be local or "site" noise caused mainly by the local operation of electrical appliances. This noise will vary widely in different localities. While in extreme cases it may be possible to have specific sources of interference suppressed, very few amateurs are blessed with anything like ideal noiseless sites. Automatic noise limiting devices fitted in a receiver can remove some of the more objectionable "peaks" of interference, but it must be accepted that in many locations and on some at least of the bands this man-made interference will limit the usable sensitivity.

(3) A certain amount of noise will be generated within the transistors or valves due to the so-called *shot effect* and in multigrid valves by *partition noise*. The noise produced in the early stages of a receiver will be amplified by all succeeding stages and so will be more important. In order to compare different valves it is usual to consider this noise in terms of an imaginary resistance in series with the grid of the valve. The equivalent value of this resistance varies widely according to the design of the

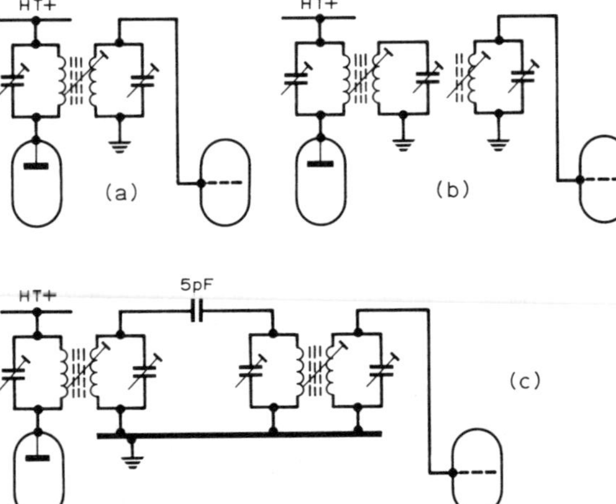


Fig 5. Inter-valve couplings for increasing i.f. selectivity. (a) The conventional double-circuit i.f. transformer. (b) The triple tuned i.f. transformer will provide a useful improvement in skirt selectivity. (c) If a triple tuned i.f. transformer is not available, two conventional i.f. transformers can be used to provide four tuned circuits per stage

valve and how it is being used. It is important to note that a transistor or valve used as a frequency changer is much noisier than when employed as a straight amplifier. A good modern rf pentode valve may have an equivalent-noise resistance as low as 1,000Ω whereas for frequency changers it is not unusual for the equivalent resistance to be higher than 200,000Ω. This gives us a clue to one of the important features of a sensitive receiver; unless it has a low-noise mixer it must have one or more carefully chosen transistors or valves operating as tuned rf stages to amplify the very weak signals *before* they pass through the frequency changing stage. If this is done, the signals will then be strong enough to over-ride the conventionally large amount of noise contributed by the frequency changer. This explains why our broadcast receivers, few of which today include an rf amplifier, are not suited to deal with very weak signals, and also explains why "straight" trf receivers, which do not have a frequency changer, can be made very sensitive, though they cannot usually be made selective enough to cope with modern conditions. Some low-noise mixers—including triodes and special beam-deflection valves and field-effect transistors—allow sensitive superhets without rf amplifiers, though an rf amplifier may have other uses.

(4) Even if we were to achieve the impossible and to eliminate all amplifier noise and then were to remove the antenna to prevent any signals from reaching the receiver, there would still be some background noise left. This would be due to what are called the thermal-agitation voltages which are always present across any impedance. In a receiver the most important source of this noise will be the first tuned circuit impedance, and a very good test for a receiver is to ascertain whether, with a resistor equal to the input impedance connected across the antenna terminals, this noise peaks up as the circuit is adjusted for resonance. The ultimate effect of this noise will depend upon the total passband of the receiver, which means that a receiver which has variable selectivity will tend to have a better signal-to-noise ratio when adjusted to its most selective position (it is for this reason that the sensitivity of a receiver is often described in terms of its *noise factor* which does not depend upon its bandwidth).

(5) Finally there may be some noise contributed by mains hum due to insufficient smoothing in the mains power supply or poor insulation of valve heaters. This form of noise is seldom a limiting factor on a good communication receiver and when encountered can be cured by more efficient smoothing or valve replacement.

Before leaving the subject of sensitivity the importance must be stressed of always presenting to the input terminals of the receiver the best possible signal, provided that this does not cause overloading of any stage. This may seem self-evident but it is not always appreciated that any "gain" that can be achieved at this end of the receiver—either by improving the antenna or its coupling (ie matching) to the first tuned circuit—will be noiseless gain, an ideal form of amplification that can never be repeated elsewhere in the receiver.

To give an idea of practical performance, any hf receiver with a 10dB signal-to-noise ratio with an input of 1–3μV is in the high-quality class, while if you can obtain this figure for a 5μV signal you will not miss many signals. At vhf the lower external noise makes it possible to use more sensitive low-noise receivers.

## Ease of tuning

For many applications, the ability to tune a station in easily and accurately, to tune away, and then to be able quickly to re-set the receiver back to the original station is an even more desirable characteristic than extreme sensitivity. For illustration, we need only to think once again of our imaginary all-band broadcast receiver; on most sets of this type the entire (20m) amateur band, although 350kHz wide, will occupy less than ¼in of a dial which has to cover all frequencies between 6MHz (50m) and 18MHz

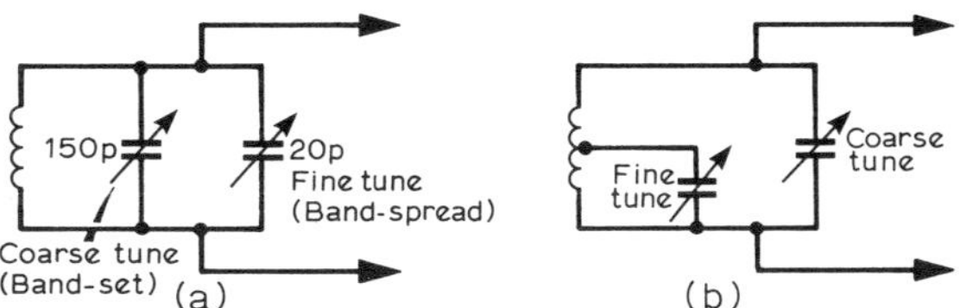


Fig 6. Alternative methods of providing electrical bandspread tuning. (a) Small capacitor connected in parallel across the main tuning capacitor. (b) Tapped coil, this system can be made to provide different degrees of bandspreading to suit individual amateur bands

(16m). It is usually almost impossible to re-set such a dial to a particular amateur station unless it has some outstanding characteristic. What is needed is to spread out the amateur band so that it occupies much more space on the dial, and this process is called bandspreading. For instance if we could arrange to eliminate all the unwanted portions of the short-wave band and leave just the 14MHz amateur band occupying almost the entire dial then it would be very much easier to tune or re-tune to a given station, more especially if the dial was accurately calibrated. In practice bandspreading can be achieved by placing across the main tuning capacitors (which may vary in capacitance anything between 150 and 500pF), much smaller tuning capacitors having a variation of about 20pF (Fig 6). Alternatively, instead of electrical bandspread—as the method just described is termed—much the same final result can be achieved by increasing the gear ratio in the slow-motion tuning mechanism and coupling this with some device to lengthen the dial itself. However if this method is adopted great care will be needed in the design and construction of the gearing to avoid "play" or "backlash" which would prevent accurate setting.

Most good communication receivers use a combination of electrical and mechanical bandspread so that a complete rotation of the tuning knob will cause the frequency to be changed by only a few kilohertz, and this permits accurate re-setting to any given frequency. Some extremely ingenious mechanisms have been developed both commercially and by amateur constructors to provide an effective scale length of the dial running to many feet.

## Stability

It is little use having an accurate dial and a high degree of selectivity unless the receiver is stable. There are several forms of instability, all of which tend to become worse as the frequency increases. First there is the steady drifting of the frequency of the local oscillator so that any given station appears to be gradually changing frequency; the set thus requires periodic retuning or else the station may be lost altogether. This drifting is caused mainly by the effect of heat on the oscillator components and can be minimized by careful layout of the receiver, by adequate ventilation and by the inclusion of compensating components chosen to balance out the changes in inductance and capacitance. However, even with the most careful design there will usually be some drift during the first 10–15min after switching on a cold receiver; but in a good design the set should then settle down and no further significant change in frequency should occur.

It should be noted that after undergoing a number of heat cycles (that is to say warming up and then cooling down several times) some components do not return precisely to their original values. This is one of the reasons why it is difficult to maintain accurate calibration of a bandspread dial over a long period. Some receivers include a crystal-controlled oscillator of high-stability providing marker signals with which the calibration can be regularly checked and adjusted. A calibration marker of this type is also extremely useful where the set has both a coarse main

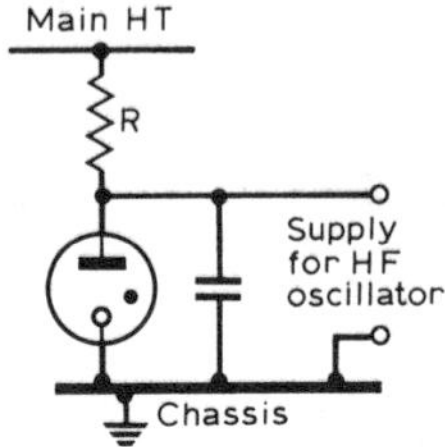


Fig 7. The use of a neon voltage regulator tube to stabilize the ht voltage to the hf oscillator stage. R must be calculated to provide a standing current of about 5mA through the regulator tube, taking into account the current drawn by the oscillator valve

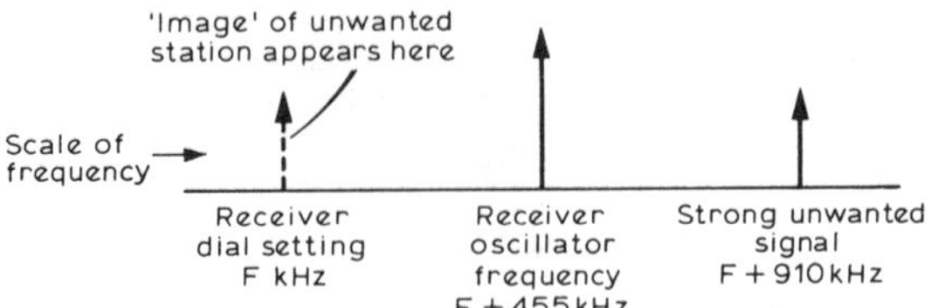


Fig 8. How a strong unwanted signal at the "image" frequency is received

(bandset) tuning knob and a fine (bandspread) tuning knob as it enables the main dial to be re-set accurately.

Another form of instability which cannot be ignored is mainly mechanical in origin. Even if you have not already done so, you are sure to come across a receiver which suddenly shifts slightly in frequency when subject to any form of mechanical shock, or even a sudden burst of electrical interference. It is difficult to eliminate completely the effects of vibration on a tuned circuit but good construction with a heavy chassis and suitable mountings will reduce this to insignificant proportions; tuning circuits must be wired solidly with no long leads left to vibrate.

A third type of instability is that of interacting controls. This usually takes the form of a gradual change in frequency when a gain control is varied, and in a valve receiver is almost always due to insufficient regulation of the ht applied to the oscillator. Fortunately this effect can be largely overcome by fitting a neon voltage regulator to stabilize the oscillator ht (Fig 7).

## Spurious signals

One of the major defects of a simple superheterodyne receiver is that the same signal can often be heard on at least two settings of the dial; one considerably stronger than the other. As the frequency increases, the relative strengths of these two signals become more nearly equal. In effect this means that loud commercial and broadcasting stations can often be heard within the amateur bands and greatly reduce the chances of hearing a weak amateur station free of interference. There are a number of causes of these spurious signals, as they are called, and these are described in detail in most standard text books. Here, however, we must confine ourselves to the most common cause; the reception of a station operating on what is called the "image" frequency.

The elementary principle of the superheterodyne receiver is that when two signals of different frequency are mixed together they combine to produce a signal on a frequency equal to the difference in frequency of the two signals. One of the original signals will be that of the required station, the other will be that produced in the local oscillator in the receiver which will be tuned to track a fixed number of kilohertz, equal to the i.f. of the receiver, away from the signal frequency. The oscillator is usually (but not always) on the hf side of the signal frequency. Supposing our local oscillator is tuned to $F1$ kilohertz and the set has an i.f. of $F2$ kilohertz, then the frequency to which the receiver will be tuned is equal to $F1 - F2$ kilohertz, and this is the frequency to which the circuits in the mixer and any rf amplifier stages will be trimmed. But unfortunately there is another frequency on which a signal would beat our local oscillator to produce a signal on $F2$ kilohertz: this is $F1 + F2$ kilohertz (since $(F1 + F2) - F1 = F2$) and a little calculation will show that this second frequency differs from that to which our receiver is tuned by $(F1 + F2) - (F1 - F2) = F1 + F2 - F1 + F2 = 2 \times F2$ kilohertz. In other words if the two signals differ by twice the i.f. of the receiver, they may both be heard. Now this means that if signals on either of these two frequencies reach the mixer valve or transistor, they will be converted to the i.f. of the set, and will be amplified in the i.f. stages without there being any way in which we can distinguish between them. It is thus vitally important to prevent the signals on $F1 + F2$ kilohertz (the "image" frequency) from appearing at the input of the mixer stage, and the only way in which we can do this is by means of tuned circuits which will accept the signals on our wanted frequency of $F1 - F2$ kilohertz while rejecting signals on our image frequency of $F1 + F2$ kilohertz. Now in the earlier discussion on selectivity it was shown that as frequencies increase, a fixed difference becomes relatively less important. For example, supposing that the set has an i.f. of 450kHz, and that when tuned to 4,000kHz the oscillator is on 4,450kHz, then the "image" frequency will be 4,900kHz: Fig 8. The pre-frequency changer tuned circuits have to discriminate between signals on 4,000 and 4,900kHz, a percentage difference of about 22·5 per cent. Should we now tune the same receiver to 28,000kHz, the image frequency will be 28,900kHz, a difference of only just over 3 per cent, so that more tuned circuits would be needed to provide equal rejection of image interference. This explains why the protection against such interference becomes progressively less in any given receiver as the frequency is increased. It can be similarly shown that if we lower the i.f., so that the signal and image frequencies become closer together, the protection also becomes less.

Here then is the dilemma in which a set designer is placed when choosing the best intermediate frequency for a receiver. In the section on selectivity it was shown that the *lower* the i.f. the easier it is to obtain high *selectivity*; but now it can be seen the *higher* the i.f. the better will be protection against *image* signals. Before explaining one method of overcoming these opposing needs, practical results in general terms are as follows.

With no rf amplification there will generally be only one tuned circuit before the mixer, and even this may be heavily damped (ie made less sharp) by the effect of the antenna; with an i.f. of 470kHz such a receiver would almost certainly suffer badly from image interference above about 3·5 or 7MHz. With one rf amplifier (two tuned circuits) there would be a great improvement, but image responses of powerful stations would almost certainly be strong enough to be a nuisance on 14MHz and above; two rf stages (three tuned circuits) accurately adjusted, would give good protection even on 30MHz. With an i.f. above say 1,600kHz good image rejection can be obtained with two tuned circuits (ie one rf amplifier), while even one well-designed circuit (no rf stage) would give reasonable protection on 14MHz and below. In general terms, the "image" response of a given signal compared with the response on its proper frequency needs to be reduced by 30dB (32 times) on 30MHz for the receiver to begin to come into the high-performance class.

## Double superheterodyne receivers

The conflicting desire to have a low i.f. for selectivity and a high i.f. for good image rejection has led to the present popularity of receivers having a double change of frequency. The incoming signals are first converted to a fairly high i.f., of the order of 3MHz or above, providing a high order of image rejection with say one stage of rf amplification, and thus avoiding the high cost of many rf tuned circuits. Then in turn these signals are changed,

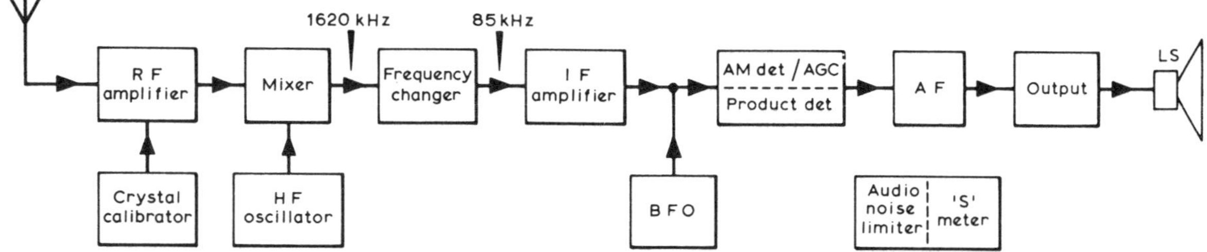


Fig 9(a) Block outline of a modern double-conversion communication receiver designed for amateur radio use

with the aid of a fixed-tuned local oscillator, to a much lower frequency, of the order of 50–100kHz (or occasionally to the standard i.f. range of 455–470kHz). It is at this second i.f. that the main amplification takes place and high selectivity can be achieved. Thus this double-conversion system (see Fig 9(a)) can offer both good image protection and high selectivity, but at the cost of added complexity in design and construction. Great care is necessary if spurious signals resulting from the harmonics of the several oscillators within the set are to be avoided completely; there is also some risk that the full benefits of the high selectivity may be lost owing to cross-modulation occurring due to overloading one of the earlier stages.

There is one interesting form of double superhet used in some of the more elaborate receivers. In this the frequency of the first local oscillator is fixed for each band, usually by a crystal, and the output from the first frequency changer represents a "spread" of frequencies. The first i.f. section is made tunable over a definite band, eg 2–3MHz and the calibration and spread for each band thus remains exactly 1MHz wide, and direct calibration can be provided with a high degree of accuracy; a very wide band such as 28–30MHz would be split into two bands each 1MHz wide. This type of design has become much more popular since the spread of ssb with its need for high stability. This principle is frequently used by amateurs when employing an hf or vhf converter in front of a lower-frequency communication receiver; the converter is fixed-tuned with the actual tuning being done on the main receiver: see Fig 10.

A common technique is to use a vhf crystal-controlled converter ahead of the hf receiver on which all tuning is carried out. This principle is outlined on p14, but for 144MHz, it is usual to tune the hf receiver over the range 28–30MHz while covering 144–146MHz. Some amateur-bands-only hf receivers make special provision for tuning over a 2MHz or 4MHz band to cover the European and American 144MHz bands when used with a converter.

Because of the lower-noise level of the vhf as compared with hf bands, considerably more attention has to be paid to designing the converter to have good low-noise characteristics.

Suitable vhf converters can be built or purchased ready built. Designs appear frequently in *Radio Communication* and full details of vhf reception are given in the RSGB *Radio Communication Handbook* and in the RSGB *VHF-UHF Manual*.

## Alignment errors

It will be readily appreciated that any receiver which has many tuned circuits in the form of i.f. transformers, and rf and oscillator tuning circuits can provide optimum performance only when all these circuits are accurately adjusted. The i.f. transformers must all be peaked to the same frequency, and if a crystal filter is fitted, the peak of the transformers *must* coincide with the filter frequency. Only sturdy construction and the use of high-grade components can bestow the ability to retain this accuracy of adjustment over long periods. For this reason regular re-alignment of a receiver will almost certainly prove beneficial.

In the front-end of a receiver, one of the main problems in superhet design is to obtain accurate tracking of the oscillator with the signal-frequency circuits, so that these always differ by the constant figure of the i.f. In practice, perfect tracking cannot be achieved over a wide band, which means that the signal-frequency circuits will be slightly off tune. In most designs exact alignment is possible only at three evenly spaced points within each waveband and a certain amount of error has to be accepted over the remaining portions of the tuning range. These errors can be minimized by keeping the total frequency coverage of each waveband as narrow as possible; the situation will be improved if the amateur bands can be made to coincide with the accurately-tuned portions. For this reason, despite the added complications in the coil assemblies, it is usually preferable that a general coverage receiver should have as many wavebands as possible, even if a good electrical or mechanical bandspreading system

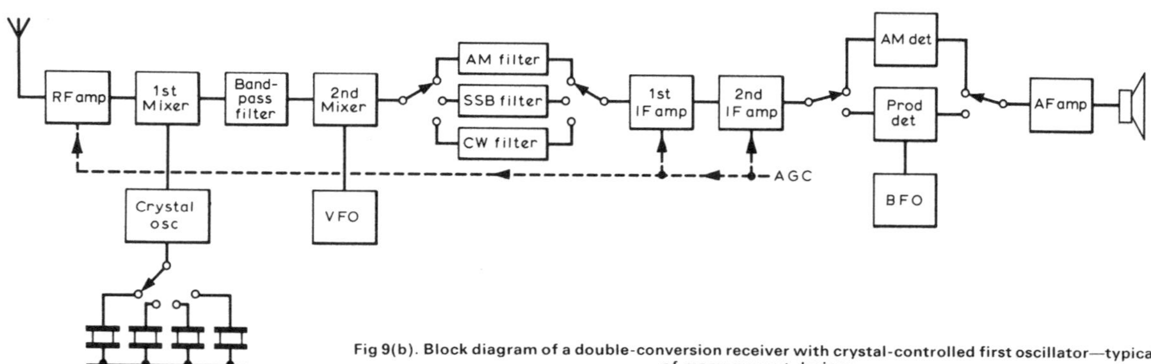


Fig 9(b). Block diagram of a double-conversion receiver with crystal-controlled first oscillator—typical of many current designs

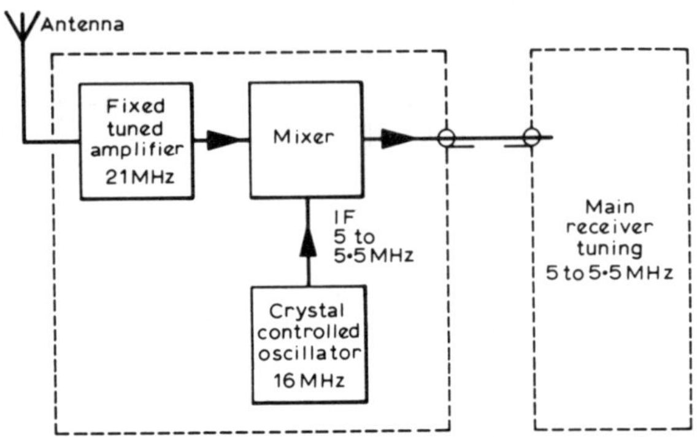


Fig 10. Showing how a fixed tuned converter is used in conjunction with a communication receiver. All tuning is done on the main receiver. A similar system is frequently used to receive 144MHz and other vhf bands

makes this unnecessary from the viewpoint of ease of tuning. Receivers which tune the amateur-bands only can keep alignment errors to a very low figure.

It has already been stated that the first tuned circuit in a receiver is of especial importance in determining the final signal-to-noise ratio; however the alignment of this circuit presents particular problems owing to the effects produced when coupling different types of antenna to the receiver. If the antennas are reactive their connection will affect the alignment of the first tuned circuit. A satisfactory method of overcoming this difficulty is to fit a panel control antenna trimmer to allow this circuit to be brought into accurate alignment by the user whenever changing bands or antennas. Alternatively, the set may be intended for use with a low-impedance coaxial cable feeder only (requiring an external antenna tuning unit for high-impedance antennas).

## Homodyne "direct-conversion" receivers

A form of communication receiver which is of increasing interest to amateurs is the homodyne, synchrodyne or "direct-conversion" receiver. This certainly offers useful advantages over superhets for the construction of high-performance, yet relatively simple hf receivers which can be realized using valves, bipolar transistors, field-effect devices or hot-carrier (Schottky) diodes for the balanced linear detectors.

The direct-conversion receiver may be considered as a superhet with an i.f. of 0kHz or alternatively as a "straight" receiver having a good linear product detector. In a superhet, the incoming signal is "mixed" with a local oscillator and the i.f. represents the difference between these two frequencies; as the two signals approach one another in frequency the i.f. is lower and lower. If this process is continued until the oscillator is precisely the same as the incoming signal, then the output will be at *audio* frequency; in effect one is using a frequency changer to demodulate the signal. But there is a snag when doing this with an a.m. double-sideband signal: strictly speaking the local oscillator needs to be *exactly* the same as the signal frequency to the degree known technically as *phase coherence*—ie the signal needs to be exact not only in frequency but also in phase, a condition which cannot be achieved with a free-running oscillator.

However, this exact phase synchronism is not at all essential for receiving ssb signals or cw, and this makes it possible to build very simple direct-conversion receivers. Furthermore, despite what we have just stated, such a receiver *can* often be used to receive amateur a.m. stations by means of what is called *exalted-carrier* detection.

Direct-conversion receivers can be built using either semiconductors or valves, and can give results which are appreciably better than simple "straight" receivers with regenerative detectors—and better than many simple superhets, although one must not expect a simple direct-conversion receiver to outperform a high-grade communication receiver.

Fig 11 shows the basic outline of a direct-conversion receiver: the incoming signal goes to a linear detector (Fig 12), either directly or via an rf amplifier, where it is mixed with the output from a *stable* local oscillator tuned to the signal frequency, and the resulting audio output passes via an audio filter to a very high-gain audio amplifier (virtually all the gain may be supplied by the audio amplifier).

Provided that the mixer-detector is truly linear (and it is often an advantage to use a *balanced* detector), it is possible to govern the selectivity of the receiver effectively by limiting the bandwidth of the af filter and amplifier. This thus provides a very much more economical approach to selectivity than using a crystal or mechanical i.f. filter. A good direct-conversion receiver means that one has found a way of providing a "straight" receiver with as high a degree of selectivity as one requires.

The newcomer may find it puzzling that this can be done with a direct-conversion receiver when apparently it cannot be done with a conventional straight receiver. The reason is to be found in the difference between the non-linear regenerative detector and the linear detector of the dc set which preserves the incoming signals, without blocking or intermodulation or cross-modulation, up to the audio filter.

A disadvantage of the simple form of dc receiver, when compared with a good superhet, is that there is no way of rejecting the "audio-image" to provide true single signal or ssb reception. However for constructors prepared to face the rather greater complexity of the "two-phase" direct-conversion receiver (which uses the phasing method of generating ssb in reverse) even this disadvantage can be overcome. The most demanding feature of the dc receiver is the need for a very stable local oscillator which is not affected by incoming signals.

Many designs for simple direct-conversion receivers have been published and there is no doubt that useful receivers, particularly for 1·8MHz and 3·5MHz where oscillator stability is less of a problem, can be constructed very inexpensively. Furthermore this is a technique that is still developing, and plenty of pioneering work remains to be done. This is not to say that the high-grade superhet is likely to give pride of place to direct conversion, but undoubtedly an amateur seeking an interesting project could do far worse than try building a direct-conversion receiver.

Because, in a direct-conversion receiver, the local oscillator is tuned to the frequency of the incoming signal, this opens the way for a further application of such receivers: this is to form the vfo for a simple transmitter (Fig 13). A number of transceivers of this type have been described, and at least one USA company has marketed a series of low-power battery-operated transceivers using this principle. It has even been shown that a simple fixed-channel hf transceiver can be built using a crystal-controlled oscillator.

Full details of a 3·5MHz dc receiver are given on pp 34–38.

## Transistorized receivers

During recent years, hf communication receivers and vhf converters based entirely upon transistors and other

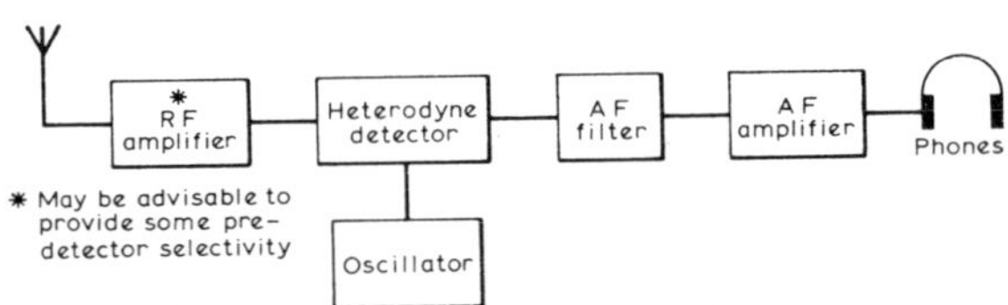


Fig 11. Outline of simple direct-conversion receiver

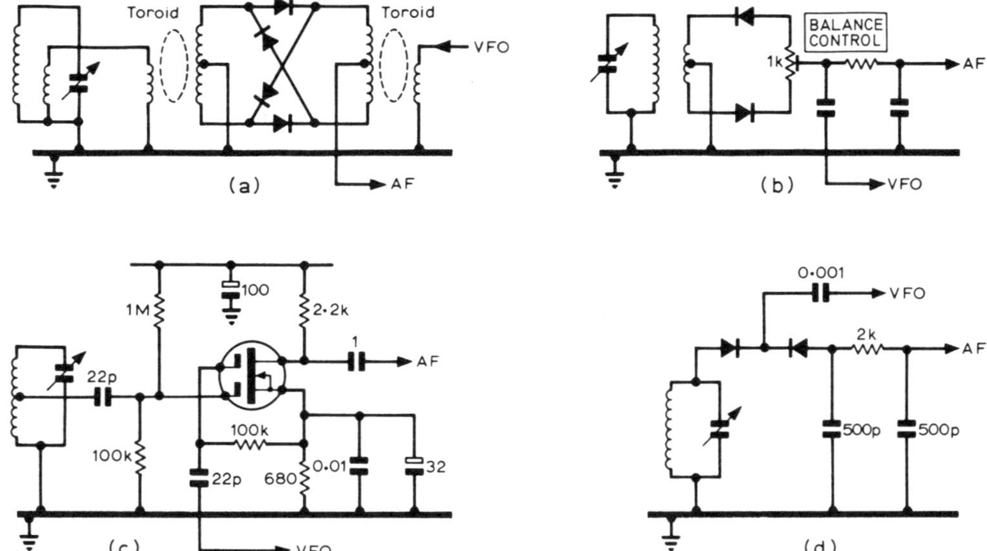


Fig 12. Linear product (heterodyne) detectors suitable for use in direct-conversion receivers (a) Double-balanced (b) Balanced (c) Dual-gate mosfet (d) Diode. For optimum results Schottky (hot-carrier) diodes should be used, but conventional germanium diodes can be used at 3·5MHz

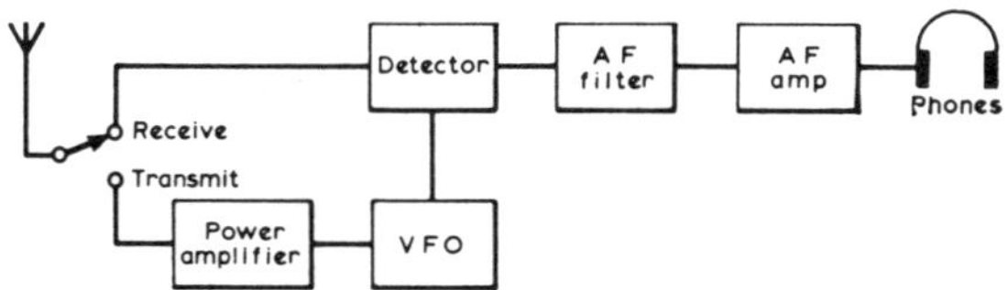


Fig 13. Showing how a direct-conversion receiver can be used as the basis of a simple transceiver

semiconductor devices have been developed both commercially and by amateur constructors. These offer the advantages of compact size, and greater reliability (provided the transistors are operated under correct conditions); and since they draw only low power, from low-voltage supplies, they can be optionally operated from dry batteries or car accumulators. Simple sets, Fig 14, can be made at considerably less cost than a mains-operated valve set of comparable performance.

Not only are semiconductors significantly more reliable than valves and less subject to ageing but modern types can have better noise-characteristics right up to uhf, and the low operating voltages and reduced heat make for more reliable and consistent performance of the associated components. The small size of the semiconductors and the complete absence of heater wiring and large holes for valveholders simplify the construction of receivers and converters; alternatively much more complex circuits can be put into a similar sized cabinet in order to improve frequency stability etc. Battery sets will usually prove a good deal cheaper due to the omission of the mains transformer and other power-supply components, especially since new transistors are available at less cost than valves. This also means that there are none of the safety problems associated with the construction and testing of a first mains-operated receiver.

Thus simple hf and vhf transistorized receivers and converters can be built easily and cheaply, and are generally used for mobile and portable work, and now often find favour for fixed station use. There are also "hybrid" designs in which some transistors and some valves are incorporated.

With the many apparent advantages of transistors it may be asked why the majority of high-grade amateur communication receivers are still based on valves, and whether this is likely to continue. Undoubtedly there will be a trend towards many more all-semiconductor receivers and converters, but it must be recognized that there are still some important problems in the design of really high-grade transistor receivers. It is only recently that designs are beginning to appear which offer performance equivalent to that of the best valve sets, and some of these are rather beyond the pocket of most amateurs.

The main problem with readily available transistors is that of

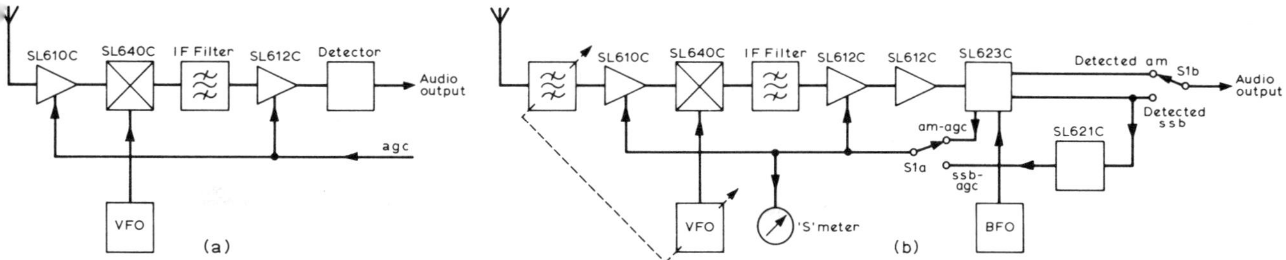


Fig 14. (a) Basic superhet receiver (b) Complete a.m. ssb superhet receiver using ICs

achieving a really wide dynamic range of the front-end of the receiver—that is to say a front-end to a receiver which will handle both very weak and very strong signals without introducing various forms of cross-modulation and blocking. This is partly because there is no near equivalent among conventional bipolar transistors to the variable-mu pentode valve and this makes it difficult to obtain really effective agc characteristics unless one uses unipolar devices such as the fet mentioned below.

Other problems include increased circuit loading due to lower input impedances; feedback capacitances that may require neutralizing; the variation of semiconductor characteristics with temperature; the fact that, unless protection circuits are fitted, transistors can be damaged by large input voltages due to static build-up on the antenna, impulse interference or a nearby transmitter; the greater spread of characteristics between devices bearing the same type number than with valves; and the need to pay greater attention to accurate impedance matching through all stages of the receiver.

This may seem a formidable list, but in fact many of these problems have been overcome by the use of low-noise "field-effect transistors"(FETS) and the special forms of these known as IGFETS (insulated-gate fet) or MOSTS (metal-oxide semiconductor transistor). These devices have extremely high input impedances and have characteristics which make them much less susceptible to cross-modulation so that they resemble vari-mu pentode valves in this respect. They can also act as extremely efficient low-noise mixers. A single fet receiver can be as effective as the once popular O-V-O. All semiconductors, of course, have to be treated with some care and the precautions listed on p17 should always be observed.

Both junction-gate (jugfet) and insulated-gate (igfet or mosfet) devices eminently suitable for use in the front-ends of hf and vhf receivers and converters are available at reasonable cost. For example, the 2N3819 and the rather more expensive 2N3823 series of junction FETS by Texas Instruments have proved most popular. In addition both single-gate and dual-gate (tetrode) MOSFETS for this application have been introduced by RCA. For example, the dual-gate 40673 can provide a cascode-type circuit with a single device, providing an hf or vhf amplifier with very good cross-modulation and also agc characteristics.

Of the various configurations, the cascode-type circuit using either two junction FETS or a single dual-gate mosfet is probably the most attractive. Since field-effect devices have characteristics and impedances closely akin to valves it has been found possible to convert some stages of valve receivers to such devices as a fairly simple modification, and sometimes without requiring full re-alignment of the receiver.

It should be noted that with some MOSFETS (IGFETS), special handling precautions are essential until after the device has been wired into circuit, to prevent static charges building up across the very high input impedance and puncturing the insulation layer. It is always advisable to keep the leads short-circuited until wired in place, and some devices are fitted with small spring clips which can then be removed when the wiring is complete. Most modern MOSFETS now have built-in diode protection against static charges.

The i.f. and af circuits of transistorized receivers will, almost certainly, increasingly use in future *integrated circuits* (ICS) in which large numbers of active transistor devices, together with resistors and capacitors and all interconnecting leads are formed on a single tiny slice of silicon. IC devices suitable for use in the linear circuits of receivers are available at reasonable cost. These devices assist the constructor since they allow much of the complex circuitry to be obtained in the form of tiny devices about the size of transistors. They offer considerable scope for the use of more complex circuitry without increasing size, constructional problems, or power drain. More predictable and reproducible designs have become possible.

The availability at reasonable prices of linear integrated circuits suitable for use in communication receivers is providing the amateur with new opportunities for the home-construction of receivers. For example, the Plessey SL600 series are suitable for high-performance receivers, each device providing sufficient gain and avoiding the need to build several separate stages in say the i.f. amplifier section. A receiver based on these devices could comprise an SL610 rf amplifier; SL641 double-balanced frequency changer; SL612 i.f. amplifier; SL623 combined a.m. and ssb demodulator; SL630 audio amplifier and SL621 "hang" type agc system based on the audio signals and suitable for ssb/cw/a.m. reception. The line-up would be completed by a vfo based on a separate transistor or fet and probably an isolator stage between this and the frequency changer. The construction of a receiver based on such a series of "modules" would normally involve much less time and effort than would be needed using many separate transistors to give comparable performance.

A much lower-cost approach would be to use one of the linear integrated circuits now being made for domestic and car broadcast receivers as the complete "heart" of a communication receiver or tunable i.f. system. Examples of these devices include the Mullard TBA570; SGS TBA651; RCA CA3088E; Toshiba TA7046P; Siemens TBA640. These are eminently suitable for use with some of the ceramic filters now available. Generally, for highest stability, it is better to use a separate device as the local oscillator even when provision is made for a "built-in" oscillator in the ic—this is because an ic contains many active devices all consuming some current and producing heat, resulting in thermal drift.

## OPERATING THE RECEIVER

Part of the interest in operating a communication receiver lies in acquiring the skill to obtain the very best results of which it is capable. No two receiver designs are exactly the same, but there are a few simple hints which are worth learning.

The basic difference between a communication receiver and a broadcast receiver is the provision of a bfo to allow reception of ssb (A3J) and cw telegraphy (A1). On a receiver without a bfo, ssb signals are completely unintelligible while cw reproduces only as a series of thumps.

For ssb, the bfo (or carrier insertion oscillator as it is sometimes called) provides a substitute carrier to allow the signal to be demodulated. To make the speech intelligible this carrier has to be within about ±100Hz and preferably ±25Hz of the original (suppressed) carrier that was generated in the

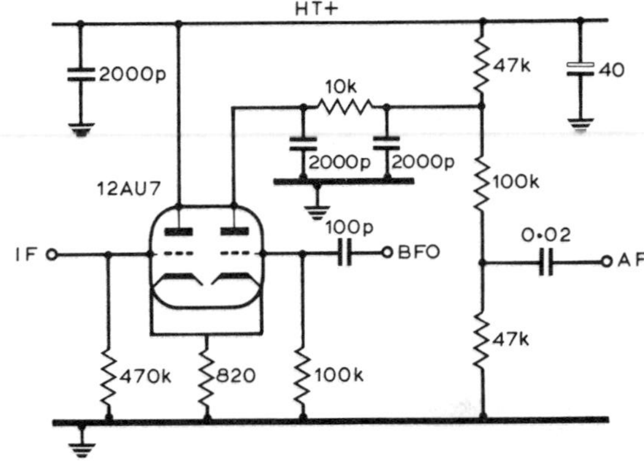


**Fig 15. The product detector has become popular for the reception of ssb and cw signals. Suitable valves are the 6SN7, 12AU7 or ECC82**

transmitter. If the receiver is fitted with a good ssb selective filter with a bandwidth of from about 2·3 to 3kHz ("nose" bandwidth) it is also necessary for the carrier to be inserted in the receiver on the correct side of the filter response, depending upon whether upper or lower sideband signals are being transmitted. Many receivers now incorporate a bfo using two crystals of appropriate frequency that can be switched to select upper or lower sideband reception.

The above requirements mean that the operator has to be able to tune his receiver very precisely to an ssb signal and that any subsequent oscillator drift in either the receiver or transmitter causes severe distortion of the signal. Except for the highest-performance receivers with precise frequency-setting arrangements, it will usually be necessary for the newcomer to obtain some experience at tuning ssb signals and cultivating the "safecracker's-fingers" that this may entail. To try out a receiver on ssb it is worth noting that amateur signals can be received at most times of the day and evening between 3,600 and 3,800kHz or, except when the band is dead, between 14,100 and 14,350kHz.

Until the relatively recent development of low-cost frequency synthesis and digital read-out systems (which can sometimes present problems if not well designed) the accuracy of tuning and the ability to set a receiver close to a required frequency depended on the degree of electrical bandspread and the provision of good mechanical reduction gearing on the tuning capacitor. For easy reception of ssb (and, with a narrow filter, cw) the tuning rate (that is the number of kHz tuned over during one turn of the knob) needs to be very low, preferably less than 25kHz or so. Tuning rates of this order are seldom found on older models and for amateur operation, where operation may be concentrated around one relatively small segment of the band, it may prove a considerable advantage to fit an additional slow-motion drive (this can often be done without drilling the receiver front panel). By this means tuning rates as low as 2–5kHz can often be achieved with very little backlash.

## Controls

Different communication receivers are fitted with different controls and facilities, but the following notes cover most of those that differ from those found on conventional broadcast receivers.

**Send/receive switch:**
This control is fitted to facilitate the use of the receiver in conjuction with an associated transmitter, rendering the receiver silent ("muted") without affecting the thermal condition of the valves or semiconductors. Modern receivers often combine this as a stand-by switch with the main "on-off" switch and a further position for calibration to bring into operation an internal crystal calibrator. In operation, send/receive switches are often not used during two-way operation, the receiver "muting" being effected automatically by means of voice-operated switching or other automatic methods.

**Gain controls**
Most communication receivers incorporate separate rf/i.f. and audio gain controls; occasionally an rf attenuator may also be fitted. The use of these controls enables a wide range of signals to be handled with minimum overloading (including the overloading of the product (synchronous) detector) while maintaining good signal/noise ratios. Normally the af control is set to provide a comfortable audio output and adjustments during tuning are carried out on the rf/i.f. control; where cross-modulation appears to be occurring in the presence of strong signals, the rf attenuator should be set to reduce the strength of signals reaching the amplifying stages of the receiver.

**Mode or filter switch**
This often provides variable selectivity based on the use of different crystal/mechanical or other forms of selective filters. A high-performance receiver might include a position for double-sideband a.m. reception (filter bandwidth 6–10kHz); upper and lower sideband reception, each with a filter bandwidth of 2·3–3·5kHz but with the injection carrier automatically positioned on the correct point of the response curve; a cw filter with a narrow bandwidth of from about 200–500Hz; and sometimes a position for nbfm reception, bringing into operation an fm discriminator detector.

**BFO control**
Where the bfo is not crystal-controlled there is usually a panel knob to tune the bfo about ±5kHz from the nominal intermediate frequency. Correct setting of this control is important since it locates the carrier to the correct side of the passband for ssb and determines the audio tone and effective single-signal reception characteristics for cw. In models fitted with the older form of single-crystal-plus-phasing-control type of filter (still an effective form of filter for cw reception), it is most important that the bfo control is adjusted carefully to precisely the correct frequency that provides a single sharply peaked af tone for cw reception.

**AGC switch**
In older receivers it is usually necessary to turn the agc off during cw/ssb reception since otherwise leakage from the bfo would greatly reduce the sensitivity of the receiver. Modern receivers, often with audio-derived agc designed primarily for ssb reception, use agc for all modes of reception, but often have provision for changing the time-constant of the system to suit cw reception. A good agc system for ssb needs a very fast "attack" time with a slow decay.

**Signal-strength meter**
This normally provides an indication of the relative strength of the incoming signals, for instance based on potentials appearing in the agc system. They are often calibrated in "S-points" and "decibels above S9" from the RST signal reporting code, but it must be appreciated that there is no universally recognized standard of S-levels and the scales adopted by individual manufacturers may vary all the way from about 3–6dB/S-unit; furthermore the S-reading will reflect the directivity and gain of the receiving antenna so that to receive a report of "S9 on my meter" cannot be considered an "absolute" signal report; nevertheless the S-meter is very useful for comparing signals from different stations or during adjustments to the antenna etc.

## Reception of cw through a crystal filter

Certain piezo-quartz crystals possess the property of resonating at a frequency which depends upon their dimensions. By means of chemical etching and "grinding" it is possible to produce crystals tuned to almost any desired frequency. The crystal may be regarded as the equivalent of a series-tuned circuit of extremely high "goodness" or *Q*. When the frequency of a crystal is made to coincide with the i.f. of a receiver it can form a very sharply tuned "acceptor" filter, with an adjacent parallel tuned rejector circuit formed by the capacitance of the metal plates in which the crystal is held. The exact setting of the rejector frequency can be adjusted by including a variable capacitor in the circuit, and this is known as a "phasing" control. The resulting i.f. response curve of a receiver with a crystal filter will be similar to that shown in Fig 16.

The degree of selectivity provided by a single crystal is often

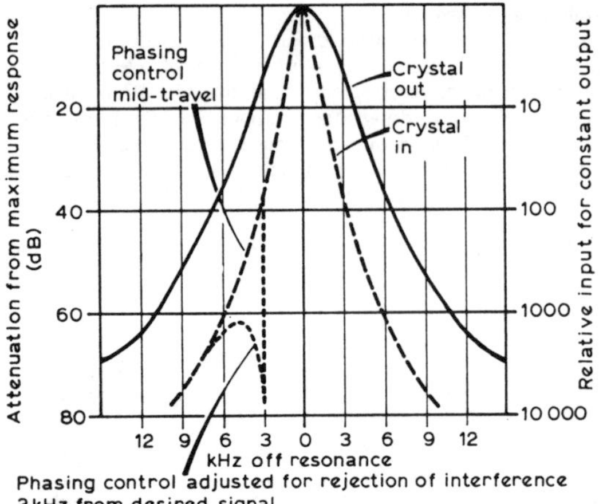


Fig 16. A graph showing the improvement in selectivity obtained by the use of a crystal filter

too sharp for distortionless telephony reception (this limitation can be overcome by using two crystals of slightly differing frequency to form a bandpass circuit or by tone correction) but is highly suitable for cw reception which requires a bandpass of only 100–200Hz.

Now although amateurs played an important role in the development and application of crystal filters, their use is not always popular, especially among newcomers and also among some operators who have not grasped fully the technical principles involved. One often hears such remarks as "When heavy interference came on, I switched in the crystal filter but it reduced the strength of the signals so much that I lost the station". Now a correctly adjusted filter will seldom reduce the strength of a cw signal by any appreciable amount but instead should considerably enhance the signal-to-noise ratio. Why then do some users so often complain of serious loss of signal strength? Simply because the bfo control has not been correctly set in advance; this has meant that the station could not be correctly tuned in with the crystal switched into circuit. The loss of strength in such circumstances is really an indication that the crystal is doing its intended job of narrowing the bandwidth.

It is essential with a crystal filter (Fig 17) that the bfo control should not be used—as it often is—as a form of tuning trimmer but should be set in advance to a frequency between 500 and 1,000Hz higher or lower than the crystal frequency. It should then be varied only when a change in the output note is required.

To set the bfo control, the phasing control should be positioned at about mid-travel and, with the bfo turned "off", a steady carrier carefully tuned in to give maximum S-meter reading (or maximum audible output if there is no meter). At this point the bfo is turned "on" and adjusted to give the beat note—usually between about 500 and 1,000Hz—most acceptable to the operator. This setting of the bfo should be carefully noted, and normally the control should not be touched again, all signals being carefully tuned on the normal tuning control for maximum output. An interfering heterodyne note can then be eliminated by adjusting the phasing control, the effect of which is to move slightly the very sharp rejection "notch" produced by the parallel resonance of the various capacitances across the crystal.

With a set in which the bfo has been correctly adjusted, the switching in of the crystal should cause little drop in signal strength, though of course if the signal has not been exactly tuned in it may be necessary to adjust the tuning slightly to restore the strength of the signal.

A "bandpass" multi-crystal filter is better for speech reception than a filter using a single crystal which is liable to cause some af distortion on account of its very narrow response. Nevertheless even the single-crystal filter is a very useful adjunct, particularly on account of its ability to reject heterodyne interference by means of the phasing control. It should be noted however that since the filter will attenuate the sidebands there will normally be a drop in audio output when the filter is switched into circuit—but the signal-to-noise ratio should improve.

Another device that has become popular is the *Q*-multiplier which depends on the very high *Q* obtainable from a conventional tuned circuit when operated near to the point of oscillation. This type of filter provides variable selectivity and also an adjustable "notch" rejection characteristic. It may be built as an external unit for improving the selectivity of almost any receiver. In operation it is much the same as a crystal filter.

With all highly selective receivers, with or without a crystal filter, there should be a noticeable difference between the strength of the signal at the two tuning points, one each side of the carrier frequency and equally spaced from it, at which the same af beat note is produced. If the selectivity is of a sufficiently high order, the strength of the signal at the second tuning point may be so reduced as to be practically inaudible except for the strongest signals. This condition is known as single-signal reception, and is a highly desirable characteristic for cw and ssb reception, since it reduces the likelihood of suffering interference by one half.

To receive speech through a selective single-crystal filter or *Q*-multiplier a treble-boosting network should be added in the audio section; this may consist of passing the af signal through a 2MΩ resistor shunted by a 200pF capacitor.

Most of the multi-crystal filters used by amateurs have, until recently, been based on the half-lattice configuration which, for bandpass units, requires the use of a number of crystals of carefully defined and different frequencies and the use of centre-tapped inductive components. With the increasing use of filters with centre frequencies between 5 and 10·7MHz, the home construction of ssb filters for transmission or reception becomes increasingly tricky and even keen constructors tend to end up by buying factory-built units and the associated bfo crystals required for upper and lower sideband generation or selection.

An alternative filter configuration— the *ladder crystal filter*— has now been shown to offer significant advantages for the home construction of ssb filters at frequencies between about 5 and 12MHz. Such filters use a series of crystals having the same resonant frequency (and similar crystals can also be used for carrier generation). The ladder filter presents relatively few problems to constructors provided that they are used with the correct impedances and are carefully screened and laid out to minimize the leakage of signals around the filter. Very successful ladder filters can be formed from three, four or more crystals and

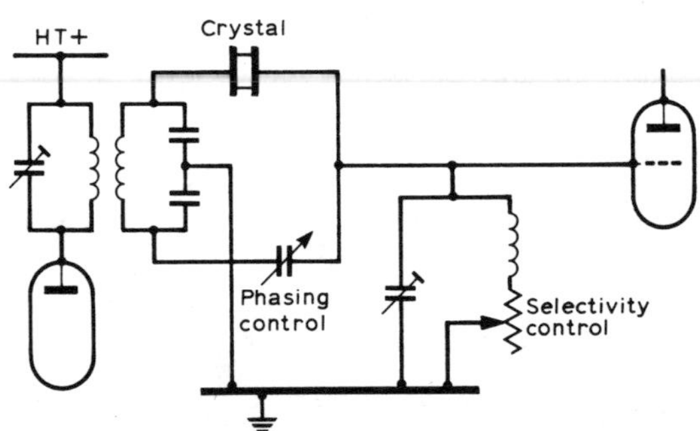


Fig 17. Variable-selectivity crystal filter. For optimum results there must be no stray coupling between the input and output circuits

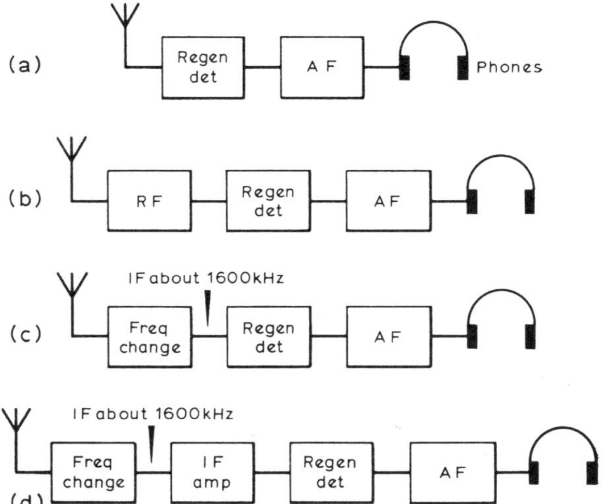


Fig 18. Simple receivers intended primarily for morse reception. (a) Two-stage "straight" receiver (the two stages may be combined in a single multiple valve, such as a double-triode). (b) The addition of a tuned rf stage will improve sensitivity and selectivity (an arrangement often known as a "trf" receiver). (c) The simplest "superhet" circuit. By using modern triode-pentode valves such as the ECF82 or 6U8 all stages can be accommodated in a two-valve receiver. (d) The addition of an i.f. stage will greatly increase the gain and make the receiver suitable for the reception of the stronger telephony stations

a handful of silver-mica capacitors. Typically they can provide an asymmetrically shaped ssb filter with a bandwidth (–3dB) of about 2kHz with low insertion loss and little ripple in the bandpass, providing very high ultimate out-of-band rejection of unwanted signals. Such ladder filters can be constructed at a fraction of the cost of a factory-made lattice ssb filter.

## PRACTICAL DESIGNS

Having read so far, the newcomer may be thinking: "Well, I can see that a good communication receiver is a fairly complex piece of equipment, but does this mean that there is no alternative to buying a ready-made set, and forgetting about constructional work?" The answer is *no*. Emphatically there is no better way than constructing a receiver—even a comparatively simple one—for finding out infinitely more about receivers than you can ever learn purely from reading theory. Even a straight regenerative receiver will teach you a good deal about the finer points to look for in tuning capacitors, slow-motion dials, careful adjustment of regeneration controls, and the like. The simplest of superhets will bring you straight up against the problems of oscillator tracking, i.f. alignment and the reduction of spurious images. And after you have gained some experience on these circuits, even the "full specification" communication receiver is by no means beyond the capabilities of an amateur having only a minimum of test equipment; this is especially true if the switched coil units are obtained as a complete assembly, or alternatively if plug-in coils are used. There are also complete kits available for building amateur-band communication receivers.

Simple superhet and even straight receivers can be quite effective, particularly for cw reception, and there is a considerable scope for modern designs, both for two-way working and for listening. Some suggestions for practical experiments are given in Fig 18. If room is left on the chassis, it is always possible to add extra stages and refinements from time to time.

Alternatively, you may buy a relatively cheap ex-government receiver such as the R1155, R107, CR100, BC348 or even one of the inexpensive "Command" series, and then set about improving its performance by adding such accessories as a crystal filter (if not already fitted), or an external rf pre-amplifier (Fig 19). Or you may extend its tuning range and improve its preformance on the higher frequencies by building an external converter unit for those bands which the original set does not cover or covers with poor sensitivity. Good converters are not difficult to build—they can be put together in an evening or two—and yet can show the constructor almost as much about the problems and principles of superhets as a complete receiver.

A set which is deficient in selectivity may be improved beyond recognition by adding an external second frequency changer coupled to the receiver i.f. strip and followed by a low i.f. amplifier (a device usually known as a *Q5-er*), or alternatively by adding a *Q*-multiplier. Then again, for cw reception it may be possible to improve selectivity by altering the af response and fitting peaked audio filters.

The keen amateur never takes the design of even an expensive receiver too much for granted. All receivers turned out on an assembly line are essentially a matter of compromise whereas amateurs tend to specialize in their interest. For example, the keen cw operator will have very different ideas from the telephony man on what constitutes an ideal receiver. The 1·8 and 3·5MHz user will seldom worry as much about extreme sensitivity as the enthusiast working 21 and 28MHz where the external noise level is much lower.

Then again the valves used in the rf amplifiers and frequency-changer stages of many of the war-time receivers are now relatively out-dated and better sensitivity can often be obtained by redesigning these early stages. Even the thorough overhaul of an older type receiver—replacing "leaky" or inefficient capacitors, checking that resistors have not changed value, and realigning—can be an interesting and instructive experience as well as a most rewarding operation from the viewpoint of improving the performance of a receiver.

This is not to suggest that the newcomer should rush into tampering haphazardly with, say, a Drake 2C, Eddystone EA12, KW201, AR88 or HRO but he should nevertheless always be ready to try a new gadget—and countless useful devices are described in the amateur journals—rather than be content to

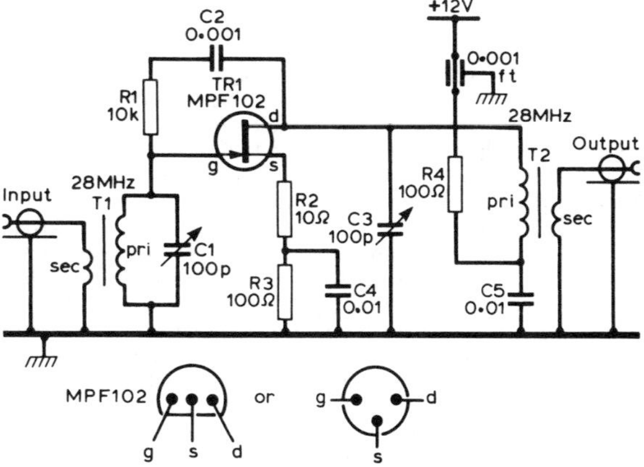


Fig 19. JFET preamplifier for 28MHz (or 21MHz) using degenerative feedback to improve stability in the high-gain common-source configuration. C1, C3: 100pF mica compression trimmers (nominally set about 45pF). T1, T2: 0·6μH with 1½-turn link windings. Typically 12 turns of No 24 enam on T50–6 powdered-iron toroid cores (for 21MHz add two turns to main windings). MPF102 or HEP F0015/HEP802, 2N4416 JFETs can be used

regard a receiver as a mysterious black box labelled "not to be opened."

## Receivers for ssb

The use of the ssb mode of transmission by amateurs—and the likelihood of a further expansion of this system—has brought about considerable re-thinking of the problems of receiver design. Many popular receiver designs of earlier years, while giving good results on A3 telephony transmissions, are not so satisfactory for ssb, usually because of the difficulty of tuning with the necessary degree of accuracy or because of frequency drift. On the other hand a receiver which has been designed to provide good performance on ssb will usually be equally good for cw and a.m. telephony. It is therefore becoming important when planning or choosing an elaborate receiver to pay particular attention to its capabilities on ssb signals.

The main requirements of a receiver for ssb are:

(1) It must be capable of being tuned very accurately. In practice this requires a very low tuning rate (see p 23). (A small handle can be fitted to the tuning knob to reduce the time needed to tune from one end of a band to the other.) Preferably the set should have the same tuning rate on all bands. A slow tuning rate enables ssb transmissions to be accurately tuned in without the need to adjust the bfo control (see p 28).

(2) It must possess the ability to stay accurately tuned to any particular frequency. In other words the drift, other than the almost inevitable "switching on" drift, must be reduced to a very low level and the receiver must not be affected by mechanical vibration etc, otherwise ssb signals even if correctly tuned to begin with will soon become distorted or unintelligible. To meet this requirement calls for all oscillators within the receiver to be highly stable. For instance to maintain a stability of 25Hz at 28MHz calls for much the same order of precision as to keep a watch accurate to within one second per week.

(3) Sufficient injection voltage must be available from the bfo and it should be possible to change the frequency of the bfo to permit rapid selection of the upper or lower sideband position (or alternatively some other means of doing this should be provided).

(4) To obtain optimum benefit from the reduced bandwidth of ssb signals, the nose selectivity of the receiver should not be greater than about 2·5kHz, and preferably, as for any other type of reception, the skirt bandwidth should be as little greater than at the nose as it is possible to achieve.

(5) It should preferably incorporate an agc system which is not affected by the bfo and which has a sufficiently long time constant to enable it to be used on ssb and cw signals.

To provide a slow tuning rate and to obtain good hf oscillator stability, it is popular to use crystal control of the first hf oscillator—a separate crystal being used for each band—and to tune the receiver by means of a variable first i.f. (this is the system outlined on p20). Where a conventional tunable hf oscillator is used, the second frequency changer oscillator is often crystal-controlled, two crystals (one above the first i.f., the other below) being fitted to provide switched sideband selection: other methods of providing rapid sideband selection include: (1) crystal-controlled bfo with one crystal above and another below the final i.f.; (2) making the oscillator of the second frequency changer or the bfo readily tunable over about 5kHz.

While factory-built receivers for ssb work must inevitably be relatively expensive, the keen and experienced constructor can obtain highly satisfactory results at much lower cost. The design, construction and alignment of such receivers, however, are likely to require several months of intensive effort.

Whether you use home-built or factory-built sets, it is worth repeating that a good understanding of the technical design of modern communication receivers will help you make the very best use of your station.

## High stability hf converter

Many of the older communication receivers provide very adequate performance on the lower-frequency bands but lack stability and/or sensitivity and good bandspread tuning on the higher-frequency bands. One satisfactory answer to this problem—enabling an additional useful lease of life to be given to receivers which can usually be acquired quite cheaply as second-hand or government surplus—is to use the receiver as a tunable i.f. on say 3·5MHz in conjunction with a sensitive crystal-controlled front-end converter.

Many suitable designs can be evolved, often with a low-noise dual-gate mosfet or double-triode rf amplifier and possibly an all-diode mixer, but most require a separate crystal for each band, or alternatively reverse the receiver tuning direction on one or more bands. However an economical and simple circuit (Fig 20) of very reasonable performance using only a single 3,500kHz crystal has been described by F. Johnson, ZL2AMJ.

This two-valve unit functions as a crystal-controlled converter on the 7, 14 and 21MHz amateur bands and as a straight pre-amplifier on 3·5MHz. The converter provides a broad-band output on the 3·5MHz band, and all tuning is carried out on the main receiver. It should prove particularly effective in conjunction with an older receiver having good bandspread tuning and stability on the 3·5MHz band. The selectivity will depend entirely upon that of the main receiver when functioning upon 3·5MHz.

One possible disadvantage is that the unit produces a strong spurious marker signal on 7,000kHz which can block the first few kilohertz of the 7MHz band, although this does not occur if the crystal is slightly lower in frequency than 3,500kHz as described later.

The converter comprises the triode section of a 6U8 (ECF82) triode-pentode as a grounded-grid amplifier. The pentode section of the 6U8 functions as mixer with a low-noise contribution. One half of a 12AT7 (ECC81) functions as crystal oscillator; the other section as a cathode-follower output stage following the mixer. There are no tuned circuits at the broadband i.f. of about 3·5MHz so that no ganged tuning circuits are involved.

For 3·5MHz reception the oscillator is switched off and the other stages act as a pre-amplifier. On other bands the crystal oscillator frequency is always 3·5MHz lower than the band being received, so that the signal frequency rises as the main receiver is tuned higher across the 3·5MHz band. For 7MHz the crystal oscillates on its fundamental frequency of 3,500kHz; for 14MHz on its third "overtone" of about 10·5MHz; and for 21MHz on its fifth "overtone" of 17·5MHz. Note that for overtone operation the output is not precisely at the exact harmonic of the crystal fundamental, although close to it, so that a small correction will be needed to the original receiver calibration. Otherwise the 3·5MHz calibration applies in terms of tens and hundreds on all bands, and the tuning rate will be the same on all bands. Thus 7·1, 14·1 or 21·1MHz signals should be received on about 3·6MHz and 7·2, 14·2 or 21·2MHz on 3·7MHz etc.

In order to maintain this approximate 3·5MHz calibration on all the other bands, the crystal needs to be fairly accurately on 3,500kHz. However if this is not regarded as vital, it is often possible to use a crystal some kilohertz lower in frequency. For example a 3,490kHz crystal would tune 7,000–7,300kHz as 3,510–3,810kHz, 14,000–14,350kHz as 3,530–3,880kHz and 21,000–21,450kHz as approximately 3,550–4,000kHz. This tolerance would allow the use of a cheap surplus crystal provided that it is sufficiently active to oscillate readily on its fifth overtone.

Note that only two adjustable coils are used to tune the four bands with L1 left in parallel with L2 on 14 and 21MHz bands.

Construction and layout are not particularly critical provided that rf leads are kept short, and output circuits placed well away from input circuits.

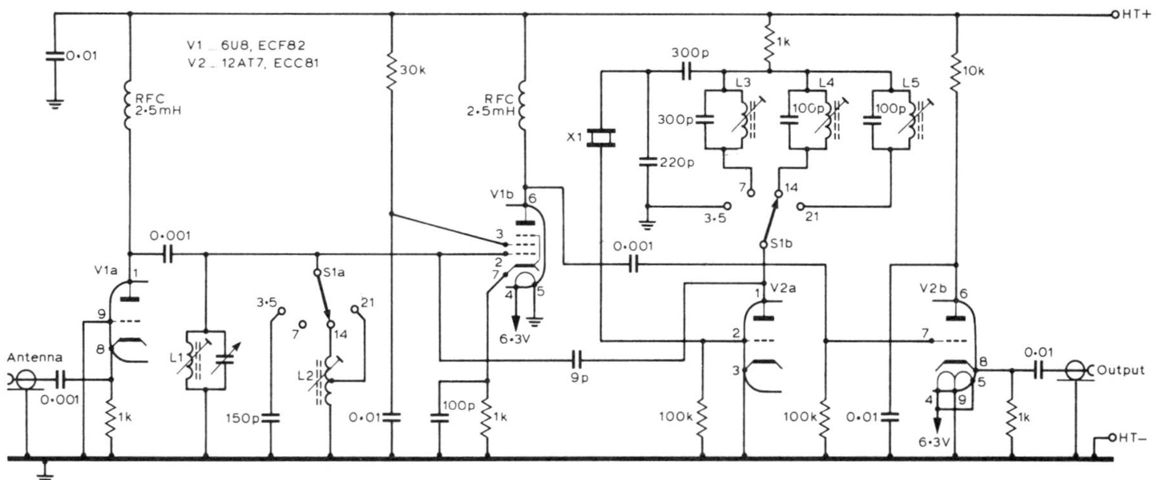


Fig 20. Circuit diagram of the high-stability converter

All coils are wound on ¼in diameter slug-tuned unshielded formers using 30swg enamelled wire: L1 42 turns; L2 26 turns, tapped at 17 turns from the "earthy" end; L3 35 turns; L4 13 turns; L5 8 turns, spaced over ⅜in. Formers could be 0·3in diameter with a slight reduction in numbers of turns—this can easily be determined if a calibrated grid dip oscillator is available. The formers can be mounted around the Yaxley-type bandswitch.

**Alignment**

After completing and checking all wiring, the oscillator circuits must be adjusted so that the crystal oscillates on the correct overtone frequencies. This can be done most readily with a gdo used as an absorption wavemeter, though it should be possible to make do by checking the output on the main receiver if this will tune to 10·5 and 17·5MHz and care is taken not to be misled by harmonic output. With the absorption device check and adjust the cores of L3, L4 and L5 until the oscillator is operating on the required frequencies.

Then set the converter to 7MHz and connect it to the main receiver which is set to 3·5MHz. Without switching on adjust L1 for coverage of 7–7·5MHz in conjunction with the 35pF tuning capacitor by using the gdo in the usual way (this can be done with incoming signals without a gdo but the process is more difficult). Switch to 3·5MHz and check that the rf tuning range now covers 3·5–4·0MHz (or just the European section of the band if this is all that is required). If the tuning range is incorrect, it may be necessary to replace or parallel the 150pF with other values. Once the 3·5MHz tuning range has been established, switch to 14MHz and adjust L2 so that the tuning range covers 14–14·5MHz. It should then be possible to tune the 21MHz band without any further adjustment (the full tuning range will probably exceed the amateur band but this is of no consequence). In practice it will probably prove necessary to change the converter tuning only when tuning right across the band; when searching around only a small section of the band the control can be left unaltered.

Once the tuning ranges have been adjusted, the unit should then be ready for use; signals can be peaked with the converter tuning control when necessary; otherwise all tuning is done on the main receiver. If the gain of the hf converter should be excessive on some signals it would be possible to fit a gain control on the cathode-follower stage.

One possible difficulty is that there may be i.f. breakthrough from strong local 3·5MHz signals when receiving on the other bands. Screening of the converter and the use of coaxial cable with screened connectors between converter and receiver may be sufficient to overcome any difficulties. Should this not be the case, it may in some cases be necessary to fit a further tuned circuit in the form of an antenna tuning unit between antenna and converter.

Provided that the main receiver is efficient on 3·5MHz this combination should provide excellent sensitivity and stability on all the bands concerned.

Fig 21 shows a solid-state version of this converter as used by a New Zealand listener, Gary Moles, ZL2125, together with the outline of a pre-selector which could be used with either version in order to minimize the problem of direct breakthrough.

The original unit is housed in an old tea-tin with the crystal oscillator in its own semi-shielded compartment. Most of the components are mounted on Veroboard, and the coils mounted on the bandchange switch, but layout is not critical.

In setting up, the first task is to ensure that the crystal oscillator is functioning correctly on 3·5MHz (for 7MHz band), 10·5MHz (for 14MHz band) and 17·5MHz (for 21MHz band), and this can be done by using a gdo as an absorption meter while adjusting the slug cores. Having set the three oscillator cores, switch to 7MHz and adjust L1 for coverage of 7 to 7·5MHz with the "rf tune" control, this time using the gdo as a gdo. Next switch to 3·5MHz and, if necessary, alter the value of C1 by substitution or paralleling additional capacitance until 3·5 to 4MHz can be peaked by the rf tune control. Switch to 14MHz and adjust L2 for 14 to 14·5MHz coverage, and similarly adjust L3 for 21 to 21·5MHz coverage when switched to the 21MHz band.

Gary Moles found the pre-selector filter necessary when using a good antenna, and his was built using a couple of toroids and a dual-gang capacitor. As this was originally developed for another application it may be that a rather simpler high-pass filter would suffice.

## A 3·5MHz direct-conversion receiver for the newcomer

A direct-conversion receiver, even in its simplest form, has many advantages over the superheterodyne, when looked at in terms of performance versus complexity and cost. This receiver was originally described in *Radio Communication* by J. Young, BRS 33339, and may be constructed using only hand tools and a small

**Fig 21(a). "The hf polished gem", a solid-state version of the ZL2AMJ "hf gem" converter. It acts as a crystal-controlled converter on 7, 14 and 21MHz, and as a preselector on 3·5MHz, providing output on 3·5–4MHz. If a 3,500kHz crystal is used, the calibration will be appropriate on all bands. All coils wound on ¼in diameter slug-tuned unshielded formers, using 30swg enamelled wire: L1, 42 turns; L2, 26 turns; L3, 17 turns; L4, 35 turns; L5, 13 turns; L6, 8 turns spaced over ⅜in**

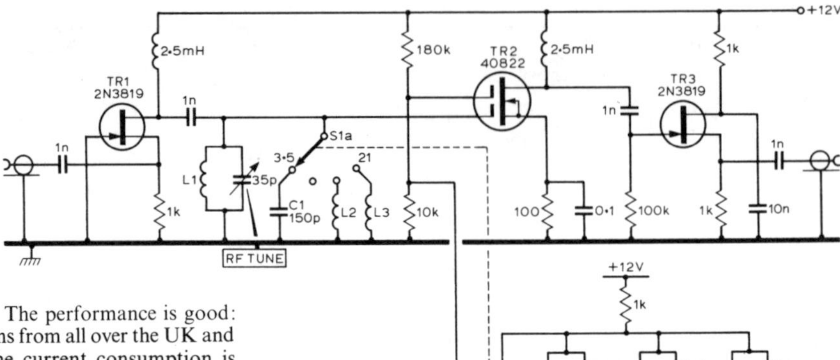


bench vice, at a cost of around £15. The performance is good: using a simple end-fed antenna, stations from all over the UK and mainland Europe can be heard. The current consumption is about 15mA excluding the power amplifier, providing many hours of operation from a set of batteries. Although intended for the reception of ssb and cw stations, a.m. can be resolved by tuning to zero beat.

The 3·5MHz (80m) band was chosen because, in the original author's opinion, it is the most interesting for the beginner. There is always something on the band, and much can be learned by listening to the technical discussions and operating procedures of the licensed radio amateurs.

The receiver should also prove an interesting project for the experienced constructor. The addition of a transistor buffer/pa driven from the oscillator would make it a very compact low-power transceiver for the licensed amateur.

**Circuit description**

A fet is used as an rf amplifier. As well as providing pre-mixer gain, this isolates the front-end tuned circuit from the antenna, thus preventing rf induced into L1 from the oscillator being radiated by the antenna.

The mixer is a balanced type. It uses two silicon diodes, feeding an m-derived, single-section, low-pass filter, with a cut-off frequency of 3kHz. The balance potentiometer is pre-set, and once set should not require adjustment unless the mixer diodes or L1 are changed for any reason.

The oscillator is a Colpitts employing a fet, followed by a buffer stage to prevent pulling of the oscillator when tuning the front-end. Both the oscillator and buffer stages are stabilized by a 5·6V zener diode. An electrolytic capacitor C11 is connected across the zener diode. If this capacitor is omitted the noise generated by the diode will be introduced into the mixer and low-level input signals will be lost.

A three-stage af amplifier provides enough gain to drive medium- to high-impedance headphones. The power amplifier is required only if low-impedance headphones or loudspeaker are to be used: it will drive loads down to 8Ω. The extra current drawn by the power amplifier will depend upon the load it is driving and the output level used; about 7mA at zero output. A

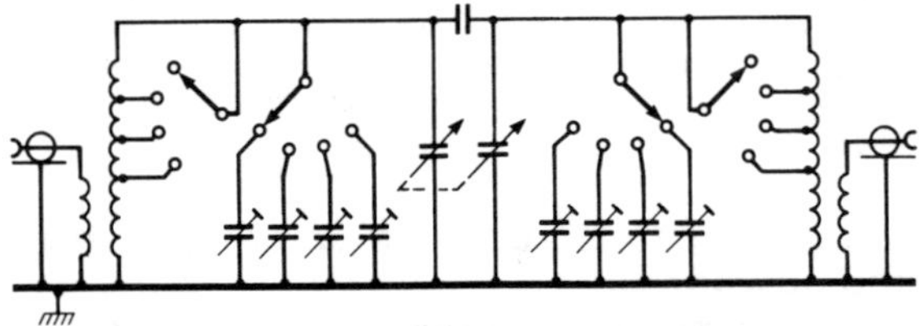

**Fig 21(b). Bandpass preselector which can be put in front of the hf converter to eliminate direct breakthrough of 3·5MHz signals. A simpler highpass filter would probably suffice, and details of this particular filter were not given by Gary Moles**

**COMPONENTS LIST FOR RECEIVER**

| | | | |
|---|---|---|---|
| **R1, 2, 4, 8, 11, 15** | 1kΩ | **C1, 4, 14, 20** | 0·01μF ceramic |
| **R3** | 390Ω | **C2, 13** | 100pF sm |
| **R5, R17, R18** | 820Ω | **C3, 5, 7, 8, 15** | 0·1μF ceramic |
| **R6** | 100kΩ | **C6** | 0·047μF ceramic |
| **R7** | 2·2kΩ | **C9, 12** | 220pF sm |
| **R9** | 120kΩ | **C10** | "470pF sm |
| **R10** | 470kΩ | **C11** | 32μF/10V |
| **R12** | 22kΩ | **C16, 21** | 10μF/16V |
| **R13** | 10kΩ | **C17, 18** | 1μF/16V |
| **R14** | 47kΩ | **C19, 22, 24** | 100μF/16V |
| **R16** | 220Ω | **C23** | 470μF/16V |
| (All above, 10 per cent ¼W) | | | |
| **R19, 20** | 2·2Ω ½W | **VC1, 2** | 50pF Jackson C804 |
| **RV1** | 1kΩ lin min pre-set | **TR1, 3** | 2N3819 fet |
| **RV2** | 10kΩ log midget control | **TR2, 6** | BC107 |
| | | **TR4, 5** | BC109 |
| | | **TR7** | 2N3704 |
| (RV1, 2 obtainable from RS Components Ltd) | | **TR8** | 2N3703 |
| | | **ZD1** | BZY88 5·6V |
| | | **D1-6** | 1N4148(1N914) |

| | |
|---|---|
| **Die-cast box** | Eddystone type 6908P |
| **Control knobs** | Eagle F19 |
| **Loudspeaker** | 8 to 35Ω, Eagle TP80G |
| **Ferrite rings** | Mullard FX1593, 2 off |
| **Ball drive** | 6:1 Jackson type F |
| **Jack plug** | Bulgin P519 |
| **Jack socket** | Bulgin J30 |
| **Sub-min toggle switch** | Eagle MTS2 |

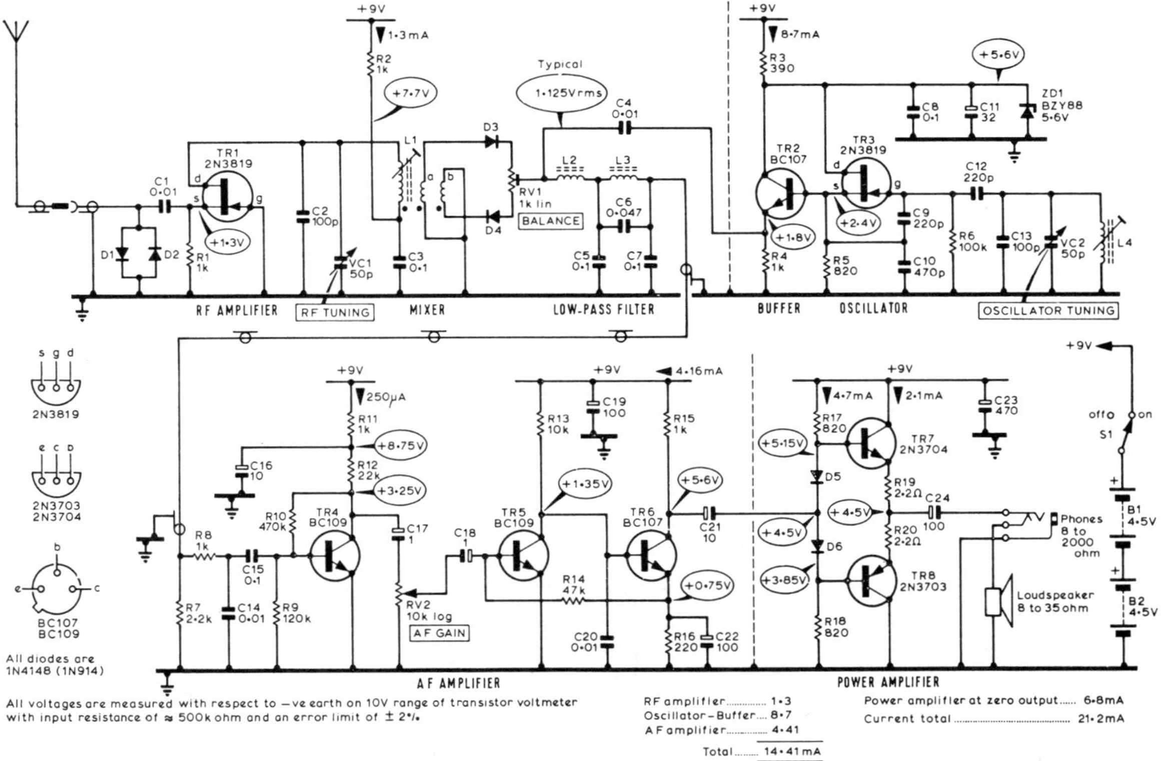


Fig 22. Circuit diagram of 3·5MHz direct-conversion receiver

35Ω loudspeaker is a good compromise between power output and current consumption.

The diodes D1 and D2 were added to the original design to protect TR1 and will also protect the receiver if it is to be used with a transmitter.

## Construction

The receiver is constructed in a 4½in by 3½in by 2in Eddystone die-cast box type 6908P. The layout and mechanical details are shown in Fig 23. To simplify construction the compartments should be assembled in the following order: oscillator, rf and mixer, af amplifier.

The oscillator tuning capacitor VC2 is mounted on a 16swg aluminium bracket, secured to the side and bottom of the case with four 6BA screws. Two screens made from 22swg tinplate or aluminium, also secured to the side and bottom of the case, but with 8BA screws, divide the case into three compartments. Extra holes are drilled in the screens for feed-through insulators. VC1 and RV2 are fixed to the front of the case. A Jackson six-to-one reduction drive is screwed to the spindle of VC2 and secured with a 6BA counter-sunk screw to the front of the case. The front of the drive should protrude ⅛in from the case to allow for fixing the tuning dial.

One word of warning, **do not** drill die-cast boxes unless they are firmly secured to the work bench.

The oscillator components are mounted on a six-way tag strip, and should be assembled before being put into the box (Fig 24). Resistor R3 is connected between the tag strip and a feed-through insulator on the screen dividing the oscillator and rf compartments. C13 is wired across VC2. A short wire from the emitter of TR2 is passed through a nylon insert in the screen and connects to the mixer board.

Both the mixer and low-pass filter are constructed on a paxolin board, fixed by 8BA screws to the bottom of the case; this is also assembled before being put into the case. The lay-out is shown in Fig 25. The mixer/filter board occupies the rear half of the centre compartment. R2 and C3 are wired to an insulated stand-off on the screen dividing the oscillator and rf compartments. TR1 and R1 are mounted on the other screen close to VC2. A short length of screened lead connects R1 to another insulated stand-off on the rear of the case, C1 is wired between the stand-off and the coaxial antenna socket. D1 and D2 are wired across the antenna socket.

A ¼in hole is drilled in the screen to take a grommet through which a screened lead is passed, connecting the low-pass filter to the af amplifier.

The af amplifier is constructed on a 1⅞ by 2⅞in paxolin board (Fig 26) and mounted on the side of the case. Small holes are drilled in the board to take the component leads. These are connected together with 24swg tinned-copper wire or $\frac{1}{16}$in wide Cir-kit strip on the underside of the board. A 16swg bus wire is used as the earth line, and is soldered to 6BA tags through which are passed the screws securing the board to the case. A screened lead from the output of the af amplifier is passed through a grommet, and connects to the headphone socket or the power amplifier.

Wiring pins are used for all leads connecting the two circuit boards to other parts of the receiver. The pa is mounted on the

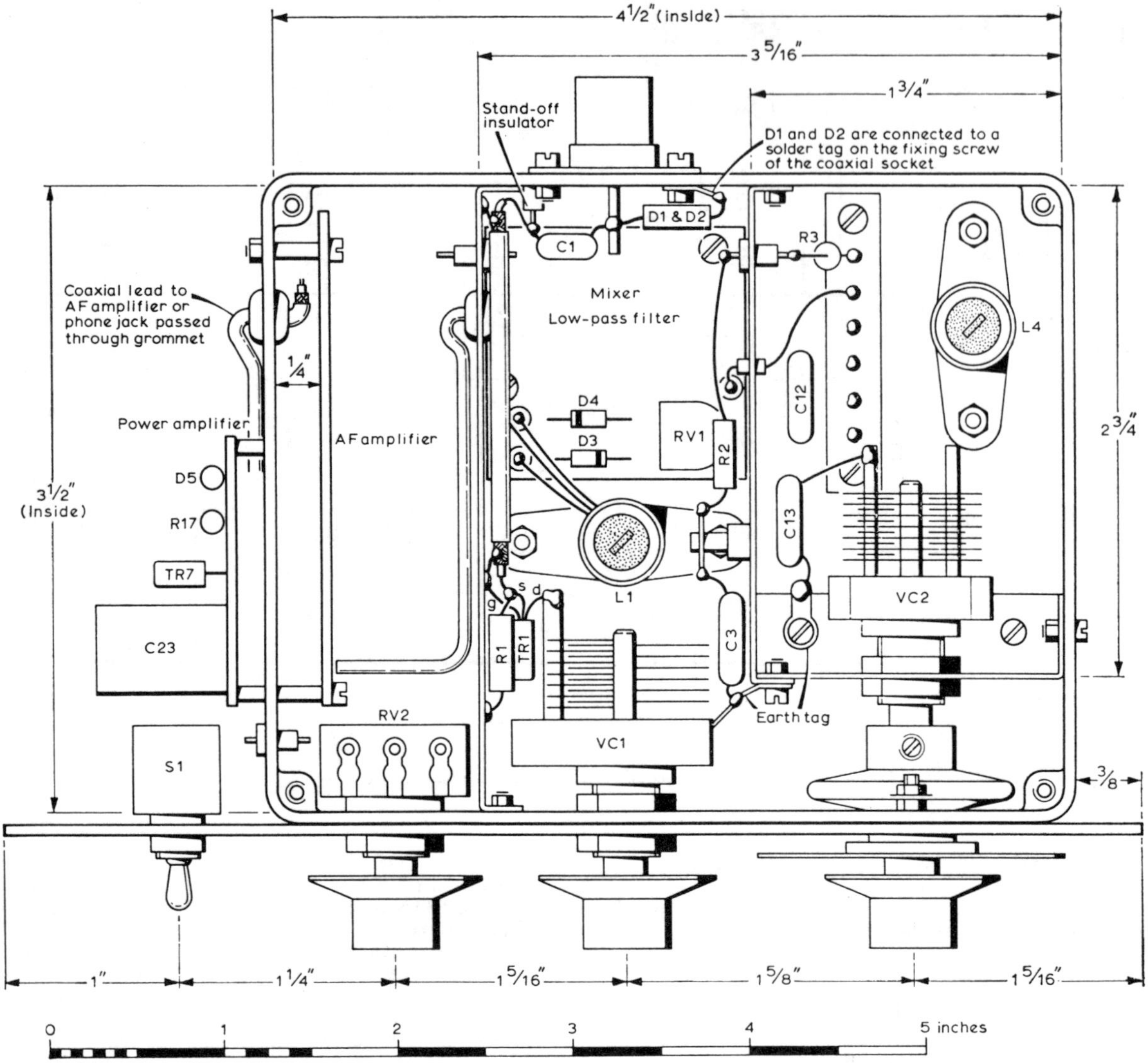

Fig 23. Top view, above; front view, below

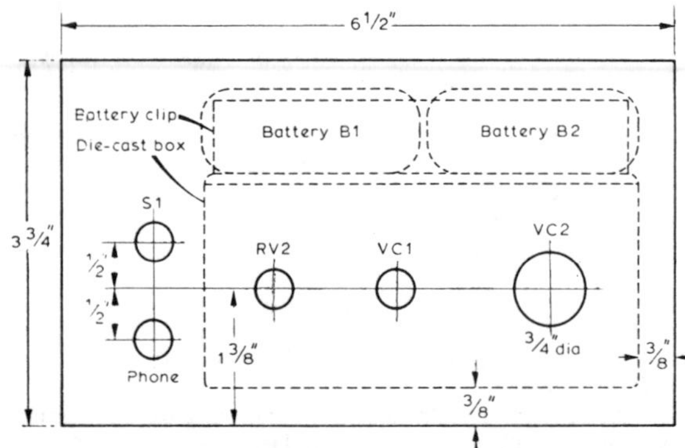


outside of the case behind the headphone socket and on/off switch.

When all the parts have been assembled and wired, the case is fixed to the front panel, secured by the lock-nuts of RV2, VC1 and two 6BA counter-sunk screws.

The batteries are held in place by two L-shaped brackets mounted on the lid of the box. The rear bracket supports a $4\frac{1}{2}$in length of $\frac{1}{8}$in paxolin strip into which round-headed screws or rivets are fixed to form the battery contacts. These contacts should be spaced in such a way as to make it impossible to connect the batteries the wrong way round. A strip of foam rubber glued to the other bracket holds the batteries against the contacts.

If the pa is not required, a PP7 battery may be used instead of the two V1289 batteries and fitted behind the front panel in place of the amplifier. The length of the front panel must be increased by $\frac{3}{8}$in and the height reduced by $\frac{7}{8}$in.

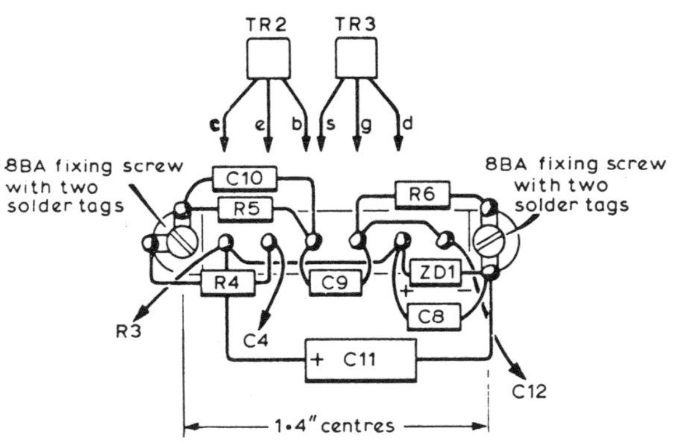


Fig 24. Oscillator component layout

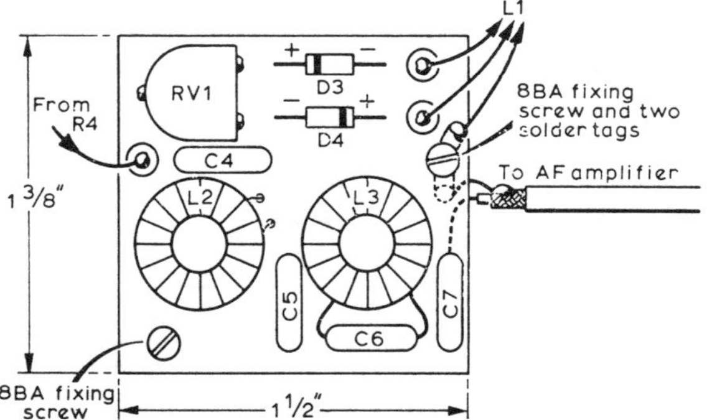


Fig 25. Mixer and low-pass filter

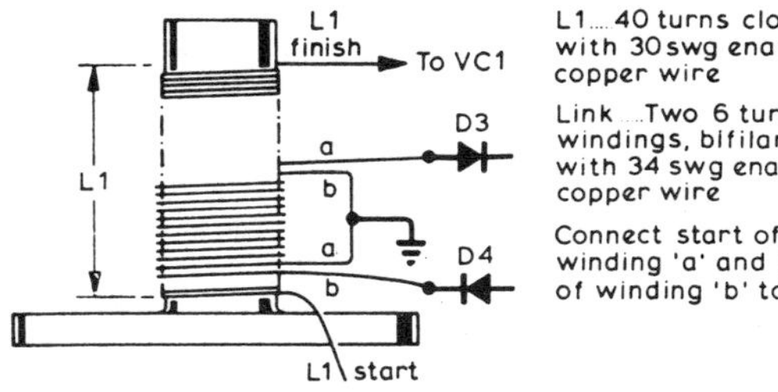


Fig 27. L1 winding details

The completed receiver is put into a wooden case with the loudspeaker.

**Coil details**

L1 and L4 are wound on $\frac{3}{8}$in diameter formers with tuning slugs. L2 and L3 are wound on Mullard ferrite-rings type FX1593.

Before winding L1 study Fig 27. L1 is wound with 40 turns of 30swg enamelled wire and given a thin coat of varnish. When dry, two lengths of 34 swg enamelled wire are wound six times over the cold end of L1. A length of sewing thread is used to secure the link windings to the coil former. All windings are close-wound in the same direction.

L4 is wound with 30 turns 30swg enamelled wire, given a coat of varnish and left to dry.

The winding of L1 and L4 can be made easier by securing the coil formers to the bench and placing drawing pins on either side to hold the ends of the windings until the varnish has dried.

L2 is wound with 40 turns of 30swg enamelled wire. A 1mH rf choke may be used in place of L2.

L3 is wound with 300 turns of 38swg enamelled wire. To do this make a shuttle from a $3\frac{1}{2}$in length of $\frac{1}{8}$in diameter knitting needle, with a $\frac{1}{4}$in slot cut in each end. Wind on 30 turns of wire, and pass it through the centre of the ring. Before winding the rings, rub them gently with fine emery cloth to remove any sharp edges. Varnish the completed coil and leave to dry. The rings are secured with sewing thread passed through small holes drilled in the board.

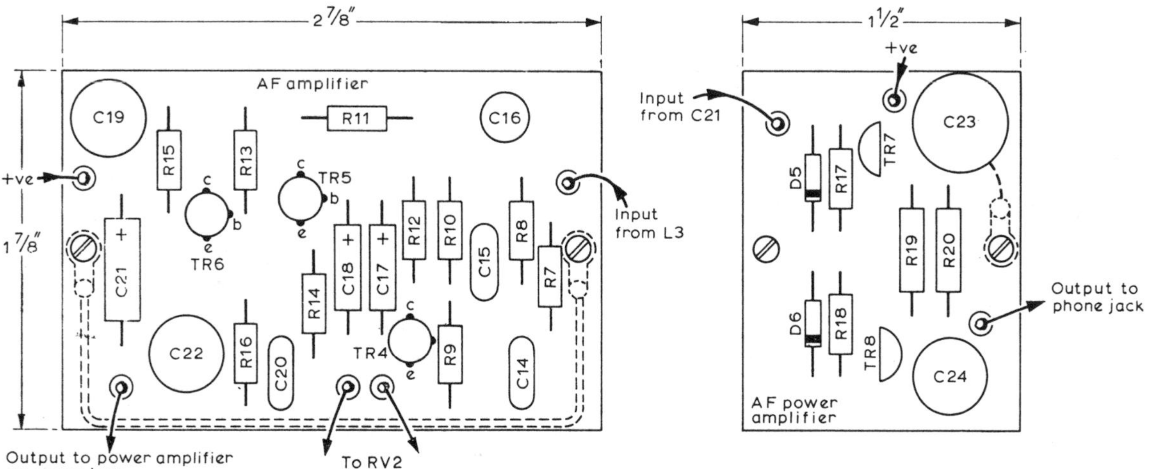


Fig 26. Amplifier component layout

**Setting up**

First make a double check of all wiring. Connect the batteries in series, negative to earth. Set the balance potentiometer RV1 to mid-way, and the oscillator tuning capacitor VC2 to full mesh. With the aid of a communication receiver (bfo on) adjust the tuning slug in L4 for zero beat at 3·5MHz. Turn VC2 to minimum capacitance. The oscillator frequency should now be between 3·8MHz and 3·9MHz. If it is below 3·8MHz, reduce C13 to 82pF and repeat the tuning procedure. The tuning can be made to cover the range 3·5MHz 4MHz, by further reducing C13. When the oscillator has been tuned, connect the antenna and a pair of headphones. Set VC1 to 90 per cent of full mesh, the oscillator to 3·5MHz, and adjust the tuning slug in L1 for maximum output in the headphones. Finally, tune the receiver to a quiet spot on the band and rotate the rf tuning capacitor VC1. Any stations heard that are independent of the oscillator frequency should be balanced out by careful adjustment of RV1.

**Conclusion**

At first sight this receiver may seem a little elaborate, but a superheterodyne with the same performance would cost more. The case, loudspeaker and FETS account for a quarter of the total cost. Most of the other components will be found in the odds and ends box. Miniature electrolytic capacitors are used in the audio amplifier, and these should be purchased new. Old or salvaged electrolytic capacitors are often faulty.

When used out of doors, the receiver works well with a $\lambda/4$ (66ft) wire antenna tied to a branch of a nearby tree. A copper earth rod will improve reception, but is unnecessary unless reception of dx stations is desired. At home even a short indoor antenna will bring in a surprising number of stations.

Should the constructor wish to make this receiver for an amateur band other than 3·5MHz, only the tuned circuits need be altered. The formulae are covered in detail in the article by the late W. H. Allen, MBE, G2UJ [1] and in the fifth edition of *Radio Communication Handbook*.

The mixer used in this design is an improvement on the type used in "The Cadet" [2] and those who have constructed it will find this a worthwhile modification.

All the components for this receiver may be purchased from Stan Reed Ltd, 109 Hillingdon Hill, Uxbridge, Middx.

**References**

[1] W. H. Allen, MBE, G2UJ. "Coils, capacitors and bandspread", *Radio Communication* November 1972.

[2] J. Young, BRS33339. "The Cadet", *Radio Communication* October 1973.

## 144MHz single-conversion superhet

This receiver, originally described by N. B. Pritchard, G8AYM, in *Radio Communication*, is suitable for the reception of amateur nbfm transmissions in the range 144–146MHz, including transmissions made through "repeater" stations operating

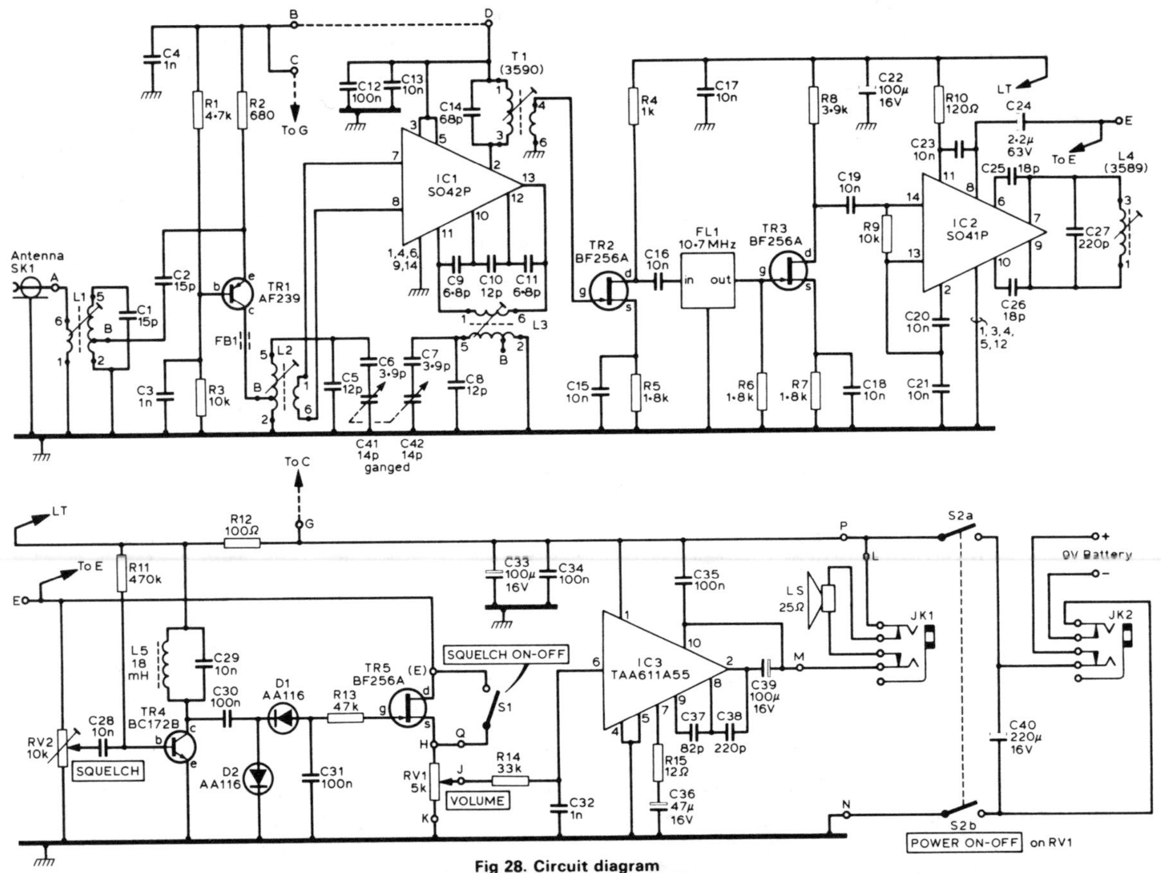


Fig 28. Circuit diagram

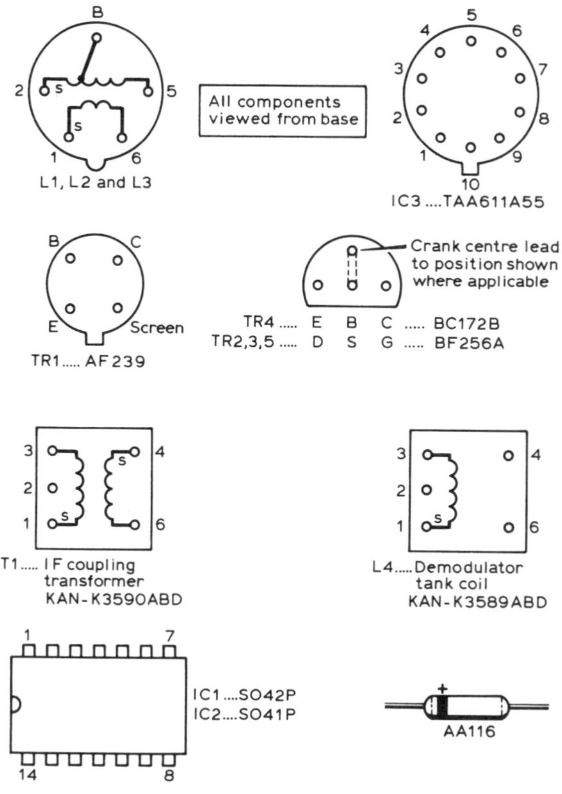


**Fig 29. Connection diagrams, bottom views**

between 145·600–145·775MHz up to distances of about 25 miles using a vertical whip antenna. A considerable number of receivers to this design have been built and tested; it is however most suitable for those with some previous experience of radio construction and with access to good test equipment.

It was developed as a club constructional kit for the benefit of members of the Echelford Amateur Radio Society. The design was based on an article in *VHF Communication* but incorporates modifications and improvements to the original design. The design is not state-of-the-art, but what was required was a cheap and simply reproducible circuit that would encourage and promote interest in construction, provide acceptable performance, and possibly allay the fears of some would-be constructors regarding the "black art" of vhf construction. A quick scan over the circuit will indicate the use of a crystal filter and, while these would normally be expensive, one club member donated a large number of surplus units for the project and, in fact, further filters were purchased as surplus for £5 each.

**Circuit description**

The antenna to be used, either a whip or Yagi for example, is connected to SK1 which links with a tuned circuit L1 and C1 resonant at 145MHz. A low impedance tap couples the signal into an rf amplifier, TR1, using an AF239 germanium planar transistor in grounded base configuration. The collector of TR1 is fed via ferrite bead FB1 to a second tuned circuit L2 which has a balanced coupling winding connected to IC1. L2 is tuned by means of fixed capacitor C5 and by the variable capacitor C41 via C6, which serves to reduce the tuning range of C41. IC1 type SO42P is a balanced mixer/oscillator, the input signal being fed into pins 7, 8 in balanced form and the oscillator being formed between pins 10, 11, 12, 13, with C9, 10, 11 forming the feed-back network. The oscillator is tuned via L3 by C42, which is ganged with C41, and fix-tuned by C8, with C7 limiting the tuning range of C42. The oscillator operates below the received signal by 10·7MHz, which is the i.f. frequency.

**Fig 30. PCB etching, underside**

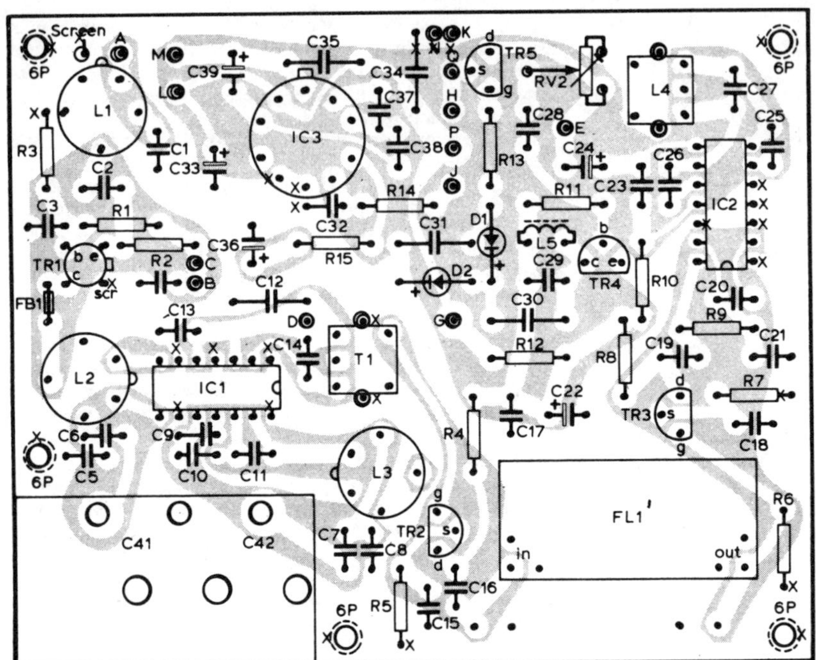


6P indicates 6BA tapped × 0·5in long pillars

Fig 31. Component layout, pcb top view

## COMPONENTS LIST

| | | | |
|---|---|---|---|
| **R1** | 4·7kΩ | **R10** | 120Ω |
| **R2** | 680Ω | **R11** | 470Ω |
| **R3, 9** | 10kΩ | **R12** | 100Ω |
| **R4** | 1kΩ | **R13** | 47kΩ |
| **R5, 6, 7** | 1·8kΩ | **R14** | 33kΩ |
| **R8** | 3·9kΩ | **R15** | 12kΩ |

All resistors 0·25W 5% carbon film

| | |
|---|---|
| **RV1** | 5kΩ carbon pot log with switch |
| **RV2** | 10kΩ carbon pot linear |
| **C1, 2** | 15pF ceramic plate ±2% |
| **C3, 4** | 1nF ceramic plate ±10% |
| **C5** | 12pF ceramic plate ±2% |
| **C6, 7** | 3·9pF ceramic plate ±2% |
| **C8, 10** | 12pF ceramic plate ±2% |
| **C9, 11** | 6·8pF ceramic plate ±2% |
| **C12** | 100nF polyester ±20% |
| **C13, 15, 16, 17, 18, 19, 20, 22, 23, 28, 29** | 10nF ceramic plate −20 +80% |
| **C14** | 68pF ceramic plate ±2% |
| **C22, 33, 39** | 100μF electrolytic pluggable 16V |
| **C24** | 2·2μF electrolytic 63V |
| **C25, 26** | 18pF ceramic plate ±2% |
| **C27, 38** | 220pF ceramic plate ±2% |
| **C30, 31, 34, 35** | 100nF polyester ±20% |
| **C32** | 1nF ceramic plate ±10% |
| **C36** | 47μF electrolytic pluggable 16V |
| **C37** | 82pF ceramic plate ±2% |
| **C40** | 220μF electrolytic (with mechanical sets only) |
| **C41, 42** | 15pF swing tuner (Wingrove & Rogers 3/CG80-03/1) |
| **L1** | Antenna coupling winding (Toko type E515HNS–210170 RIE9) |
| **L2** | Mixer tuning coil (Toko type E515HNS–210170 RIE9) |
| **L3** | Local oscillator coil (Toko type E515HNS–210170 RIE9) |
| **L4** | Demodulator tank coil Toko type KAN–K3589ABD |
| **L5** | Squelch inductor 18mH Toko type 187LY–183 |
| **T1** | I.F. coupling transformer Toko type KAN–K3590ABD |
| **FB1** | Ferrite bead |
| **FL1** | Crystal filter 10·7MHz bw ±7·5kHz, ITT type 024BG/923B (or suitable surplus unit) |
| **D1, D2** | Germanium diode type AA116 noise demodulator |
| **TR1** | AF239 Germanium pnp rf amplifier |
| **TR2, 3** | BF256A jfet i.f. amp/filter matching |
| **TR4** | BC172B silicon npn squelch amplifier |
| **TR5** | BF256A jfet squelch switch/gate (identified with white spot) |
| **IC1** | SO42P balanced mixer/oscillator (Siemens) |
| **IC2** | SO41P limiting amplifier/demodulator (Siemens) |
| **IC3** | TAA611A55 af output amplifier |
| **LS1** | Internal loudspeaker 25Ω/1W Mullard type AD2071/Z25 |
| **SK1** | Belling-Lee antenna socket |
| **JK1** | 0·25in headphone jack |
| **JK2** | 3·5mm power jack |
| **S1** | SPST squelch on-off |
| **S2a, b** | Power on-off dpst (on RV1) |

It is recommended that a 6:1 reduction drive is used with the tuning gang. Also, some experiments with the removal of rotor vanes would produce a more expanded scale if less coverage was required.

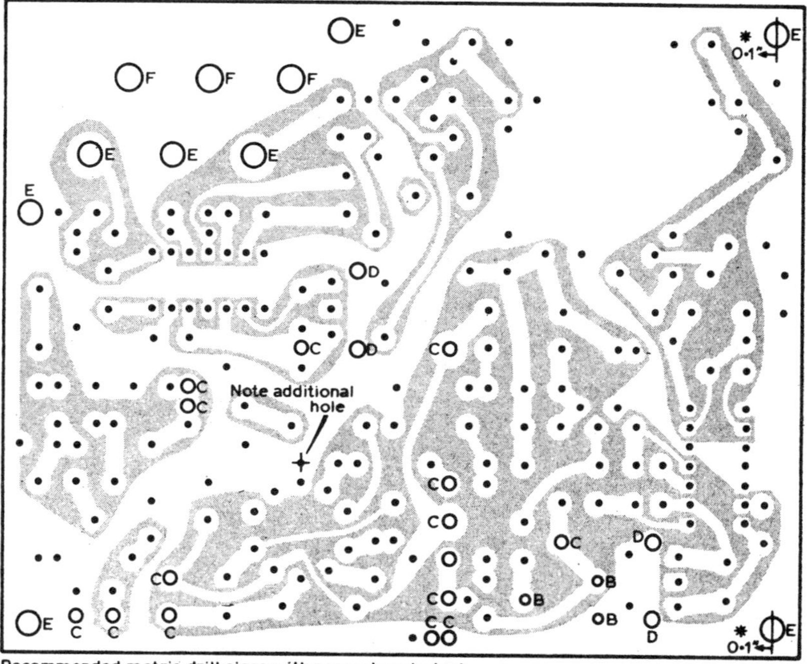


Fig 32. PCB drilling guide. (*For those building the receiver in a die-cast box, these two mounting holes should be relocated upwards by 0·1in to allow the corners to be removed to clear screw bosses in the box)

Recommended metric drill sizes with nearest equivalent

| Holes | Metric | Drill No | Inches |
|---|---|---|---|
| 'A'• | 1mm dia | No60 | 0·040" |
| 'B' | 1·2mm dia | No 56 | 0·047" |
| 'C' | 1·5mm dia | No 53 | 0·059" |
| 'D' | 1·8mm dia | No50 | 0·070" |
| 'E' | 3mm dia | No31 | 0·118" |
| 'F' | 3·3mm dia | No30 | 0·129" |

The i.f. appears at pin 2 of IC1 and is coupled via a tuned circuit, C14/T1, to the gate of TR2 which is an N-channel fet type BF256A operating in grounded source. TR2 acts as an i.f. amplifier, and via its drain resistor R4 presents the correct source impedance for the quartz crystal filter FL1 which has a bandwidth of ±7·5kHz at the −3dB points. The filter is terminated by R6 in parallel with the gate impedance of TR3, another BF256A in grounded source configuration again used as an i.f. amplifier. The amplified signal is taken from the drain of TR3 and couples into IC2 via C19.

IC2 type SO41P is a six-stage limiting amplifier/fm demodulator, the tank tuning circuit being formed by C25, 26, 27, and L4 resonant at 10·7MHz. The audio recovered from the demodulator is partially de-emphasized via C23 and couples via C24 to the audio circuits. The audio signal passes via TR5, BF256A, which acts as a squelch gate, to the volume control RV1 and R14, C32 (giving final de-emphasis) to the audio amplifier IC3.

IC3 type TAA611A55 raises the level of the audio signal enough to provide a maximum of about 1W into an 8Ω loudspeaker. R15, C36, 37, 38 tailor the audio response to be 120Hz to 4kHz −3dB relative to 1kHz with a 22Ω load. An external load such as a low-impedance loudspeaker (4Ω minimum) or a pair of headphones may be connected via JK1 which disconnects the internal loudspeaker. External power may be applied to the receiver via JK2 which disconnects the internal battery. In the absence of an rf carrier, only noise will exist at point "E". This will be passed via RV2 and C28 to the noise amplifier TR4, a BC172B. The collector load is an LC circuit tuned to about 10kHz and is coupled via C30 to a diode detector/doubler. The negative-going output of this detector turns off TR5 (the squelch gate) and hence prevents the noise from being passed to the audio amplifier. If the receiver is now tuned to a carrier, stronger levels of rf signal will progressively quieten the limiting rf amplifier, reduce the negative bias applied to the squelch gate, and therefore allow any fm modulation on the carrier to be passed to the audio amplifier. Thus, by varying the setting of RV2 the sensitivity of the squelch circuit may be suited to any appropriate rf level, eg adjusted so that signal levels below, say, 1μV may be effectively cut off and only signals above that level may be resolved.

A degree of power supply flexibility has been built in. For a basic circuit not associated with a transmitter, the rf amplifier, mixer/osc and i.f./audio stages may be run from the same power supply. To effect this, link points B to D and C to G. However, when used with a companion transmitter circuit it is desirable to shut down the receiver, except the oscillator, in the transmit mode in order to maintain frequency stability. The mixer/osc may be powered continuously via point D, preferably by a stabilized 7V

**COIL WINDING DETAILS**

| Coil | Toko part No | Function | Winding (0·7mm wire) 1 | 2 | 3 |
|---|---|---|---|---|---|
| | | | Pins 2–5 | Pins 2–8 | Pins 1–8 |
| L1 | E515HNS-210170 RIE9 | Antenna coupling | 2½ turns | ¾ turn | 2⅙ turns |
| L2 | E515HNS-210170 RIE9 | Mixer tuning | 2½ turns | ¾ turn | 2⅙ turns |
| L3 | E515HNS-210170 RIE9 | Local oscillator | 2½ turns | ¾ turn | 2⅙ turns |
| T1 | KAN K3590 ABD | 1st i.f. | Pins 1–3 15 turns | Pins 4–6 8 turns | — |
| L4 | KAN K3589 ABD | Demodulator tank coil | Pins 1–3 7 turns | — | — |

All wound on ferrite cores. (Original coils custom-made by Toko.)

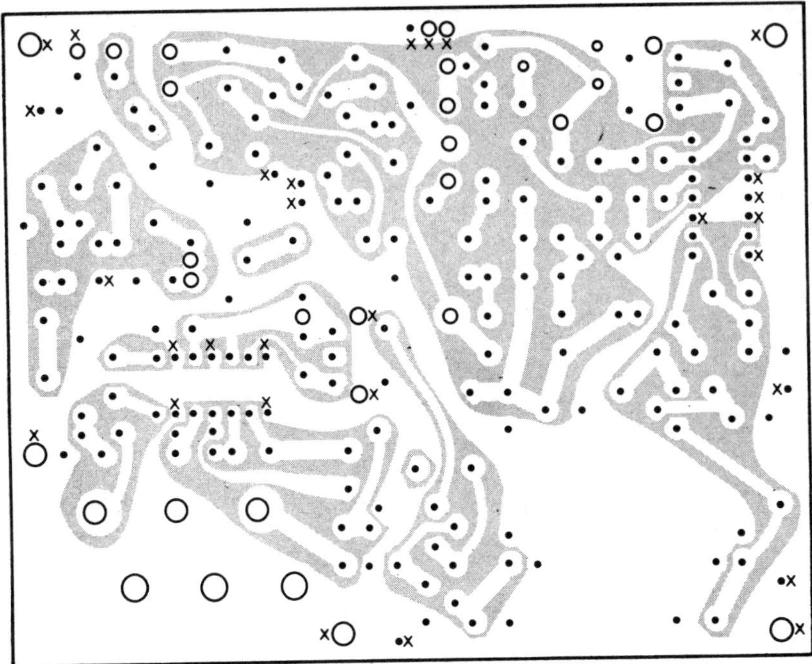


Fig 33. Top side counter-boring guide. File excess glass fibre at circumference to match the underside copper pattern. If building receiver in die-cast box, corners must be removed to clear box bosses. Crop front end corners at 45° by about 0·1in and then crop rear corners to give good fit in the box. It may be necessary to file away the base of the mounting bosses in the box due to their large taper. The holes marked "X" should NOT be counter-bored, but the drill burrs should be removed carefully. The remaining holes should be counter-bored to expose the glass fibre to a diameter of about 0·1in

**RECEIVER SPECIFICATIONS**
(based on two different receivers)

| | |
|---|---|
| **Frequency coverage** | 144–146MHz |
| **Image response** | –34dB relative to 145MHz |
| **Sensitivity** | Low level detectability 0·25μV<br>⩽1μV for 10dB quieting<br>1·5μV for 20dB quieting<br>Start of quieting 0·5μV |
| **Oscillator drift** | **Variation with supply**<br>11V + 15kHz<br>10V + 8kHz<br>9V 0<br>8V – 8kHz<br>7V – 15kHz<br>**Switch-on drift after 1min from switch-on**<br>+5min ............ – 5kHz<br>+30min ............ 11kHz<br>**Oscillator detuning with input level** (detuning starts at 5–10mV)<br>10mV ............ –11kHz<br>100mV ............ –70kHz<br>1,000mV ............ –31kHz |
| **I.F. bandwidth** | ±7·5kHz at 10·7MHz |
| **Overall af bandwidth** | 100Hz–3kHz at –3dB |
| **Output versus deviation** | 3kHz peak deviation at 1kHz gives ⩾250mW into 25Ω at maximum volume |
| **Quiescent current consumption** | 20mA at +9V |

*Note.* RF measurements were made with a Hewlett-Packard 8640B signal generator. RF voltages are rms levels assuming a 50Ω load, ie the emf levels are twice those indicated.

supply but, with points C–G linked, the main power to the circuit may be switched off in the transmit mode at point P.

**Construction**

The circuit is constructed on a double-sided pcb, the top side of the board being plain copper to provide an earth plane. All holes are spot drilled with a 1mm drill, and opened out according to the drilling guide. Most of the holes on the top side have to be counter-bored in order to isolate component leads from earth. The counter-boring guide shows which holes should *not* be counter-bored, all others being counter-bored with either a Vero track cutter or a suitable drill to produce a copperless area of about 0·1in diameter. Finally de-burr all holes slightly with a drill, removing as small an amount of copper as possible around the hole.

**Assembly**

Assembly is generally straightforward, but care should be taken to mount all components well down on to the pcb, not forgetting to check for inadvertent shorts to the top side earth plane. Wherever possible, earthed component leads should be soldered to the top side. Care should be taken over soldering, since the components will have short leads, particularly the ICs.

**Alignment**

Turn RV1 and RV2 to minimum and switch off the squelch (switch shorting) and the loudspeaker load to M and L. Make connections C–G and B–D and ensure that the power lines to P and N are decoupled with 220μF if testing with a battery, otherwise af instability may result due to high dynamic source impedance.

Inject to pin 14 of IC2, via a 10nF capacitor, a signal at 10·7MHz with a deviation of 3kHz peak at about 1mV amplitude and 1kHz modulation. Adjust L4 for maximum recovered audio at H or M.

Inject at point H a sine wave of 1kHz and 10mV p-p, and check for undistorted output at M with an oscilloscope. Output will be approximately 1W (8Ω) or 250mW (25Ω) for about 20mV p-p at H with maximum volume.

Adjust C41/42 to mid-travel and inject a sweep generator into the antenna socket, taking the output from IC2 pin 14 to the equipment demodulator via a 10nF capacitor. Locate the 145MHz marker on the sweep generator and align L3 to produce a response on this marker. Then adjust L1,2 and T1 to produce maximum amplitude, being careful not to overload the receiver as this will mask the alignment. Check that by adjusting C41/42 that the whole of 144–146MHz is tunable with about 0·5MHz extra at either end.

Remove all input signals, switch on the squelch (S1 open) and check that the noise at the audio output is killed when RV2 is adjusted anticlockwise between 40 and 75 per cent of travel. The receiver is now ready to use.

**Conclusion**

In a few cases the local oscillator would not operate satisfactorily, and this has been cured by the addition of a 1kΩ resistor to ground on each of pins 10 and 12 of IC1. This increases the transconductance of the mixer-oscillator and maintains the oscillation correctly. This problem has been the only one encountered, and in all other respects the circuit has been found to operate satisfactorily.

In spite of its simplicity, the receiver performs well compared with some well-known "black boxes", although one could find several areas where improvements could be made.

All in all, the receiver has proved to be an interesting little project to design and build.

## COMPONENT COLOUR CODES

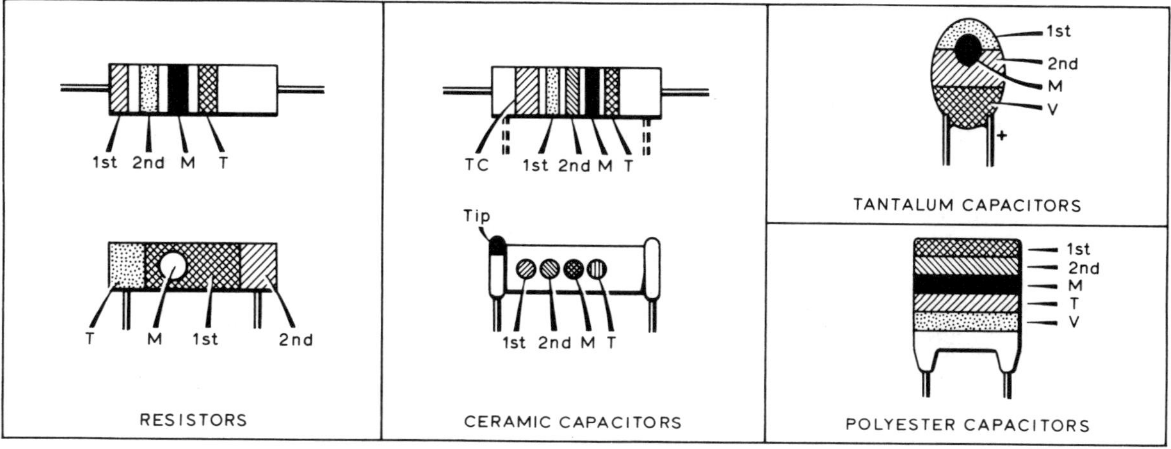


| Colour | Significant figure (1st, 2nd) | Decimal multiplier (M) | Tolerance (T) (per cent) | Temp coeff (TC) (parts/10⁶/°C) | Voltage (V) (tantalum cap) | Voltage (V) (polyester cap) |
|---|---|---|---|---|---|---|
| Black | 0 | 1 | ±20 | 0 | 10 | — |
| Brown | 1 | 10 | ±1 | −30 | — | 100 |
| Red | 2 | 100 | ±2 | −80 | — | 250 |
| Orange | 3 | 1,000 | ±3 | −150 | — | — |
| Yellow | 4 | 10,000 | +100, −0 | −220 | 6·3 | 400 |
| Green | 5 | 100,000 | ±5 | −330 | 16 | — |
| Blue | 6 | 1,000,000 | ±6 | −470 | 20 | — |
| Violet | 7 | 10,000,000 | — | −750 | — | — |
| Grey | 8 | 100,000,000 | — | +30 | 25 | — |
| White | 9 | 1,000,000,000 | ±10 | +100 to −750 | 3 | — |
| Gold | – | — | ±5 | — | — | — |
| Silver | – | — | ±10 | — | — | — |
| Pink | – | — | — | — | 35 | — |
| No colour | – | — | ±20 | — | — | — |

Units used are ohms for resistors, picofarads for ceramic and polyester capacitors, and microfarads for tantalum capacitors.

CHAPTER 4

# Transmitters

For many years amateur hf and vhf transmitters were relatively simple to build compared with communication receivers and were based almost entirely upon valves. Mostly they comprised a low-power *oscillator* (crystal-controlled or variable frequency) followed by a number of intermediate power stages operated as *buffer amplifiers* (to isolate the oscillator from the succeeding stages) or as *frequency doublers* or *frequency multipliers* to obtain outputs on several harmonically-related bands (eg 3·5, 7, 14, 21MHz etc) or to raise the frequency to, say, 144MHz. In hf practice the final stage of this *exciter* acted as a *driver* to provide sufficient rf drive for the final *power amplifier* which was normally biased to beyond cut-off during operation either by a bias supply or by bias derived from the rf drive; in the latter case current in the power amplifier was reduced in the absence of drive by means of a *clamp* valve which automatically reduced the screen voltage applied to the stage, and so eliminated the need for a bias supply. The high-power stage forming the power amplifier (pa) might be "single-ended" (using either a single high-power valve or several medium-power valves in parallel) or alternatively in "push-pull" configuration with two valves in a balanced form.

For amplitude-modulated (A3) telephony transmitters, high-level modulation techniques were most commonly used; this meant that modulation was applied to the anode and screen grid of the power amplifier only, using a *speech amplifier* and high-power *modulator* stage capable of providing an audio output power of roughly half the dc input of the transmitter power amplifier. Often various forms of speech processing in the form of peak clipping or speech compression were used to increase the average *talk power* of the transmitter.

Such an approach requires the use of bulky, well-smoothed power supplies capable of providing the full carrier power continuously and able to cope with the higher peak requirements during modulation; a large generously-rated modulation transformer was also needed for the usual "plate and screen" modulation system. The final rf and af valves needed to be capable of dissipating high power.

Transmitters designed in this way for a.m. (A3) operation are still very well suited for cw (A1), nbfm (F3) and fsk (rtty) applications since they are capable of providing the total carrier power (up to 150W dc input) over periods of say 10 minutes, and also coping with the higher-power speech peaks. All these factors mean that a 150W a.m. transmitter, although basically quite simple, requires the use of fairly expensive rf and af valves (for example the 813 tetrode), hefty power supplies and large "iron" components such as swinging and smoothing chokes, mains transformers and modulation transformers, with high-grade and generously rated components. Frequency stability requirements however are fairly relaxed, and are governed primarily by the desire to produce good clean tone for cw operation.

The gradual take-over of the single-sideband mode (A3J) for hf operation during the period 1950–1970 brought in trail many significant changes in transmitter design (and the emergence of the transceiver). However it should be stressed that the traditional approach remains, even now, entirely suitable for transmitters intended for cw/fsk/nbfm operation. Such designs tend to result in rugged transmitters that are less critical of operation into mismatched loads, are less susceptible to abuse during tune-up and, because of the Class C power amplifier, often provide considerably more rf output for a given dc input power than a transmitter intended primarily for ssb operation.

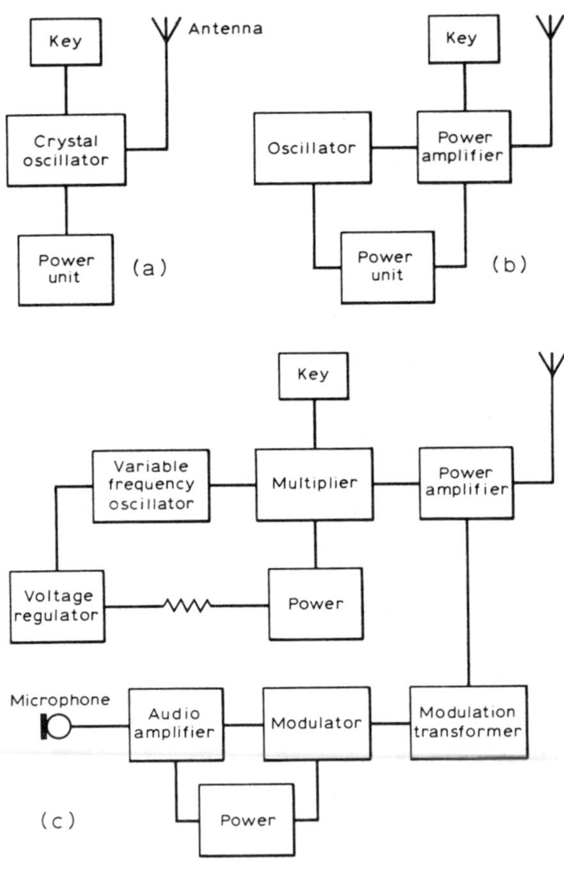


Fig 1. Block outlines of simple transmitters. (a) The simplest telegraphy transmitter is a single valve or transistor oscillator. This is seldom suitable for more than 15–20W input and will have low efficiency. (b) A simple two-stage transmitter usually provides a much cleaner and more potent signal; the input power will be governed by the type of valve or transistor(s) used as power amplifier and the voltage and current available from the power supply unit. (c) The incorporation of a multiplier stage makes it easier to operate on a number of different amateur bands while the audio-frequency amplifier and modulator stages permit the use of a.m. telephony as well as telegraphy

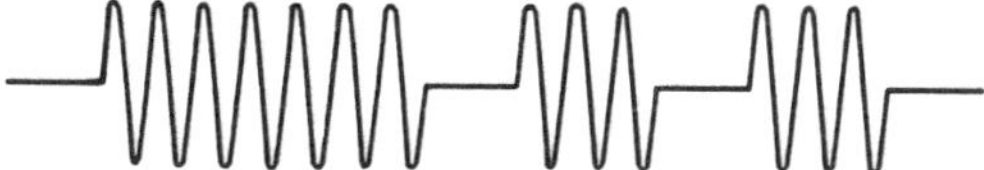

Fig 2. Keyed rf signal for cw operation

Since ssb transmitters are designed for low *duty cycles* (that is the percentage of time during which full output has to be delivered on voice peaks) they may use transformers and valves that can cope with very high peaks of current for short periods of time. However, considerable care has to be taken when tuning up or when operating with a mismatched load; for example, it is usually not possible to provide "key down" operation into a dummy load for more than a few seconds unless the input power is reduced. Since no high-power audio amplifier or modulator transformer is required for ssb, the overall cost of an ssb transmitter, in spite of greater complexity, may be significantly lower than for an a.m. (A3) transmitter of equivalent rating.

It is in the low-power stages that the ssb transmitter tends to be more complex, often with many transistor or valve stages being used for ssb generation and frequency conversion. The output from this section of the transmitter is then fed to a linear *driver* stage and *linear power amplifier* (often called a *linear*) where the power level is built up to the required level with valves operated in Class A, AB or B (modes which require less drive power than Class C which cannot be used to provide linear amplification). This question is discussed later in this chapter.

Modulation is carried out at a very low power level using a *balanced modulator* immediately after the carrier is generated: Fig 3. A balanced or double-balanced modulator closely resembles the balanced mixers discussed in the receiver section; such a stage suppresses the carrier frequency but leaves two sidebands, one of which is then filtered or phased out in the manner described later. We now have a very small amount of ssb at a fixed frequency; such a signal cannot be amplified without very serious distortion by a Class C stage. This means that all subsequent stages, whether for amplification or frequency conversion, have to operate in a linear manner in so far as the input signal is concerned; in other words the output of each stage must be a faithful replica of the input signal, although it may be at a greater power level or on a different frequency.

Any non-linearity in these later stages will introduce distortion in the form of *flat-topping* (ie the stage will not provide the desired peaks of output but will flatten the peaks) and in the process produce unwanted intermodulation products and spurii. A stage which is linear at the correct input power will inevitably become non-linear if driven too hard so that power levels need to be specified for a given degree of distortion. Peak envelope power (p.e.p.) of an ssb transmitter can be defined as the power integrated over an rf cycle at the crest of a two-tone wave when the transmitter is generating a power output conforming to the two-tone spurious specification.

In fact the specification for an ssb transmitter should indicate the degree of carrier suppression, the degree of unwanted sideband suppression, and the peak envelope power output for a specified degree of intermodulation distortion (imd).

From the above it will be appreciated that no Class C frequency multiplying stages can be used in an ssb transmitter; instead frequency conversion involves (as in superhet receivers) the use of the basic mixing or heterodyning techniques. Unfortunately most forms of linear amplifiers and frequency converters are less efficient than the Class C stages that can be used in a.m., fm, fsk and cw operation. This is of practical importance mostly in the final power amplifier. Mixer stages produce unwanted as well as wanted output signals, and the wanted and unwanted products may be quite close together in terms of frequency. Such unwanted spurious outputs must always be suppressed to the maximum possible extent, using selective filters (ie often additional tuned circuits of high *Q*) before the transmitter output is fed to the antenna.

The average power output of an ssb transmitter is usually quite low because of the extremely peaky nature of the human speech waveform (for example the amplitude level is likely to be less than 20 per cent of the peak level for some 80 per cent of the time) and this has encouraged the use of various forms of speech processing designed to raise the average output power. To do this effectively, and without introducing unacceptable distortion, is more difficult to achieve with ssb than with a.m. or nbfm signals, and usually involves low-level clipping of the rf signal (*rf speech clipping*); such facilities may be built into modern ssb transmitters and transceivers or as optional extras, or as separate external units. It should be noted that since raising the average power has the effect of raising the duty cycle of the power amplifier, it may, for example, be necessary to install a fan to reduce the bulb temperature of the output valves when installing a speech processor: bulb temperature is often the limiting factor in the power-handling capabilities of modern high-perveance rf power valves.

## Oscillators

The frequency stability of any transmitter is governed by that of its oscillator(s). In absolute terms the frequency variations become proportionally greater when the low-level output of an oscillator stage is multiplied upwards in frequency through one or more frequency doublers, triplers etc. This does not happen when the frequency is changed upwards by means of a heterodyning mixing process, where the output frequency will have a frequency stability determined by both oscillators, and since one at least of these is likely to be crystal-controlled, the stability may at the higher frequency be virtually that of the basic vfo at its fundamental frequency (thus representing an *improvement* in terms of frequency tolerance in percentage terms). The heterodyne vfo, either as a "stand-alone" unit or as part of the transmitter exciter, has thus brought about a marked improvement in vhf transmitter stability. It is one of the factors that has made it possible to use ssb with variable-frequency operation (although it should be noted that the heterodyne form of vfo can result in spurious outputs in non-amateur frequency bands).

The simplest method of generating stable rf energy is to use a quartz crystal in place of a conventional tuned circuit: Fig 4.

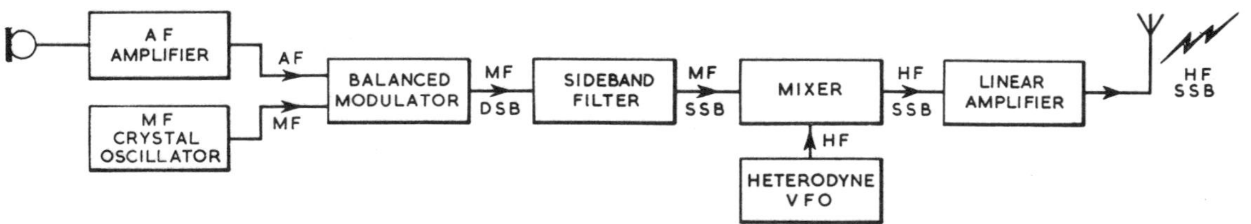


Fig 3. Basic arrangement of filter type ssb transmitter. To facilitate band switching more than one frequency conversion stage would usually be incorporated, and the signal is now usually originated at hf rather than mf

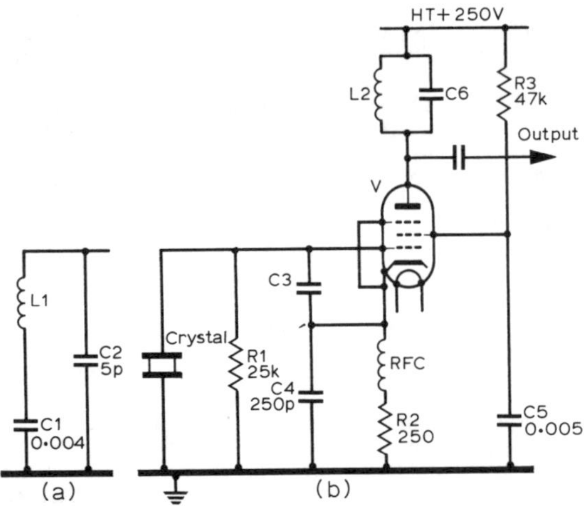


Fig 4. Crystal oscillator. (a) Approximate equivalent of 1·9MHz crystal. C1, 0·004pF; C2, 5pF; L1, 2H. (b) Typical crystal oscillator for 1·9MHz; C3, 75pF; C4, 250pF; C5 0·005μF; R1, 25kΩ; R2, 250Ω; R3, 47kΩ; V, 6AG7

Quartz is a piezo-electric material; this means that when an electrical potential is placed across a suitably cut thin plate of quartz the plate will be deformed and, conversely, if the plate is deformed an electrical potential will be produced across the two faces. If a mechanical vibration is set up within the quartz a corresponding alternating electrical voltage will appear across the two faces; normally, this vibration would soon die out but, by using a semiconductor or valve amplifier to supply suitably timed electrical impulses to the plates, the oscillation can be maintained. The crystal is thus equivalent to a conventional LC tuned circuit but the resonant frequency depends on the physical dimensions of the quartz and its mode of oscillation. It thus provides a precise generator of spot frequencies (although these can be shifted very slightly by externally loading the crystal with capacitance or inductance). Crystal frequency will depend to an extent upon temperature, although the quartz is so cut as to result in a zero temperature coefficient at the usual operating temperature.

Over the years a large number of oscillator circuits both for crystals and variable LC tuned circuits have been developed from the basic Colpitts, Hartley and Miller oscillators. Crystal arrangements include fundamental, harmonic and "overtone" circuits designed to provide output on the fundamental crystal frequency, on one of its harmonics (eg tritet oscillator) or to induce oscillation at a crystal "overtone" (usually very close to the third, fifth or seventh harmonic of the fundamental). It should be noted that the overtone frequency is not precisely the harmonic frequency of the fundamental, and that in such an arrangement there is *no* output at the fundamental.

A crystal frequency can be "pulled" slightly as noted above by external inductance and capacitance and this characteristic is used in a class of oscillators known as variable crystal oscillators (vxo), tunable over a small fraction of one per cent of the crystal frequency and having a stability rather less than that of a "spot" crystal oscillator but usually rather better than that of an LC oscillator. More complex are the various forms of *frequency synthesizers* in which a large number of output frequencies in small fixed increments are derived from one or more stable crystal oscillators, using mixing or more commonly phase-locked-loop techniques.

For LC-type variable-frequency oscillators (vfo) the more popular circuits include: Colpitts, Vackar (Tesla), Clapp-Gouriet, electron-coupled oscillator, Franklin etc. All these circuits are capable of excellent stability if they are correctly designed, well constructed and take into account mechanical and thermal stability as well as good electrical design. The components forming part of the frequency-determining resonant circuit (tank circuit) require particular care and the vfo should be mounted in an enclosed metal box (as a sub-assembly if within the main transmitter enclosure); mounting and wiring of components must be mechanically rigid and of good quality; the tuning capacitor should preferably have good bearings at each end. To enable the oscillator to be accurately reset to a given frequency the tuning should be large and clearly marked (it is possible to provide a "digital" read-out by incorporating a frequency counter as part of the unit); the tuning mechanism should be free from backlash and it may be advisable to use a flexible coupler between the drive and the capacitor to prevent undue strain on the bearings due to imperfect alignment.

Any fixed capacitors associated with the tank circuit must have high stability and will normally be of silver mica type. Paper dielectric, high-K ceramic and moulded mica types should be avoided; all resistors should be very generously rated to avoid a significant rise in operating temperatures. The use of a coil former having a low coefficient of expansion will help reduce inductance changes with temperature; a silica former with the wire wound in grooves while hot and allowed to contract has a low temperature coefficient. It is possible, with care, to use negative-temperature-coefficient capacitors to compensate for the positive coefficient of the other components, and various ways of doing this very accurately have been developed. However the modern semiconductor oscillator dissipates very little power so that unless in thermal contact with hot-running stages of the transmitter it is possible to limit temperature drift, even without compensation, to a very low figure after the initial switch-on changes, and even these can be kept to a few tens of hertz in a good design.

## Stability

For effective ssb operation it is essential that the frequency of the transmission remains stable to within a few hertz during the entire contact. The vfo control needs to be capable of setting the transmitter frequency very precisely in order to allow perfect "netting" with the other station(s) so that, for instance, an ssb "net" of several stations can all be received by all operators without having to keep retuning their receivers. With transceivers it is important that these are accurately set up so that the "receive" and "transmit" frequencies are normally identical.

The modern hf transmitter thus requires an extremely stable oscillator, whether this is vfo, vxo or one of the many forms of mixer-vfo or frequency synthesizer. Where a vfo is used, it is normally designed to operate over a fixed and limited tuning

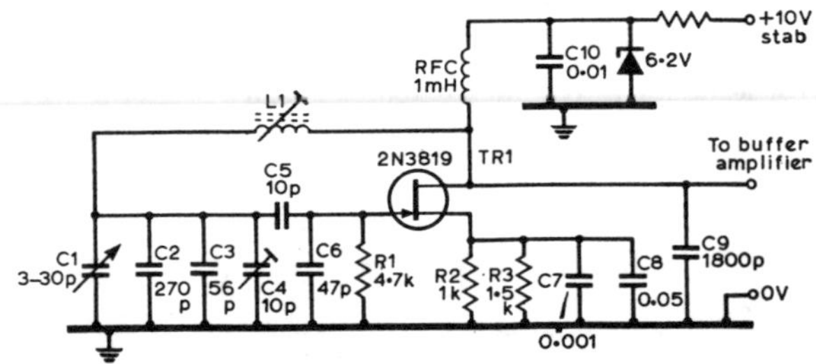


Fig 5. A 4·2–4·4MHz fet Vackar vfo with source-bias components. L1, 17t 1in dia, length 1·6in, 20swg; C1, 3–30pF; C2, 270pF sm; C3, 56pF sm; C4, 6·5pF (Tetfer trimmer); C5, 10pF sm; C7, 0·001pF; C8, 0·05μF; C9, 1,800pF sm; C10, 0·01μF; R1, 4·7kΩ 2W, hi-stab; R2, 1kΩ; R3, 1·5kΩ; TR1, 2N3819. Note that source bias components C7 and C8 can be combined as single 0·05pF capacitor and R2 and R3 combined as single 600Ω resistor

range (for example 5,000 to 5,500kHz); mixer techniques are then used with crystal-controlled oscillators to frequency convert the low-level signal to the required amateur bands before final amplification: Fig 6.

For a vfo there is no one circuit arrangement that ensures adequate stability or is superior to all others: in practice good stability depends also on the mechanical and thermal arrangements. Any mechanical instability of the frequency-determining resonant tuned circuit (tank circuit) will be reflected in the electrical output: a vfo needs to be constructed in a very rigid enclosure, using high-grade variable capacitors (or tuning diodes) with all associated fixed capacitors of the silver mica type and with any resistors extremely generously rated (to ensure that they do not heat up); ceramic coil formers are to be preferred, iron dust cores may be used but avoid ferrite cores; rigid interconnecting wires are similarly essential. Temperature variations are usually the limiting factor of a well-constructed vfo and a degree of temperature-compensation or temperature-stabilization is most desirable. It is possible to achieve stabilities of the order of better than 5Hz per 30min (after the initial warm-up period) at a fundamental frequency of 5–6MHz although with a free-running vfo this requires great care and is reached by very few designs.

It should be appreciated that to achieve good transmitter stability all oscillators within the system must be extremely stable. For this reason crystal control is invariably used for oscillators in the frequency conversion processes.

Note that during mobile and portable operation the range of ambient temperatures over which a transmitter may be expected to operate will be significantly greater than for normal domestic operation. Even when used as a fixed station, care should be taken to ensure that the transmitter is well ventilated. Ventilation should not be restricted by placing the transmitter in confined spaces.

## Power amplification

A stable vfo may provide only a few milliwatts of rf power over a single span of frequencies. It needs to be associated with other stages within the exciter section of the transmitter in order to provide perhaps 1W or so output at all the required frequencies. The exciter output must then finally be brought up to the full output power in one or usually two further stages, the driver and power amplifier (pa).

In the case of ssb and nbfm operation it is customary to carry out modulation at low power; cw keying is often at low power. Amplitude modulation in amateur transmitters was formerly left until the final power amplifier stage, but with modern multi-mode transmitters may be carried out at a similar low level to ssb. With low-level modulation a.m. is similar to ssb in requiring linear amplification for all stages subsequent to the modulator.

A linear amplifier is one in which the output waveform is an exact (or near exact) replica of the input waveform, which may consist of a *sideband spectrum* containing many different signals or tones at slightly different frequencies; non-linearity with such an input signal will result in severe intermodulation between the different signals, resulting in very heavy distortion.

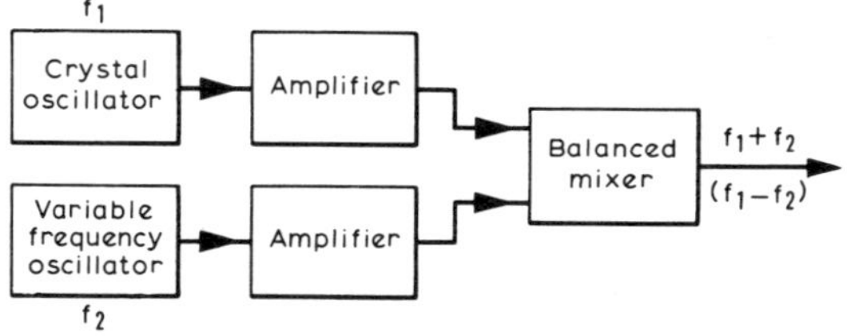


Fig 6. Block diagram of mixer vfo

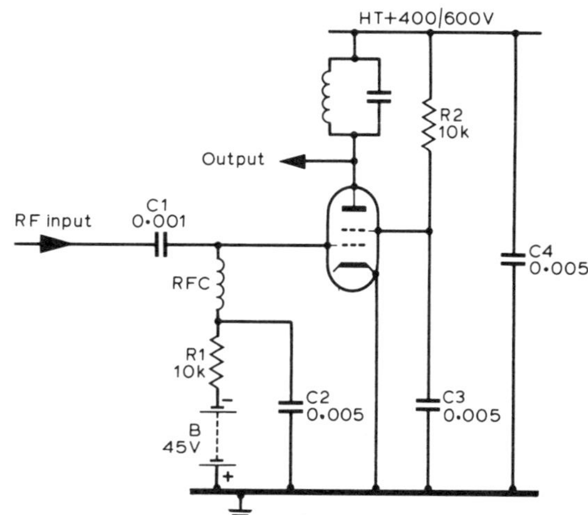


Fig 7. A typical rf power amplifier. Some representative component values are shown

A linear amplifier thus differs from stages in which the bias conditions are such that the valve or transistor conducts during only part of the input cycle (eg Class C), producing pulses of power which is then restored to sine-wave form by the "flywheel" effect of a resonant tank circuit. Newcomers are often surprised that any rf amplifiers should deliberately be operated in a non-linear mode that will handle only one sine-wave frequency at a time; such amplifiers however can be made appreciably more efficient than a linear amplifier, producing more rf output for a given dc input to the stage, although requiring more rf input power.

To understand some of the difference between linear amplifiers and the Class C type of non-linear rf amplifier, it is useful to consider a typical tetrode valve rf amplifier (Fig 7) in some detail. It should be appreciated that the same classes of power amplification apply also to transistors, although these are current-operated rather than voltage-operated devices. As we note later, the use of transistors for rf power amplification offers some additional problems.

The operating characteristic of a typical tetrode is shown in Fig 8(a). If the valve is operated under Class A conditions the grid would be biased to about −15V. Then, provided that its amplitude is not excessive, a sine-wave input voltage would produce a corresponding sine-wave anode current as shown in Fig 8(b).

If the grid bias were increased to −30V the anode current, with no rf input applied, would be practically zero and the bias would be said to be at the cut-off point. Under these conditions, when a sine-wave voltage is applied to the grid, anode current flows only for half of every cycle. We can now drive the grid positive for part of the cycle and obtain higher peaks of anode current for the same average anode current as we obtained in Fig 8(b). Greater efficiency has been achieved in the sense that more rf output power is being obtained for a given power consumption from the ht supply. It should be noted, however, that a much larger rf input is required. Not only must the amplitude of the rf voltage be greater, but the grid will have a low resistance to earth when it is positive and power will be required to drive it above earth on the positive peaks. Used under these conditions the amplifier is said to be operating in Class B.

If the grid bias is increased still further, say to about two or three times the cut-off value, the anode will conduct for considerably less than half a cycle, as shown in Fig 8(d). Under these conditions the grid can be driven even more positive before the average value of the anode current becomes excessive, and an even greater rf output power can be obtained for a given ht input power. This fact makes Class C operation, as it is called, particularly attractive for an amateur transmitter where the power limitation is imposed on the dc input to the anode of the final power amplifier.

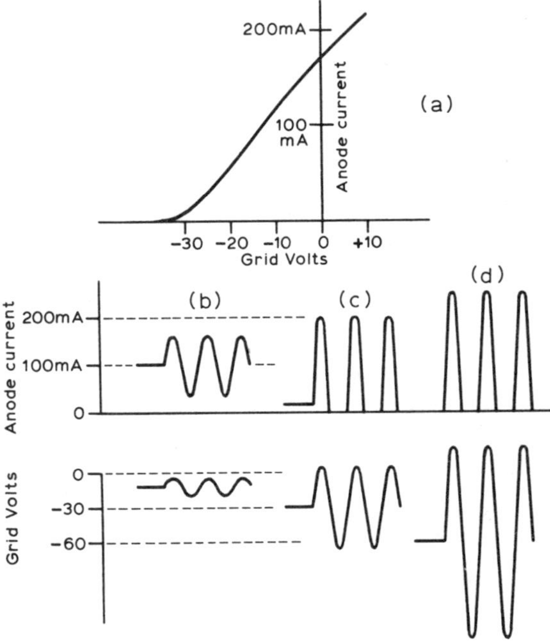


Fig 8(a) Characteristics of a typical transmitting tetrode. (b) Class A operation. (c) Class B operation. (d) Class C operation

Class B is, however, occasionally used in preference to Class C, because it has the advantage that its output can be made to be directly proportional to the rf input. This, as we have seen, is extremely useful for ssb transmission.

With a maximum input power of 150W a typical Class C amplifier might give 110W of rf power. In Class B the output would be less than 100W and in Class A less than 50W.

## Anode current

It has already been shown that when rf excitation is applied to the grid the anode will take pulses of current. If the anode circuit is off-tune the reading given by a dc milliammeter connected in the ht supply to the anode will indicate a high current. If the anode circuit is tuned to resonance a large sine-wave voltage will appear across it. The phase of this voltage will be such that the anode potential will be very high when the grid is very negative and nearly zero when the grid is positive. In consequence, the average anode current will be very low.

If a load such as an antenna or a succeeding power amplifier is coupled to the anode circuit, energy will be drawn from the tuned circuit. This will damp the oscillatory current there and the rf voltage across the coil will fall. The anode will consequently be more positive at the instant the grid potential is high and more anode current will flow.

It is quite easy therefore to bring the anode circuit into resonance by tuning for minimum anode current. When an external load is coupled to the anode the increase in anode current will give an indication of the amount of energy being extracted. Ideally the effect of the external circuit should be the same as if a resistance had been introduced into the tuned circuit. In practice some capacitance or inductance may also be introduced and a slight readjustment of the tuning may be required to compensate for this condition.

If there is no load on the anode circuit it is inadvisable to run an amplifier with full anode and screen voltages because of the high rf voltage which may be developed. Furthermore, because the anode potential will be so low during the time that the grid is positive, most of the electrons from the cathode will flow to the screen and as a consequence the screen current may become excessive. When first adjusting a power amplifier the ht voltage should be reduced to at most 50 per cent of its normal value. With a tetrode it is sufficient to lower the screen voltage only.

When adjusting the anode circuit a useful indication of the presence of an rf voltage can be obtained by a small neon bulb. When held in the hand near a point of high rf potential the small rf current which passes through the capacitance of the air gap, via the neon and thence through the body to earth, is sufficient to make the neon glow.

**TABLE 1**

**Typical valves used in amateur transmitters**

**RF amplifiers (Class C)**

| Valve | Max ht (volts) | Max dc input (watts) | Max freq at full power (MHz) |
|---|---|---|---|
| 6AG7 | 375 | 11·5 (A1) | 10 |
| 5763, QV03-12 | 375 | 15 (A1) | 175 |
| 6V6 | 350 | 16·5 (A1) | 10 |
| 6AQ5 | 350 | 16·5 (A1) | 54 |
| 6146, QV06-20 | 750 | 90 (A1) | 60 |
| | 600 | 67 (A3) | 60 |
| 6146B | 750 | 120 (A1) | 60 |
| | 600 | 84 (A3) | 60 |
| 807 | 750 | 75 (A1) | 60 |
| | 600 | 60 (A3) | 60 |
| TT21, TT22 | 1,250 | 160 (A1, A3) | 30 |
| 4-65A | 3,000 | 345 (A1) | 30 |
| | 2,500 | 270 (A3) | 160 |
| 813 | 2,250 | 500 (A1) | 30 |
| | 2,000 | 400 (A3) | 30 |

**Notes:** Figures given for maximum power input apply to a single valve, and may be doubled if two valves are used in parallel or push-pull. Valves may be operated at higher frequencies than those shown provided the power ratings are reduced. There are also certain American colour television line output valves such as the 6HF5, 6KD6 and 6LF6 popular for use in linear amplifiers for ssb operation. Typically a single 6HF5 can operate in Class C up to 115W input with an ht of 500V, or can give over 55W p.e.p. output in linear operation in Class AB1, up to 30MHz. Up to 175W p.e.p. output can be obtained from a single 6LF6.

**Typical efficiencies** (per cent)

| | |
|---|---|
| Tripler stage | 20–25 |
| Doubler stage | 30–40 |
| Class A amplifier | 10–50 |
| Class AB amplifier | 50–60 |
| Class B amplifier | 55–70 |
| Class C amplifier | 65–80 |

## Feedback

It must not be forgotten that a power amplifier has its grid and anode circuits tuned to the same frequency and that stray coupling between the two may cause oscillation. Most power tetrodes designed for rf use have a top-cap anode and in such cases it is sound practice to mount all the anode circuit above and all the grid circuit beneath the chassis. It is advisable also to place an earthed cylindrical metal screen around the lower half of the valve to shield the grid assembly from the anode circuit. To prevent overheating of the valve, allowance should be made for air from beneath the chassis to pass up between the valve and the cylinder. All leads should be kept short and direct, earth connections for any one stage being taken to points on the chassis which are as close to each other as is practicable. On no account should a length of wire be used to serve as the common earth lead for the grid and anode circuits. Such a lead would have some inductance and the rf voltage developed across it by the anode circuit, although very small, might introduce sufficient feedback into the grid circuit to cause trouble.

On very high frequencies, with a poor transmitter layout or with tetrodes having very poor internal screening, the stray anode to grid capacitance may be high enough to cause undesirable interaction between the anode and grid circuits, if not actual oscillation. It will then be necessary to neutralize this capacitance. If, as was common some years ago, a triode is used, it will be essential to take this step.

## Neutralizing

A popular neutralizing circuit is shown in Fig 9. In this arrangement the anode coil is centre-tapped and tuned by a split-stator capacitor. The voltage at the upper end of the anode coil is 180° out of phase with the anode and the feedback through Cn can therefore be made to cancel out the anode-grid capacitance. The required value of Cn will be approximately equal to the anode-grid capacitance. With a triode this might be of the order of 5pF; with a tetrode it will be of the order of 0·1pF and will require nothing more than the suitable positioning of a short piece of stiff wire.

Neutralizing is adjusted with all ht removed from the amplifier. Grid excitation is applied and Cn set so that tuning the anode circuit through resonance produces no kick on the grid current meter.

## Grid bias

There are various ways of providing grid bias but that shown in Fig 7 is one of the best for Class C operation. The battery (or small power supply) B provides about $1\frac{1}{2}$ times the cut-off

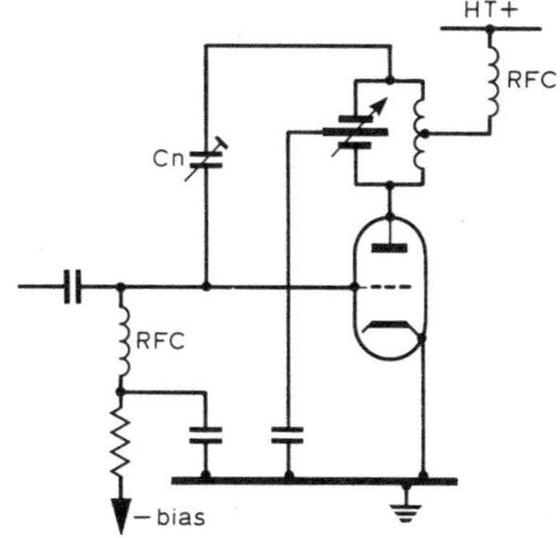

Fig 9. Neutralization of a triode valve

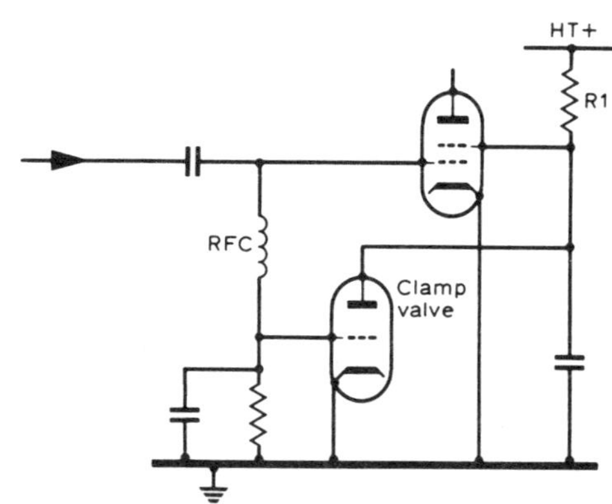

Fig 10. Clamp valve circuit

voltage; the extra voltage required is produced by the rf input itself. For a short period every cycle, the grid is driven positive and during this time it draws electrons from the cathode. These then return to earth through the resistor R1 and the battery. The current flow through R1 is smoothed out by the capacitors C1 and C2 so that there is a steady voltage developed across R1 which provides the extra bias. Since this voltage is dependent on the rf excitation it is self-adjusting, becoming high for over-excitation and low for under-excitation.

It is possible to dispense with the battery and use grid current bias entirely but this introduces the danger that if, for any reason, there is no excitation the bias will disappear and the valve will draw excessive anode current. A popular method of preventing this happening is to use a second valve—commonly called a clamp valve—to prevent excessive anode dissipation in the absence of drive by reducing the screen voltage of the pa valve to a low value. The arrangement is illustrated in Fig 10.

During normal operation the pa grid current bias causes the clamp valve to be cut off and to have no effect. Should the excitation fail, the grid of the clamp valve will rise to earth potential. The clamp valve will then conduct heavily and the resulting voltage drop across R1 will reduce the screen voltage of the tetrode, thus preventing the latter from taking excessive anode current.

The grid current of a pa gives a very useful indication of the amplitude of the excitation being applied to the grid and thus it is common practice to make provision for connecting a dc milliammeter in series with the bottom of the grid resistor. There is no advantage in increasing the current beyond the figure recommended by the valve manufacturer; it will give no worthwhile increase in efficiency and is quite liable, by overheating the grid and liberating gas, to have the reverse effect.

## Interstage coupling

It has been assumed so far that the anode of the driver valve would be tuned and would be capacitance coupled to the grid of the amplifier as shown in Fig 11(a). The capacitor C1 isolates the grid from the ht supply but provides a low-impedance path for the transfer of rf energy from the driving anode. The rf choke RFC gives a dc path for the grid bias but its high rf impedance prevents it from absorbing any appreciable rf energy.

An alternative system which is very popular is that shown in Fig 11(b). In this arrangement the grid circuit also is tuned and is inductively coupled to the anode of the driver. One advantage of this method is that by suitably adjusting the coupling between the two coils it is possible to obtain a bandpass characteristic. This means that it is possible to obtain an efficient transfer of energy

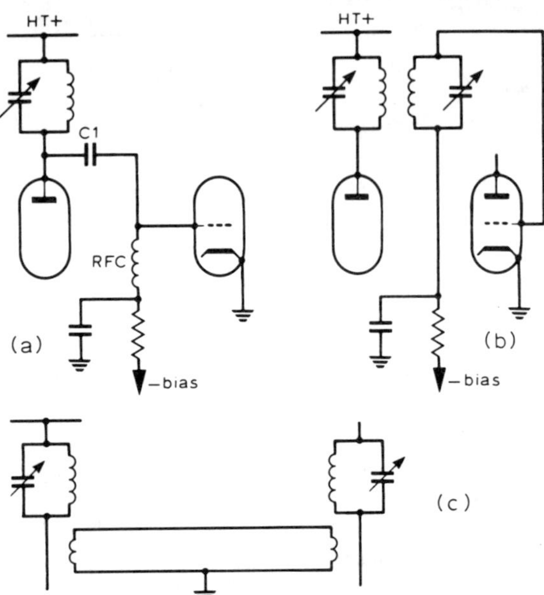


Fig 11. Interstage coupling. (a) Capacitive coupling. (b) Inductive coupling. (c) Link coupling

over the whole frequency range of one amateur band without any retuning being necessary.

A modified version of inductive coupling is the link coupling method shown in Fig 11(c).

## Frequency multipliers

The anode circuit can, of course, be tuned to a multiple of the excitation frequency. It will not then receive a driving pulse during every cycle of its resonant frequency but its own pendulum action will maintain a circulating sine-wave current. This arrangement is most commonly used to give twice or three times the input frequency. The amplifier is then said to be a *doubler* or a *tripler* and the output frequency the second or the third harmonic of the input frequency. Because the anode and grid are tuned to different frequencies there is less danger of feedback and no need for neutralization.

A frequency multiplier is less efficient than a straight rf amplifier and is generally only used as an intermediate stage between the master oscillator and the final pa. Such frequency multipliers are not linear and cannot be used for ssb.

Since many of the amateur bands are harmonically related to each other it is possible to obtain outputs on a number of bands by using one master oscillator followed by different arrangements of frequency multipliers. The more ambitious transmitters, for example, often have a vfo giving an output in the 3·5MHz band followed by switchable frequency multipliers which enable the input to the final amplifier to be in the 3·5, 7, 14, 21 or 28MHz bands.

## Output networks

Most transmitters are designed to provide a low-impedance output of between 50–75Ω although with valves the tank circuit must provide a high-impedance load (with transistors this is not the case since rf power transistors usually require a low-impedance collector load).

One way of connecting the tank circuit to the output lead is to join the latter to a small coupling winding wound around the "cold" (or low rf potential) end of the tuned circuit as in Fig 12(a). The tank circuit then becomes an rf transformer and the coupling can be regulated by varying the number of turns on the output winding or by adjusting its proximity to the main winding.

Several other possible arrangements are illustrated. In Fig 12(b) the ht has been removed from the tuned circuit by the buffer capacitor C3 and an rf choke has been added to give a dc connection from the anode to the ht supply. Instead of using a coupling winding the output lead has been tapped into the coil.

A more elegant arrangement is to tap the output lead into the capacitance as in Fig 12(c). If both C1 and C2 are variable we can use them not only to alter the tuning but also to adjust the effective tapping point. In order that the rotors of both capacitors can be earthed the connections may be modified to those shown in Fig 12(d).

The circuit in Fig 12(d) is extremely valuable. It is usually drawn as in Fig 12(e) and, because of the resemblance of the network C1-L-C2 to the Greek letter $\pi$, is known as a *pi-network tank circuit*. Its great merit is that it is far more successful than most other circuits in suppressing harmonics. C1 is effectively from anode to earth and provides a fairly low impedance at harmonic frequencies; L is between the anode and the output socket and presents a high impedance; C2 provides a low-impedance shunt right across the output socket. The suppression of the second harmonic is, in fact, four times better than with the circuit in Fig 12(a). With the higher harmonics the improvement is greater still and the suppression of the seventh harmonic, for example, is 49 times better. In the interests of reducing television interference, therefore, the use of the pi-network circuit is to be strongly recommended for 7, 14, 21 and 28MHz—and for 3·5MHz also if the input power exceeds 10W.

Assuming that the network is to feed from the anode of a tetrode into 72Ω, and that the $Q$ is to be 15, the correct values for C1, C2 and L will be approximately as given in Table 2. A typical pi-network tank circuit derived from this table is shown in Fig 13.

The adjustments of C1 and C2 will be to some extent interdependent but C1 can be considered as the tuning control and C2 as the loading control. The tuning procedure is first to set C2 to maximum capacitance (to give minimum loading) and then to tune C1 for the anode current dip. The meter reading should then be low, indicating insufficient loading. The value of C2 is therefore decreased a little and C1 retuned, the process being

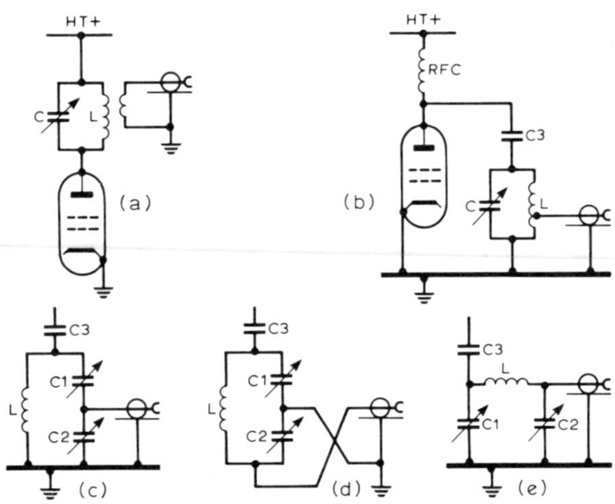


Fig 12. Methods of feeding output to a coaxial cable. The pi-network circuit (e) is recommended because of its low harmonic output

**TABLE 2**

**Correct values for C1, C2 and L to give Q of 15 in a pi-network tank circuit feeding from a typical beam tetrode into a 72Ω load.**

| Va/Ia | C1* (pF) | C2* (pF) | L* (μH) |
|---|---|---|---|
| 4 | 140 | 730 | 4·3 |
| 6 | 100 | 640 | 6·1 |
| 8 | 75 | 560 | 7·8 |
| 10 | 61 | 510 | 9·5 |

*Values are for 7MHz. Multiply by 4 for 1·8MHz, 2 for 3·5MHz, ½ for 14MHz, ⅓ for 21MHz and ¼ for 28MHz.

repeated until the anode current at the dip has been brought to the desired value.

Additional harmonic suppression can be achieved by the use of a further coil connected between C2 and RFC2; the circuit is then known as a *pi-L network*.

## Harmonic suppression

The mere use of a pi-network tank circuit alone does not guarantee that there will be negligible harmonic radiation in the television bands and it is still important to pay attention to a number of practical points. Among the most important are: (a) enclosing the entire rf section of the transmitter in a metal box, ensuring that all parts of the box make good electrical contact to each other; (b) taking all earth connections associated with the pa stage by the shortest possible length of heavy-gauge wire to one common point on the chassis close to the cathode pin; (c) using screened wire for all leads carrying ht, grid bias or heater supplies; (d) fitting bypass capacitors of 0·001 to 0·005μF from all unearthed heater pins and from each side of the mains input to chassis; (e) avoiding excessive rf input to the grid of the pa.

Provided that sufficient care is taken with the above points there should be no appreciable vhf radiation from the transmitter itself or from the mains lead. There may, however, still be a large enough harmonic content in the output to cause troublesome radiation from the antenna and it may be necessary to connect a filter between the transmitter and the antenna tuner. If this filter is to give severe attenuation of frequencies above 40MHz and yet have negligible effect on frequencies in any amateur band from 1·8 to 30MHz it must be carefully designed.

Spurious signals from a transmitter need not necessarily be caused by harmonics. They can also be due to parasitic oscillations occurring in one of the rf amplifiers or even in the oscillator. Sometimes they make themselves apparent by unstable or inefficient operation of the transmitter but sometimes are only noticeable as additional signals, usually rough in tone, picked up on a receiver.

Parasitic oscillations in the vhf range can often be traced to poor screening or long wiring. Lengths of wire in the grid, anode, cathode and screen connections can act as tuned circuits which resonate at vhf and, even though their *Q* may be low, the gain of a modern tetrode is so high that oscillation can easily result. Oscillations at vhf are particularly liable when two or more valves are connected in parallel. It is then always advisable to include a parasitic stopper in the lead to each grid and to each anode. The stopper should be wired direct to the valveholder and can be made by winding 5-10 turns of about 20swg wire around an insulated type 100Ω ½W resistor, connecting the coil in parallel with the resistor. The optimum number of turns may have to be found by experiment.

If any stage has grid and anode chokes resonating at the same

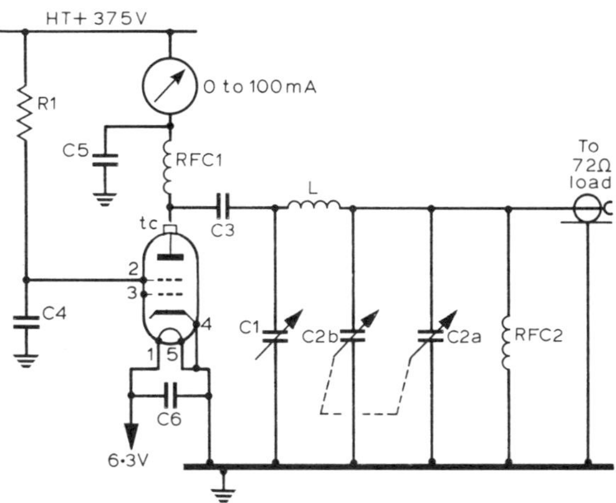


Fig 13. A typical pi-network circuit for a 25W 14MHz transmitter. Assuming the anode current to be 62·5mA with the transmitter loaded, Va/Ia = 6, the correct values (Q = 15) for C1, C2 and L are 50pF, 320pF and 3·05μH respectively. Practical component values are C1, 100pF variable; C2, 350 + 350pF receiver type; C3, 0·002μF mica 500V wkg; C4, C5, 0·001μF 500 wkg; C6, 0·001μF 350V wkg; L, 13t 14swg enam, 1in dia, 1¾in long, self-supporting; R1, 6·8kΩ; RFC, 1mH; RFC2, 2·5mH is to prevent ht from reaching the antenna if C3 fails

frequency, the feedback through the inter-electrode capacitance of the valve may occasionally give rise to spurious oscillation. The frequency will then usually be lower than the transmitter frequency. The remedy in such cases is to use a different type of choke in one position or to replace the grid choke by a resistor. Ferrite beads can also be used to prevent parasitics in rf and af amplifiers.

## Television interference (tvi)

The most difficult single problem arising in modern transmitters is to avoid causing interference to nearby television receivers. Unless interference can be avoided, or is not the fault of the transmitter, operation has to be restricted to those hours when the television stations are not on the air. No amateur would pretend that this is a simple problem or that there is any one all-embracing remedy. However, experience has shown that it is possible to avoid interference by effective screening and filtering of the exciter and pa, and of all leads before they emerge from the metal screening cabinets. There are many amateurs who have to all intents and purposes completely solved their interference problems and can operate their stations at all times even with television receivers running in the same room—though the occasional unexpected complaint may still turn up from time to time to test their ingenuity in devising further means of protection. Unfortunately television receivers are often overloaded by strong local signals, even when these are on frequencies well removed from the television bands.

Provided that the transmitter is itself effectively screened and filtered so that no harmonics can leak out through the cabinet or along the power supply or auxiliary leads, then any harmonics generated by the transmitter can only be radiated if they reach the antenna system. One of the most successful methods of preventing such harmonic radiation thus consists of placing a harmonic filter in the low-impedance coaxial transmission line used to feed the transmitter power to the antenna system. For an hf transmitter this is often a *low-pass filter* (lpf) designed to reduce (attenuate) to a high degree all output from the

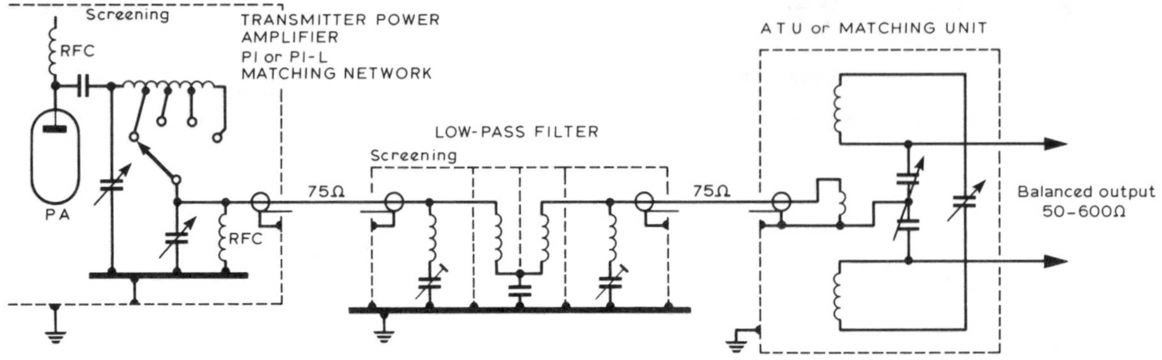

Fig 14. Typical anti-tvi precautions to prevent the radiation of transmitter harmonics. The harmonic signals are reduced first by the pi-network output circuit, then by the low-pass filter, and finally by the antenna tuning unit

transmitter above say 30MHz while having little effect upon signals below this frequency. The construction and alignment of such filters are described in the standard amateur handbooks. A filter of this type can simply be left in position on all bands below 30MHz.

An alternative arrangement which does not have this advantage, but which is somewhat simpler to build and adjust—and is quite suitable where most activity is on only a few bands—consists of a series of separate *bandpass filters*, each designed to pass signals on only one amateur band and to attenuate all other frequencies. Harmonic suppression may not be as good with this type of unit as with a correctly adjusted low-pass filter, but impedance matching problems are less likely to arise.

The simplest form of bandpass filter takes the form of a good *antenna tuning unit* (atu) whose main purpose is often to match the low-impedance coaxial output from the transmitter into the antenna system. A good atu combined with a pi-network in the pa tank circuit (or preferably the slightly more effective "pi-L" network) may well provide all that is required to eliminate tvi in areas where there are strong television signals. An atu may, of course, be used in combination with a low-pass or bandpass filter since the attenuation of a series of filters will normally add together, each reducing any harmonic content still further: Fig 14.

## A low-pass tvi filter

A form of low-pass tvi filter which is rather easier to adjust than some others is shown in Fig 15. This type of filter offers several advantages: it uses two less coils for equivalent stop-band attenuation; it provides relative freedom in the selection of frequencies of maximum attenuation; and it can be tuned with the aid of only a grid-dip meter. The example shown has a 30MHz cut-off, with tuned circuits at 78·1MHz, 40·5MHz and 47·3MHz resulting in the attenuation curve shown in Fig 16. As with all such filters it is preferable for the filter to be built in a small screened enclosure with additional internal screening to separate the various sections and to prevent leakage of vhf harmonics through the filter. Coils can be made from 14swg enamelled copper wire formed on a ½in diameter mandrel and then self-supporting: L1 8 turns; L2, L3, 6 turns. When the coils have been formed the capacitors C2, C4 and C6 are soldered across them and initially tuned to resonance by adjusting the spacing of the turns until a gdo indicates resonance at the appropriate frequencies. The coil/capacitor assemblies are then mounted in the filter box individually and again checked for resonance. Finally the shunt capacitors C1, C3, C5 and C7 are soldered into the filter. The prototype unit was in a 5in by 3in by 2in aluminium box, with aluminium shields providing isolation between the three filter sections. Each shield should be carefully secured at eight points to assure good isolation, and preferably angle brackets used to reduce leakage from the enclosure with any paint removed from the edges of the cover to ensure good metallic contact between overlapping flanges when the unit is assembled. These precautions are important in preventing harmonic currents from reaching the outer surface of the housing and so bypassing the filter.

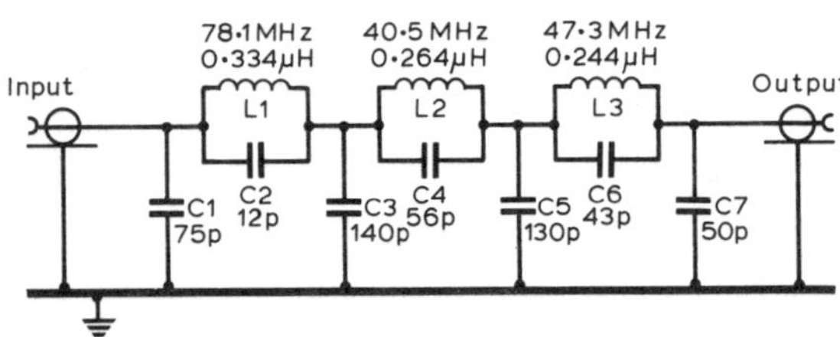


Fig 15. A low-pass filter providing attenuation exceeding 50dB above 40MHz with easy adjustment

## TV, radio and audio breakthrough

The adoption in the UK of the uhf Bands IV and V as the main tv broadcasting bands has decreased the problem of tvi caused by harmonic radiation from hf transmitters. Unfortunately, however, many solid-state television sets are susceptible to high rf fields from nearby hf, vhf and uhf transmitters even when these are on frequencies far removed from the tv frequencies and when

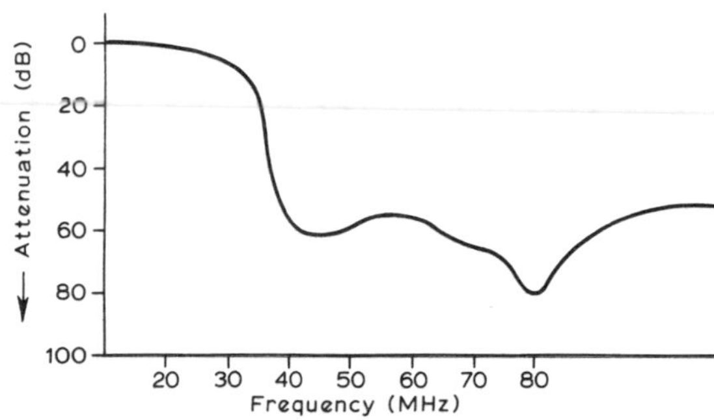


Fig 16. Attenuation of the low-pass filter; points of maximum attenuation depend on the resonant frequencies

the outputs have been carefully filtered to remove any spurious outputs.

The early stages of the tv sets may thus be overloaded by the local, strong signals of the amateur, or his signals may be rectified directly in the af stages of radio and tv sets; similarly, interference may be experienced by the users of hi-fi audio and stereo equipment, tape recorders, electronic organs or even hearing aids. In most of these cases, it is unlikely that any additional precautions or modifications to the transmitting station will have much effect, other perhaps than a very significant reduction of power, or placing the transmitting antenna as far as possible from the domestic equipment, or the use of a constant carrier system such as nbfm or fsk rather than ssb or a.m.

A number of effective techniques for dealing with such types of breakthrough interference have been developed, aimed at reducing the rf entering the domestic equipment via antennas, mains leads or loudspeaker cables. For example a *high-pass filter* or "*braid-breaker*" type of filter may be placed in the antenna feeder of the affected tv or vhf/fm receiver, or a mains filter connected in the mains lead. Bypass capacitors or rf chokes may be required across loudspeaker leads in audio equipment, and occasionally bypass capacitors have to be connected within the audio stages. Where a simple cure cannot be effected, the owner of the equipment should be encouraged to report the problem to the manufacturer, importer or retailer of the equipment concerned. The more reputable firms accept that no responsibility for this type of interference can be ascribed to the amateur station.

Fig 17 shows a simple uhf high-pass filter made from double-sided copper laminate printed circuit board. In more difficult situations a double-section filter may be necessary.

## Keying

The purpose of any radio transmitter is to convey information to the distant listener; this may be in the form of speech, visual images, teleprinter messages or, the simplest way of all, by

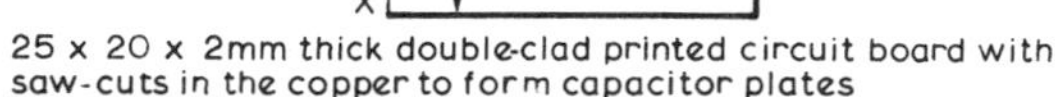


**Fig 17. Method of constructing uhf high-pass filters using double-sided copper laminate board. Single-section filter suitable for most cases of tvi**

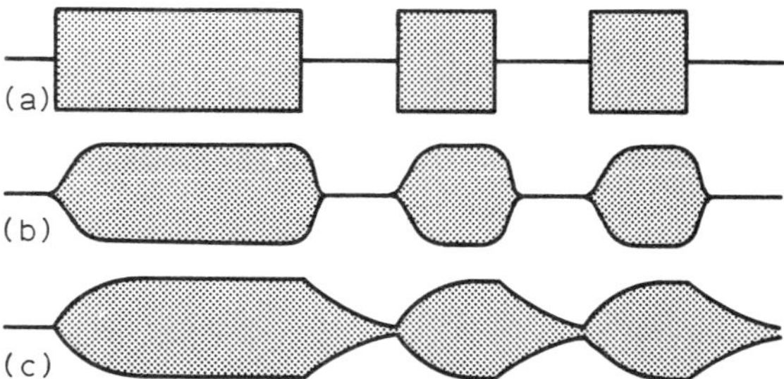


**Fig 18. Envelope of a cw signal. (a) No click filter. (b) Adequate click filter. (c) Excessive click filter**

signalling in morse code (cw or A1). For cw transmission we need to control the rf output of the transmitter by means of a morse key or keyer, changing from zero output to full output and back to zero output several times in a second. A morse key is simply a convenient form of electrical switch and we can arrange to switch a transmitter on and off in a number of ways, in any of its stages. However care is necessary if this is to be done effectively, without causing interference to local radio and television receivers, yet providing a clean, stable, pleasant-sounding signal to the distant listener.

In a cw transmission, over which no care has been taken, the envelope of the rf waveform might be as in Fig 18(a). The modulation waveform would in effect consist of square waves and if analysed would be found to be composed of a large number of sine waves, the frequencies of which might extend to as high as 100kHz. The sidebands of such a transmission might therefore extend over an entire amateur band and be apparent to listeners as annoying clicks. To prevent this happening all the high frequency components of the modulating waveform must be severely attenuated. In practice this means that we should prevent any sudden changes in amplitude of the rf signal, and produce rounded dots and dashes such as in Fig 18(b). With suitable shaping it is possible to reduce the frequency band covered by a cw signal to considerably less than 1kHz.

One means of shaping the output waveform is the key click filter: Fig 19. When the key contacts close, the inductance of the iron-cored choke L prevents the anode current rising too suddenly. When the contacts are broken the capacitor C is initially in a discharged state and the anode current continues for a brief period, dying away as C is charged up to the voltage across the key contacts when open. When the key contacts close again they will short-circuit C, but the presence of the resistor R prevents the discharge current being excessive. The larger the values of L and C the more gradual will be the rise and fall of the envelope of the signal, and this must not be too gradual or it will resemble that of Fig 18(c), making the characters indistinct and difficult to copy. The correct value of L and C depend upon the current being keyed and are usually found by experiment.

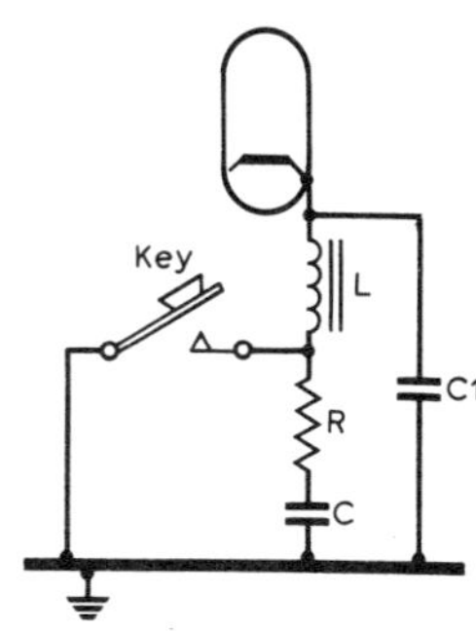


**Fig 19. Cathode keying. C, L and R are the click filter. C1 is for rf bypassing and should be about 0·005μF**

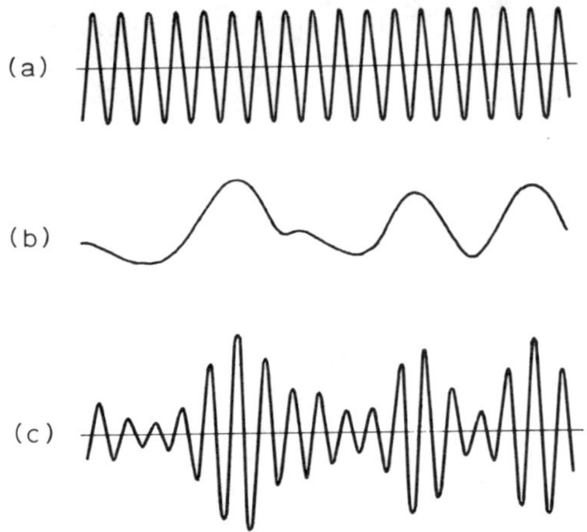

Fig 20. Amplitude modulated telephony. (a) Unmodulated rf signal. (b) Typical speech waveform. (c) Speech-modulated rf signal

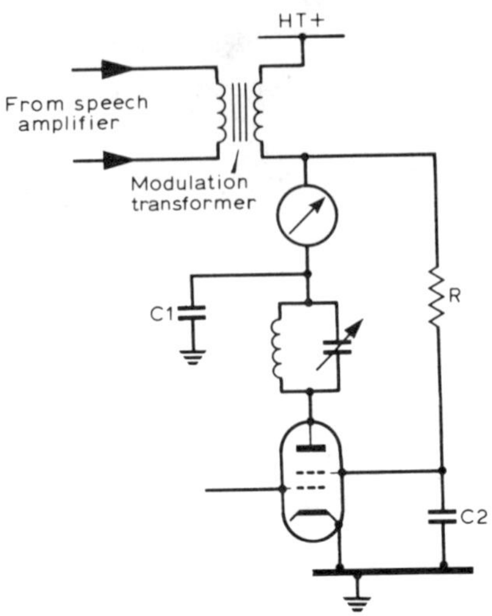

Fig 21. Method of applying anode modulation to the pa valve. With a triode, R and C2 would be omitted. With a tetrode, the system is more correctly described as anode and screen modulation. The values of R, C1 and C2 can be the same as for an unmodulated stage except that C2 should not be greater than about 0·001μF

Valves and transistors may be keyed in any of several different ways, and often a fairly early stage of the transmitter is keyed so that the currents and voltages will be lower than when the output stage is keyed. It is very convenient if we can arrange the keying so that not even a weak signal is picked up in the local receiver in the key-up condition, and this suggests that the most logical stage to key would be the oscillator. However it is not easy to key an oscillator without introducing "chirp" into the signal, caused by the slight change of frequency that occurs with most oscillators when they go into or cease oscillation. One method of overcoming this problem is to use *differential keying* in which two stages (one of them the oscillator) are keyed simultaneously but arranging that the vfo is swiched on a tiny fraction of a second before the later stages of the transmitter and switched off a fraction of a second later.

Some cw operators like to use "break-in" techniques. That is to say that they have both transmitter and receiver arranged so that there is no need for manual operation of switches other than the morse key when going from transmit to receiver or vice versa, with the receiver muted only when the key is actually depressed. This means that they can receive incoming signals at any time when the key is "up" so that the distant operator can interrupt immediately if required, or it can be heard if the station called has in fact come back to someone else. In effect to do this successfully involves arranging for automatic switching of the antenna between receiver and transmitter and de-sensitizing the receiver when the key is down.

## Amplitude modulation

The easiest way of modulating a transmitter is amplitude modulation of anode and screen electrodes of a tetrode power amplifier. During the years when this was the usual system for amateur phone transmission, many variations to this approach were developed, but are no longer often used.

Amplitude modulation implies the use of speech waveform to control the amplitude of the rf signal envelope. One way of achieving this is to amplify the audio frequency voltage which appears across the microphone and to superimpose this amplified voltage on to the ht voltage of the rf power amplifier by means of a transformer as shown in Fig 21. It is a useful characteristic of a Class C amplifier that the output voltage is directly proportional to the ht voltage and the shape of the rf envelope is, therefore, a faithful replica of the af waveform.

When, as in Fig 22(b), the modulated ht supply swings between zero and double its unmodulated level the modulation is said to be 100 per cent. If the modulating voltage is greater, as in Fig 22(c), the anode of the rf valve will periodically be driven negative. The abrupt cuts in the rf output which this causes will produce modulation frequencies much higher than those present in the original speech waveform and these will cause "splatter" to be radiated perhaps as much as 50kHz either side of the signal. The aim, therefore, with an amplitude-modulated transmitter is to maintain the modulation percentage as high as possible without running the risk that it will exceed 100 per cent. For the

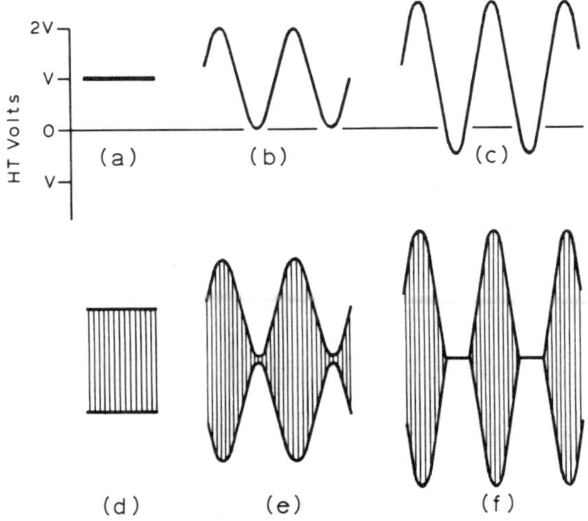

Fig 22. Anode modulation. (a) Unmodulated ht. (b) HT with superimposed af to give 100 per cent modulation. (c) HT with superimposed af to give 150 per cent modulation. (d), (e) and (f) show the waveforms of the resultant rf outputs

most satisfactory results, it is advisable to have some form of modulation monitor.

When amplitude modulation is applied to the pa the mean rf output power is increased, and in the case of anode modulation this extra power is supplied by the speech amplifier. For 100 per cent modulation the speech amplifier must supply an af power equal to about 50 per cent of the total dc input power to the pa valve. Thus, in the example shown in Fig 21, the total of the anode and screen currents might be 0·08A at 350V. The dc input power would then be 0·08 × 350 or 28W and thus the required af power would be about 14W.

A high-power a.m. transmitter thus requires a high-power audio amplifier and for input powers much above 100W, amplitude modulation in this form, although simpler technically, is a more costly approach than ssb. To the cost of the af amplifier must be added the cost of a modulation transformer of suitable impedance matching. Various lower-cost approaches to a.m. are possible but result in somewhat lower efficiencies.

## Frequency modulation

Frequency modulation, like amplitude modulation, is a means of conveying intelligence by varying the characteristics of a transmitted carrier wave in accordance with changes in the audio waveform. But instead of varying the transmitted *power* by the amplitude of the audio signal, the *frequency* of the carrier wave is changed. In accordance with the loudness of the speech the carrier is deviated away from its nominal frequency, and the louder the speech peaks the more the frequency is deviated. An exaggerated representation of frequency modulation (fm) is shown in Fig 23. It will be seen that when a modulating voltage is applied the carrier frequency is increased during one half-cycle of the modulating waveform and reduced during the other half. The change in carrier frequency (ie frequency deviation) is made proportional to the instantaneous amplitude of the modulating signal and will be greatest on peaks of the modulating waveform.

A signal having a large frequency deviation, as found in fm broadcasting, has important advantages over one that is amplitude modulated: it provides a much improved signal-to-noise ratio for a given transmitter power and is much less susceptible to interference of an impulsive form, such as car ignition or static. But we obtain these advantages only at the cost of using more bandwidth for the transmission. In fm broadcasting the signal when modulated fully has a deviation of ±75kHz and requires a transmission channel of some 200kHz. In the full 2MHz bandwidth of the 144MHz band there would be room for only 10 wideband fm transmissions of this type. The full advantages of this form of fm transmission can be achieved only when the signal being received is above a fairly sharply defined "threshold" which is determined by the fm demodulator in the receiver. On weak signals, quality and signal-to-noise degrades very rapidly indeed below the threshold.

For vhf communication, wideband fm is not permitted and in practice two-way communication is limited to a deviation of less than ±15kHz; for amateur radio the signals should not be deviated by more than ±3kHz, so that they occupy roughly the same amount of bandwidth as an a.m. signal. Such systems are termed *narrow-band frequency modulation* (nbfm). With nbfm it is not possible to achieve the improved signal-to-noise ratios of wideband broadcast fm and in terms of communication efficiency, nbfm depends very much on the form of demodulator used in the receiver. Generally overall efficiency will be rather lower than for a.m. and considerably lower than for ssb.

Nevertheless there are several reasons why nbfm has become popular on vhf and uhf amateur bands. For instance the constant power output of an fm transmitter tends to cause less interference to local television and radio receivers; fm is more suited to the limiting characteristics of rf power transistors and so can be more readily implemented in solid-state transmitters; it is also a technically simple and inexpensive form of modulation, although it is rather more difficult to check that the signal is not being over-modulated (ie its deviation is kept within ±3kHz). Since the sensitivity of receiver discriminators tends to be low on nbfm there is an in-built tendency to over-deviate the transmission to make it sound louder, but producing an unduly broad signal.

Frequency modulation can be produced by reactance modulation of an oscillator, in practice using a form of diode, the capacitance of which varies according to the reverse voltage applied across it. Very little audio amplification is needed.

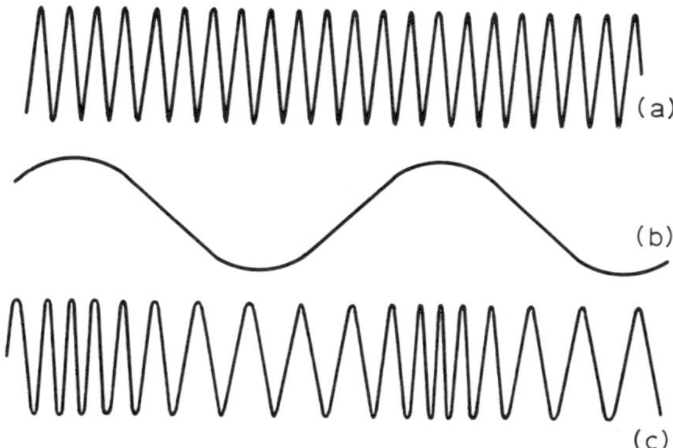


Fig 23. Frequency modulation. (a) Unmodulated rf signal. (b) AF waveform. (c) Modulated rf signals

However, nbfm is often in the form of *phase modulation* (pm) and achieved by varying the phase in a tuned circuit in one of the stages subsequent to the oscillator, without attempting to vary the frequency of the oscillator itself. With fm, deviation is directly proportional to modulation amplitude, whereas, with phase modulation, the deviation is proportional to modulation amplitude multiplied by the modulation frequency. Phase modulation can be converted to true frequency modulation by inserting a network having an attenuation of 6dB per octave with increasing frequency in the audio amplifier. Such a network, of course, helps to limit the audio frequency spectrum. Fig 24 shows one form of phase modulator for 144MHz nbfm operation.

It is possible to receive nbfm signals on a conventional a.m. envelope detector by slightly off-tuning the receiver so that the signal is received on the "slope" of the i.f. response curve, automatically providing the necessary conversion from fm to a.m. Better results can be achieved by using an fm discriminator type of demodulator such as the ratio detector or the Foster-Seeley discriminator. However it should be noted that a high sensitivity discriminator is needed for nbfm capable of producing reasonably strong audio output from small deviations of frequency. To achieve this, many amateur equipments demodulate nbfm by means of double conversion to 455kHz, rather than using the wideband fm standard i.f. of 10·7MHz. There are however some forms of discriminators, such as those using quartz crystals or phase-locked-loop (pll) demodulators, which can be used very successfully at 10·7MHz.

It is also possible to receive nbfm signals on a product detector, treating the signals exactly as though they were ssb signals and using the bfo or carrier-insertion oscillator: this is possible because nbfm can be thought of as an ssb signal with the carrier re-inserted 90° out of phase; the locally re-inserted carrier then over-rides the transmitted carrier to provide a form of "exalted carrier" reception.

We have noted that, with fm, the louder the audio signal, the

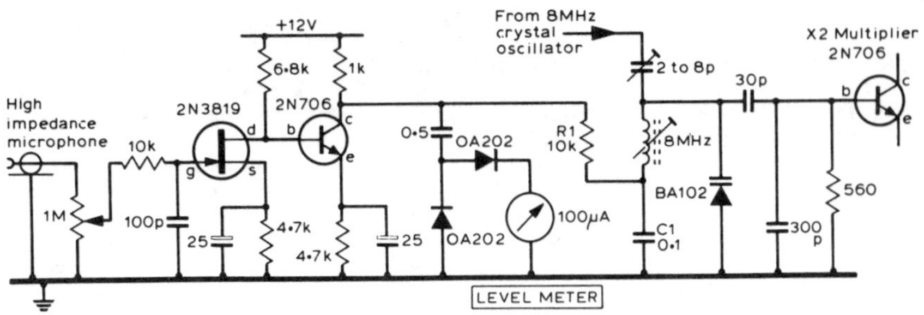


**Fig 24. Phase modulator for 144MHz. R1 is the 10kΩ resistor connected to the 8MHz coil; C1 is the 0·1μF capacitor also connected to this coil**

greater the frequency deviation of the transmitter; the *rate* at which the frequency changes about its nominal carrier frequency represents the audio *frequency* transmitted. As with a.m. this process results in a carrier wave accompanied by sidebands; but because the total radiated power of an fm signal remains constant, there is a continuously varying and rather complex relationship between the power in the carrier and the total sideband energy. In theory (but fortunately not in practice) the bandwidth of any fm signal is infinite and there are certain relationships at which the power in the carrier wave is actually zero. These principles are quite difficult to grasp but luckily it is not really necessary for a newcomer to analyse mathematically just how nbfm works, and indeed most amateurs think of nbfm as a continuous wave signal varying about a nominal frequency. The question of sidebands however is of some practical significance: it means that an fm signal actually requires more channel space than is represented purely by the maximum deviation.

## Single-sideband equipment

In amplitude-modulated (a.m.) transmission the radiated power is contained in a steady carrier (representing the basic unmodulated output of the power amplifier) plus, whenever there is speech, in two sets of sidebands equally placed on either side of the carrier (see Fig 25(c)). For a long time it has been appreciated that all the power represented by the steady carrier is wasted, in the sense that it conveys none of the actual speech information. The sole use of the carrier is its value in making it easy for the receiver to recover the original speech information from the incoming signals. Yet during typical a.m. telephony, some 80 per cent of the power is in the carrier.

By eliminating the basic carrier from the transmission (and replacing it locally in the receiver), all the transmitted power can be used to convey the actual speech information. We can do this by using what is termed a *balanced* modulator; such a stage, when fed with an rf carrier and an af signal, produces as output the amplified sidebands of a normal modulator but balances out or suppresses the carrier. Such an output signal, if radiated, is known as *double sideband suppressed-carrier transmission*, or more simply as dsb (although this abbreviation is used by some professional radio engineers to describe an ordinary double sideband a.m. signal with carrier). This type of transmission is used by a number of amateurs, and has the advantage that a conventional a.m. transmitter can often be altered very easily to provide dsb—but for various reasons is much less popular than *single sideband* (ssb) transmission.

In ssb we go one stage further than in dsb and eliminate not only the carrier but also one or other of the two similar sets of sidebands which result from a balanced modulator. Since each set of sidebands is merely a mirror image of the other, clearly any information contained in them both is available from only one. Thus for ssb we concentrate the entire output power of the transmitter into one set of sidebands, covering a bandwidth equal to the audio bandwidth which we are transmitting, preferably for speech about 200 to 3,500Hz. The ssb signal thus has the same benefit over a.m. of dsb in that all power is put into useful information, but has the advantage that this full power gain can be utilized at the receiver without the special synchronous demodulation techniques which would be required for optimum dsb reception. And also—most important—ssb reduces the total bandwidth of the radiated signal, when correctly adjusted, to rather less than half that required for either a.m. or dsb. With so many amateurs packed into relatively-narrow amateur bands this is a most valuable feature of ssb. Furthermore there are no carriers to produce the unpleasant heterodyne interference that occurs with closely spaced a.m. signals.

But how can we eliminate the unwanted sideband from the output of a balanced modulator? In practice the balanced

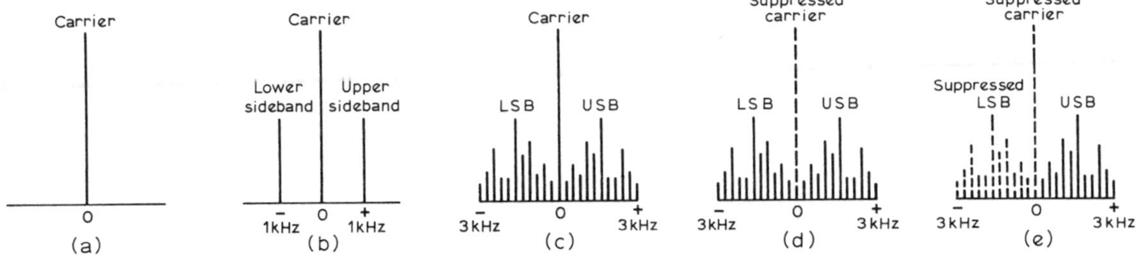


**Fig 25. RF spectrum diagrams of various forms of transmission. (a) Basic unmodulated carrier, occupying only bandwidth of carrier. (b) Same carrier fully modulated with a 1kHz audio sine wave. Note that two exactly similar sideband signals are generated 1kHz away from carrier frequency. (c) A.M. speech transmissions contain many audio frequencies simultaneously, up to about 3kHz for communications, and corresponding sidebands appear on either side of the carrier frequency, each set of upper and lower sidebands forming a mirror image of the other. The total rf bandwidth is thus at least 6kHz. (d) DSB speech transmission is similar to a.m. except that the carrier is suppressed (unless received as "ssb" by filtering in the receiver, a synchronous demodulator must be used). (e) An ssb speech transmission with the carrier and the lower sideband suppressed. Note that the total bandwidth occupied is now less than half that for (c) or (d)**

modulator works at low power and the unwanted sidebands are then *filtered* or *phased* out. Although several different phasing methods have been fairly widely used by amateurs on the grounds of their economy and their ability to generate an ssb signal directly at a high frequency, the tendency in recent years has been much more towards systems using *sideband filters*. Such filters must have rigorous selectivity characteristics to allow them to pass the wanted sidebands while eliminating the unwanted ones, which will be spaced only some 200–2,000Hz away. This calls for a very good filter with a pass bandwidth of roughly 2,600Hz and a high degree of rejection of signals outside these limits. At one time it was only possible to design filters meeting these requirements with a very low passband centre frequency of say around 100kHz, using high-quality inductors and capacitors. Nowadays however suitable filters can be achieved with *half-lattice crystal filters* using some four or more crystals with carefully staggered resonant frequencies, or *ladder crystal filters* using crystals of the same frequency or alternatively with *mechanical filters*. Most of the mechanical filters have a resonant frequency (passband centre frequency) of about 455kHz; some crystal filters are also designed for use at about this frequency, but more recently crystal filters up to 5 or 9MHz have become available with the necessary selectivity characteristics, though at some cost. The higher the frequency at which the initial ssb signal can be generated the less complex will be the later frequency conversion arrangements, but on the other hand the more difficult it is to achieve high selectivity in the sideband filter. A good ssb filter may cost £20–£35 or even more.

At present, by using "surplus" crystals and either by carefully selecting them, or by "etching" them to the desired frequencies, the amateur can build his own sideband filter, usually around 450kHz, inexpensively and without undue difficulty, provided that a certain amount of test equipment is available.

It is more difficult to build good ssb filters with the half-lattice network at hf. For filter frequencies from say 3–12MHz a more effective approach is the crystal ladder filter. This can use crystals of the same, or slightly different frequency, and does not require the use of carefully etched crystals or special input and output transformers. Several articles in *Radio Communication* have provided the necessary guidance for building effective ssb filters for the cost of four or five "surplus" crystals and a few fixed capacitors.

With any ssb crystal or mechanical filter it is most important that the carrier frequency is carefully selected or trimmed to match the filter response characteristics: this point is further described later in this chapter.

The initial carrier (that is rf drive) fed to the balanced modulator is crystal controlled. Once the dsb output from the balanced modulator has been converted to ssb by passing it through the filter, it can be amplified and heterodyned (in one or more steps) to the required transmitter output frequency by mixing it with appropriate signals. The hf heterodyne oscillator providing one of these signals can be made variably tuned; the transmitter output can then be tuned to the required position in the band by adjusting the vfo; any variable oscillator must be of very high stability.

At the receiver, a steady carrier is needed to assist in the recovery in the detector or demodulator of the original audio signals. This is generated in a low-power oscillator in the receiver itself, usually at the i.f., and re-inserted alongside the incoming signals in the correct relationship. Often the same oscillator is used as bfo for cw reception. This re-insertion process has to be done accurately, for unless the local signal is within about 25Hz of the original (suppressed) carrier frequency, there will be distortion. If the frequencies differ by appreciably more than this figure or if no locally generated carrier is available, the ssb transmissions remain completely unintelligible.

This indicates clearly that one of the most important requirements for both transmitters and receivers for ssb working is a very high order indeed of frequency stability. Since even a drift of 25 or 50Hz will cause distortion we must try to ensure that the equipment stays well within such a tolerance over quite long periods. A tolerance of 25Hz is very little when we are thinking in terms of transmitting frequencies of say 21,200,000Hz.

Apart from the actual generation of the ssb signal, an ssb transmitter differs in several respects from those designed solely for a.m. Normally in amateur a.m. practice, only the final pa stage is modulated whereas, as we have shown, in ssb this is always done at a very early stage. This has the big advantage of eliminating completely the need for high-power audio amplifiers

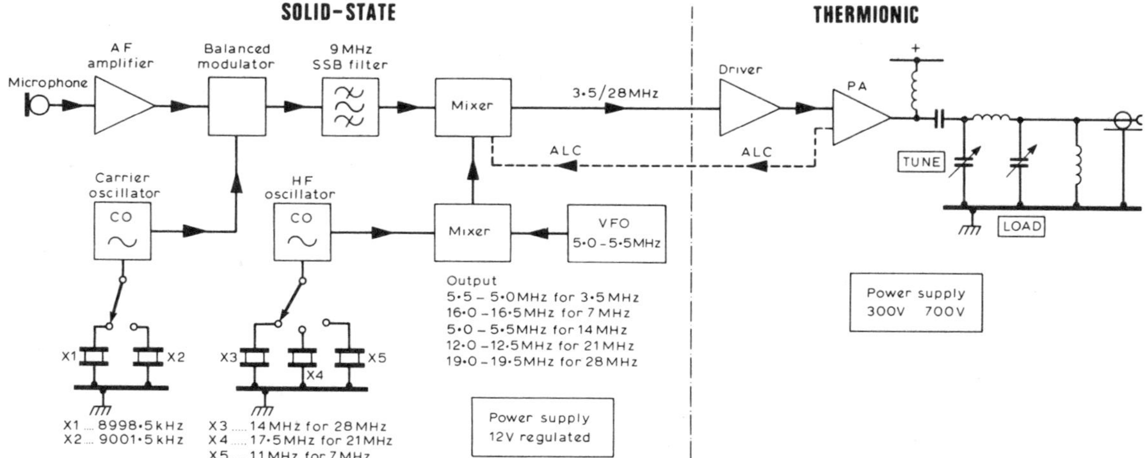


**Fig 26. Typical ssb/cw hf transmitter for power of about 100W p.e.p. The frequency conversion indicated requires relatively few crystals but results in sideband inversion on 7MHz and on 21MHz tuning inversion. Diode switching of crystals would be convenient and typically the pa could use two 6146B valves, with 6GK6 or 12BY7 driver. Balanced modulator might be MC1496G ic or similar. Mixers could be dual-gate MOSFETs, and addition matching and buffer semiconductor stages would normally be incorporated. For high-power operation such a transmitter might be used with an external "linear" amplifier**

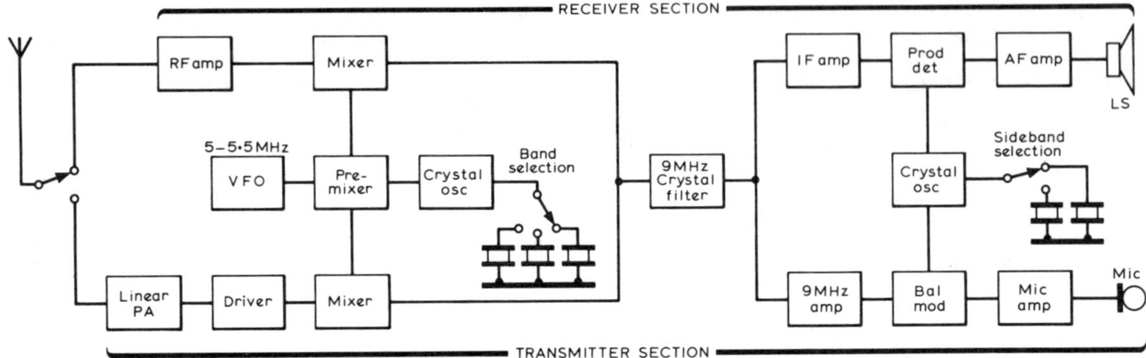


Fig 27. Block diagram of a modern ssb transceiver

for use as a modulator, but it does mean that all subsequent frequency conversion and amplifying stages must be *linear*. The pa for an ssb transmitter is thus called a *linear amplifier*, or simply a *linear*.

It was soon appreciated that many of the more expensive parts of an ssb transmitter, such as the sideband filter, were equally valuable for high-performance receivers, in this case as an i.f. selectivity filter. The same high-stability oscillators and other stages can be used. Today there are many compact ssb/cw multi-band transceivers (often with some facilities also for a.m. transmission) which are often cheaper to build or buy than would be separate units of comparable performance. Frequently, these equipments are built for use with alternative power supply units; one for normal home operation from domestic ac mains supplies; the other for 12V dc supplies (using transistor *inverters* for converting the dc to ac to allow mobile operation). Even where separate receivers and transmitters are used it is becoming common for a single vfo to be used to tune both together.

## The phasing methods

Almost all the factory-built ssb transmitters and transceivers use the filter method of ssb generation, and modern filters enable very good suppression of the unwanted sidebands to be achieved. With the availability of good filters working at frequencies up to 9MHz, less stages of frequency conversion are now needed to provide outputs on the amateur hf bands. But a high-quality ssb filter is still a relatively expensive component.

By using the phasing method of ssb generation which requires no filter an amateur constructor can build a complete ssb exciter for probably less cost than a good ssb filter—though he is unlikely to achieve such a high order of carrier and sideband suppression—so that the phasing method may still be a preferred choice for a newcomer wishing to build a rig at low cost.

At one time, phasing systems usually generated the ssb signal directly in the 3·5MHz band, and this sometimes led to difficulties when attempts were made to change the transmitter frequency. However, modern phasing exciters frequently generate the signal at 9MHz and then heterodyne the signal to the required band using a vfo to permit changes of frequency, along similar lines to filter exciters.

The phasing system depends upon the use of 90° phase shift networks for both rf and af signals. While the production of two rf signals 90° out of phase is fairly simple, an effective phase-shift network which will function effectively over the frequency range of about 300 to 3,000Hz calls for the use of close tolerance resistors and capacitors of very precise values—and this can call for very careful checking of components; nevertheless networks have been designed using standard values of one per cent tolerance fixed resistors and capacitors. Fig 28 shows the basic arrangement. The af and rf signals at low level are passed through the phase-shift networks and then fed to a balanced modulator; simultaneously the original signals are fed to a second balanced modulator. When the outputs from these two modulators are combined, both the carrier and the upper sideband are suppressed; by changing over the two af signals fed to the two balanced modulators, the lower sideband can be suppressed. Fairly simple and quite effective phasing type exciters can be built using bipolar and field-effect transistors.

The phasing technique can also be applied in reverse for reception, though relatively little use has been made of this arrangement. In conjunction with direct-conversion receivers the "two-phase" system will provide rejection of one set of sidebands of a double-sideband signal to provide effective single-sideband reception.

The conventional phasing-type ssb generator requires the use of accurate 90° phasing networks for both the af and rf signals. In particular the broadband af network calls for the use of very precise and non-standard resistor and capacitor values; the rf phasing network presents fewer difficulties.

In recent years, several techniques have come into use which simplify the building of low-cost phasing-type ssb generators and allow standard components to be used successfully.

For example digital rf 90° (quadrature) phase shifting can be accomplished by using a standard ttl ic (for example a 7473) if the device is fed with a signal at four times the required frequency. Broadband af phase shifting can be tackled by using the "third method" with four ICs as the four balanced modulators; digital phase shifting the rf and af carrier signals and then using fairly simple audio low-pass filters.

A recent further variation on phasing techniques is the use of *polyphase* networks which are much more tolerant of component values (and can be constructed entirely from preferred-value resistors and capacitors), together with four balanced modulators. Such ssb generators, at 9 or 10·7MHz, have been very successfully used by amateurs in low-cost hf and vhf ssb transmitters. The polyphase approach to ssb generation (or demodulation) was originally developed by M. J. Gingell of STL in the UK.

With all ssb generators, care has to be taken that the balanced and double-balanced modulators do not rapidly drift out of balance with changes of temperature. This can be a problem with ic devices: the high-density packing of the tiny active devices tends to result in internal heating, which can cause drift.

## VHF transmitters

A common arrangement for a.m./fm/cw vhf transmitters is to use a crystal-controlled oscillator (or vfo or vxo) at about 8MHz; and then to multiply the carrier frequency up to 70, 144 or

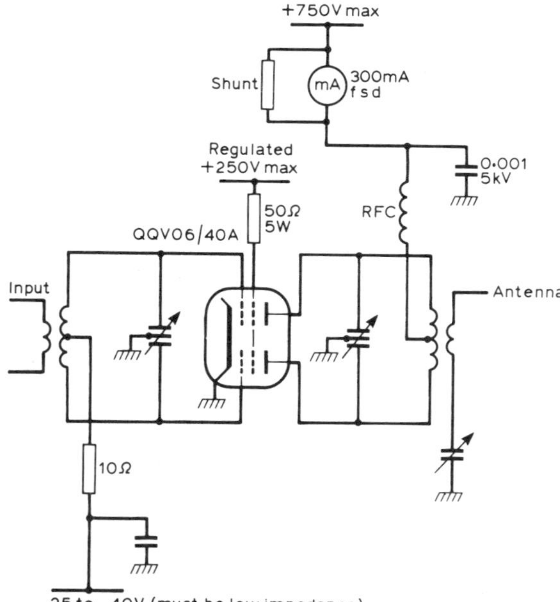


**Fig 28. 144MHz linear amplifier using QQV06/40A. Screen voltage 250V regulated. Low impedance bias supply (maximum source impedance 5kΩ) and bias adjusted for "just a sniff of grid current at maximum drive"**

432MHz in a chain of Class C frequency multiplying stages. For example an 8MHz oscillator, tripling to 24MHz, doubling to 48MHz and tripling to 144MHz. Such multiplier chains may use either transistors or valves with the trend now very much towards the use of transistors. The crystal oscillator is often an *overtone oscillator* in which the oscillator output is at an odd multiple of the fundamental frequency of the crystal; for example 24MHz from an 8MHz crystal. Where the overtone technique is used it should be noted that the crystal is made to oscillate at the higher frequency and there is no output at the fundamental frequency. This reduces the likelihood of spurious signals since the spurious outputs are further apart in frequency; however, the overtone form of oscillator is rather less stable than when a crystal is used at its fundamental resonance. Increasingly, vhf transmitters are using all-solid-state techniques with rf bipolar transistors in the power amplifier at input power levels of about 15 to 25W. For higher powers it is more usual to have thermionic valves in the final stages.

For nbfm telephony, modulation occurs at low-level; often directly at the oscillator stage by using a variable-capacitance diode across the crystal; or alternatively by phase modulating the output of the oscillator before it passes to the frequency multiplier chain. It should be appreciated that the process of frequency multiplication automatically increases the deviation of an fm signal, so that for nbfm an 8 or 24MHz oscillator would require very little deviation. On the other hand, frequency conversion by the mixing process does *not* increase deviation.

There is often some confusion between *frequency* and *phase* modulation since both techniques are widely used for nbfm; an important advantage of phase modulation is that there is less chance of producing some of the rather rough signals that stem from non-linear reactance modulation of oscillators.

The increasing use of ssb at vhf has meant that the basic design of vhf transmitters has changed; so that the ssb signal passes only through linear stages with frequency conversion by means of mixing techniques. It is common practice to achieve ssb generation at hf (eg 9 or 10·7MHz) and for the signal then to be converted to the vhf band by heterodyne mixing. Although transmitters designed specifically for vhf ssb operation are now common, considerable use is made of *transverters* in conjunction with hf ssb/cw transceivers. A transverter accepts an ssb signal (typically at 28MHz) and includes the necessary frequency converters and vhf power amplifier; it also often contains a similar frequency converter for vhf *reception*, designed to operate in conjunction with the receiving section of the hf transceiver. This, in effect, converts an hf installation into a vhf station. Rather less common is for amateurs who have an ssb vhf transceiver to provide similar transverting techniques to give them hf facilities; however the power levels used at hf may require the use of a high-power linear amplifier.

## Speech processing

The frequencies of male speech range from about 100Hz to about 7,000Hz and to reproduce a faithful copy of the original all these frequencies would have to be transmitted. However to convey intelligence we can restrict the bandwidth to about 300–3,300Hz and by doing so we not only reduce the bandwidth of the transmission but we also use our available power more effectively. With ssb equipment using the filter process, the filter is arranged to restrict speech to the required frequencies, but it is also normal practice for all modes to compress the frequency band initially in the speech amplifier. For example suppression of bass frequencies can easily be achieved by using low-value coupling capacitors; both high and low frequencies can be reduced by means of an audio bandpass filter or a low-pass af filter in conjunction with low-value coupling capacitors.

The audio waveform is of a spiky intermittent nature, with amplitudes exceeding 50 per cent of peak amplitude for less than 10 per cent of time. Again very little intelligence is conveyed by these short high-amplitude peaks and we can "clip" them off without losing intelligence. This allows the average modulation level of the transmitter to be greatly increased, and this again applies to a.m./nbfm/ssb. With a.m. or nbfm it is useful to limit peaks of audio by means of back-to-back diode pairs, followed by a low-pass filter to remove the harmonics generated in the process. This simple system is much less satisfactory for ssb where it is advisable to carry out clipping at rf in order to ensure that the harmonics produced in the clipping process are at a sufficiently high frequency to be well outside the basic af spectrum. It is possible to do this either within the low-level rf section of the transmitter or by means of an adaptor in the af input, provided that this unit translates the af to rf and then back to af; such systems are generally known as rf peak clippers. Care should be taken when adding such facilities to an existing transmitter to ensure that the extra duty cycle at which the linear amplifier will operate is within the capability of the output stage and the power supply; it is not uncommon for a cooling fan to have to be added.

The dynamic range of the af signal may also be compressed by means of a feedback loop rather like the agc system in a receiver and a popular form of this system with ssb transmitters is known as *automatic level control*.

In any telephony transmitter, some care needs to be taken to ensure that rf signals do not affect the speech amplifier, producing howl-round or other forms of distortion. This may call for screening of microphone leads and careful bypassing of the speech amplifier to rf signals.

It should be appreciated that these various forms of speech processing can result in very potent signals, fully equivalent to unprocessed transmissions from transmitters many times more powerful, but equally it should be recognized that transmitters used in this way need very careful design and operation or they may result in unpleasant sounding as well as splatter-prone transmissions. For example, heavy clipping or compression can result in an increase in background noise during speech pauses giving rise to a form of "pumping".

## Transistor power amplifiers

It is today possible to build all-solid-state hf and vhf transmitters at power levels of up to hundreds of watts, although in practice this approach is used by only a small minority of amateurs at power levels above about 15–25W. For high-power operation it is still common to use hybrid systems, using solid-state devices (bipolars and some field-effect devices) in the low-level stages such as ssb generators, heterodyne band-switched frequency conversion stages and the like, with thermionic devices still popular for the pa stage and (usually) the driver stage. However, the development of broadband solid-state power amplifiers, eliminating the need to "tune up" when band changing, is making the all-solid-state hf transmitter very attractive.

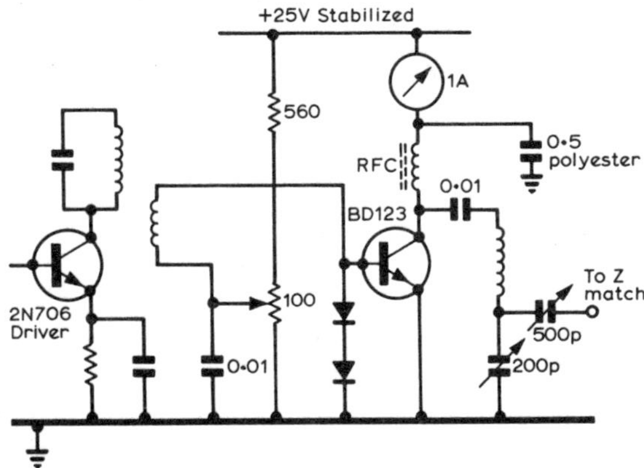


Fig 29. Linear transistor amplifier in 3·5MHz mobile ssb transmitter, providing 14W p.e.p. output. RFC is 50t 22swg enam, wound on ferrite rod. The standing current is set to 50mA by forward bias on the base, and this can be readily talked up to an indicated 500mA using a 2N706 running 10mA at 12V. The silicon diodes across the base of the BD123 prevent destruction of the base-emitter junction by overbiasing or overdriving. With series tuning the pa current rises on resonance and the load pulls it down. Good regulation and decoupling are necessary with the peak currents of around 1A and ordinary paper tubular capacitors are useless for the collector decoupling capacitor

An increasing selection of rf power transistors is available at reasonably low cost for use up to the 5W or so level, and some for the 10–20W level, and these can often be operated directly from 12V vehicle batteries. Virtually all recent designs for mobile vhf operation and some for hf operation are all-solid-state (Fig 29).

One advantage that transistors have over valves (apart from small size and absence of heater supplies) is that they operate as Class C stages with an extremely simple bias arrangement; if, in the common-emitter arrangement, the base is returned to the emitter through an rf choke, with no other form of biasing, then the stage automatically functions in Class C, requiring a positive drive voltage of roughly 0·3V or more for a germanium device and 0·6V or more for a silicon device, before collector current flows. For Class A and Class B operation, suitable forward bias (derived usually from the collector supply line) must be applied.

With bipolar transistors, it is most necessary for optimum results to design carefully both input and output circuits to present approximately correct matching; these matching impedances are usually much lower than for valves, and series-resonant tuned circuits are often used; or alternatively parallel resonant circuits with impedance taps.

Great care is needed when operating a transistor pa at anywhere near its maximum ratings; for example amplifiers often need protection circuits unless great care is taken to ensure that even momentarily they are never operated without a correctly-matched output load. It will be noted from Fig 29 that it is common practice to series-tune the tank-circuit to match the low collector impedance of rf power transistors.

Unlike valves, the factors which limit power output from a transistor are less average power dissipation than *peak* collector voltages and currents; this is the reason why a transistor power amplifier can be run at significantly higher power for nbfm or fsk operation, than for a.m. or ssb.

Unfortunately, it is very easy to destroy rf power transistors during the adjustment or operation of a solid-state transmitter unless the dangers are understood and avoided, or the stage has been designed to incorporate various forms of protection. Many of the practical problems, which can lead to the destruction of an expensive transistor in a matter of microseconds, stem from various forms of instability to which transistor power amplifiers are prone when operated near to their maximum ratings. For instance such instability (often in the form of low-frequency or vhf parasitic oscillation) can arise from ineffective bypassing or when the tank circuit is detuned; peak voltages on a collector may occur as a result of a transmitter being operated without a load, or with an incorrectly matched load (for instance into a transmission line having significant swr).

Several transistors may be used in parallel, and high-power amplifiers often comprise a number of modules, each of which may comprise several transistors in parallel or push-pull configuration. In such circumstances care should be taken to ensure that the drive power to the individual transistors is roughly similar. Neutralization of a power amplifier may be necessary when a bipolar transistor is operated at more than about one-third of its alpha cut-off frequency, particularly with common-emitter circuits.

A recent development is the V-mos form of power device which has a high input impedance (and thus simplifies input matching) and presents fewer problems, including *secondary breakdown*, to which bipolar devices are prone; however the rf V-mos devices are still limited to relatively low power and at present are expensive for the amateur.

In general, as the cost of rf power transistors has come down, it has become less essential to operate them close to their limits, with consequent reduction in the problems of building and adjusting solid-state linear and Class C amplifiers.

## Transistor power amplifier hints

Despite the steady improvement in transistors, it is still necessary to be rather careful if one is to avoid littering the pathway to success with discarded and destroyed devices when building rf power amplifiers. However, not only is the price of rf power transistors coming down but the price of power valves is rising quite rapidly.

Remember that the bipolar transistor is a *current*-operated device with low impedances; this implies the use of heavy-gauge conductors and high-value capacitors. A collector tank coil may be carrying both dc and rf currents and needs special attention. If shunt feed is used watch the wire size in the rf choke. The heavy peak currents should also be reflected in the power supply cabling etc.

"Mode jumping" in linear amplifiers is generally due to a tuned tank circuit having a different resonant frequency for a strong drive signal than for a weak one, and this presents problems with the very peaky nature of an ssb drive signal. Precautions include carefully choosing bias values, correct grounding, and using only transistors with low values of parasitic capacitance and inductance (usually met in recent rf power devices but not always in the older types).

Make sure there is sufficient drive; these days, particularly for vhf mobile applications, manufacturers often offer a "family" of devices each capable of providing sufficient drive for the next higher-power stage.

Use triple bypassing for af, hf and vhf signal components; electrolytic capacitors for af bypassing should preferably be of the heavier tantalum type; hf bypass capacitors can be ceramic feedthrough or (second choice) disc ceramic types; vhf bypass capacitors should be silver-mica button types with shortest possible leads.

A number of these suggestions are indicated in Fig 30. The biasing circuit uses a stud-type silicon power diode bolted to the same heat sink as the rf power transistor(s) as near as possible to the transistor or between a pair of transistors. Any increased heat then lowers the diode resistance, thus helping to maintain a safe dissipation level in the transistor(s). A temperature-sensitive diode can be selected using an ohmmeter and soldering iron: measure the drop in diode forward resistance after touching the hot iron to the stud for a given number of seconds, selecting the diode with the fastest thermal response.

The following are precautions to be taken during initial tune-up of a medium-power transistor pa:

(1) Always carry out initial checks with a low supply voltage and low drive, gradually increasing both together.

(2) It may be advisable to connect temporarily an inexpensive power transistor (eg 2N697) in order to gain the "feel" of tuning-up before inserting the expensive device.

(3) Emitter resistors offer useful protection during preliminary checks. Several 1W resistors in parallel will present lower impedance than ½W types.

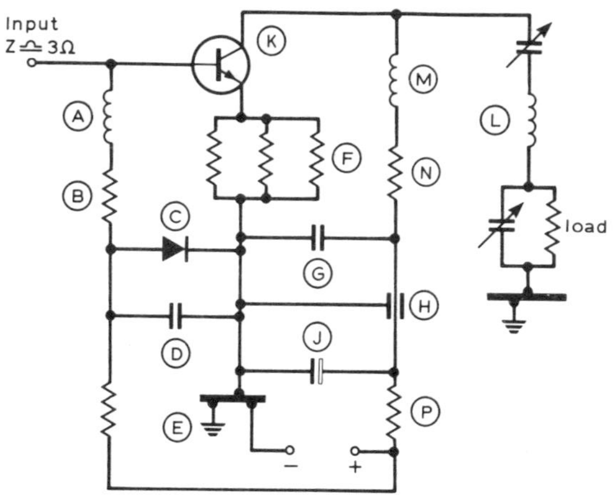


Fig 30. Typical rf power transistor amplifier for medium power illustrating safeguards and suggestions made by W4ATE
A RFC must be low-Q (high-Q causes lf oscillation).
B Resistor lowers Q of rfc (typically 100Ω).
C Bias stabilization diode (see text).
D Diode bypass (0·001μF).
E Group all earth leads near emitter earth connection using short leads.
F Low emitter resistor helps prevent secondary breakdown.
G 0·001μF button vhf type.
H 0·1μF feedthrough capacitor (hf bypass).
J 10μF tantalum af bypass capacitor.
K High-level harmonics are generated due to non-linear characteristics of the transistor plus large dynamic voltage and current swings.
L Network for load matching and reduction of harmonics.
M RFC tunes out reactive component of admittance.
N Low-value resistor lowers Q of the self-resonance of the rfc.
P Decoupling resistor, typically 12Ω.

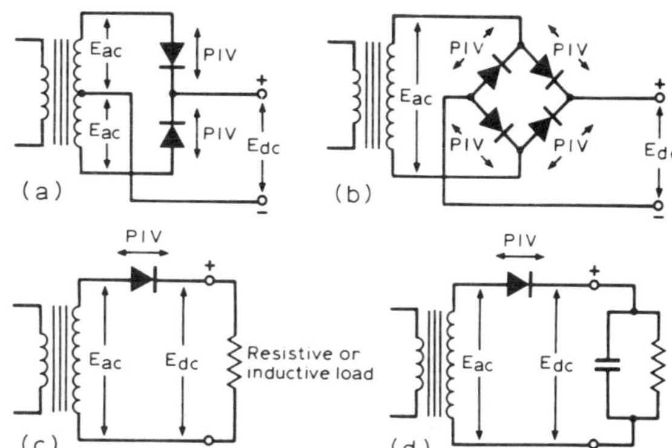


Fig 31. Fundamental power supply relationships. (a) Full-wave rectifier. PIV about 2·8 times Eac or 3·14 times Edc. Edc is about 0·9 times Eac. With a capacitive load the actual value of dc output voltage will be affected by the value of the reservoir capacitor in relation to the dc load. (b) Full-wave bridge rectifier. PIV (each diode section) about 1·4 times Eac or 1·57 times Edc. Edc is about 0·9 times Eac but note in (a) also applies. (c) Half-wave rectifier with inductive or resistive load. PIV about 1·4 times Eac. Edc is 0·45 times Eac. (d) Half-wave rectifier with capacitive load. PIV about 2·8 times Eac. Edc is 0·9 times Eac and note in (a) applies

(4) Use generous heat sinking. During alignment it is advisable to keep a constant check on the temperature of the power transistor, using a finger, or a thermometer attached to the device with putty.

(5) Never operate a transistor pa without a dummy load or matched antenna.

(6) Monitor collector current continuously; a climbing current is the earliest warning of junction heating. Remember that maximum rf current output does *not* occur exactly at the collector current dip.

A tunable rf indicator is very necessary since, because of the unusual (to the valve man) L-C ratios it is easy to tune to an incorrect frequency. The indicator will also help check spurious and harmonic outputs. Use an indicator and not collector dip to tune to maximum rf output.

Harmonic output is reduced by using higher C in tank circuits, tapping coils closer to rf "earth" and careful attention to the bias operating point. But learn to expect and live with higher harmonic content than from valve amplifiers: at the present state of the art a low-pass filter between a transistor pa and the antenna is more or less essential.

## Power supplies

One of the main units, both from the viewpoint of cost and also its importance in determining the remainder of the design, is the main ht power supply. For medium power, 500V at about 150mA may be ample for both pa and the exciter; and this can be obtained from a conventional type full-wave rectifier circuit, see Fig 31(b). For higher powers there will be need for a heavy-duty supply at either 750, 1,000 or even 1,500V and rated at from 100–300mA. Power units of this voltage call for careful choice of components with ample tolerances, care in construction and in operation. The failure of a second-grade or over-run smoothing capacitor can easily lead to the loss of high-voltage rectifiers and an expensive transformer, and with such lethal voltages involved *full safety precautions are essential.*

A point not always appreciated by newcomers to power supply

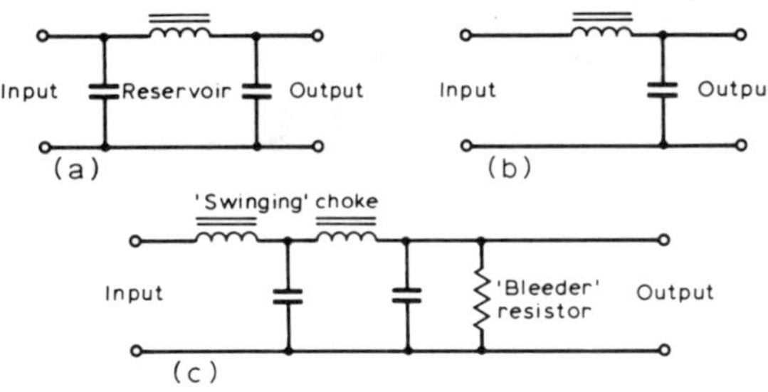


Fig 32. Basic ripple filters to reduce the "hum" content of power supplies. (a) A conventional capacitor-input filter provides effective smoothing but has poor voltage regulating qualities when used with varying loads. It also imposes a considerable strain on the rectifier valve. (b) A simple choke-input filter. This provides less smoothing but gives better voltage regulation and imposes less strain on the valve. (c) A double-section filter combines the advantages of both choke- and capacitor-input filters, and is commonly used in power supplies for modulators. In all these circuits, capacitors should be at least 4μF and generously rated—particularly the reservoir capacitor in a capacitor-input filter

construction is that the peak voltages present may be considerably above the transformer ratings. For example, with a 500-0-500V transformer and conventional full-wave rectification feeding a cw transmitter, the voltages when the supply is fully loaded by the transmitter will be of the order of 500V, but when the load is removed, as will usually be the case whenever the morse key is not depressed, the dc voltages will rise to about $1{\cdot}4 \times 500 = 700$V. All capacitors and components in such a unit would have to be capable of withstanding this higher voltage; in practice it would be better to add an extra safety margin to take care of voltage surges and to fit capacitors rated at a minimum of 1,000V. It is the peaks, surges and the combination of rf with dc voltages that cause flash-overs and break-downs. Closer regulation of voltage can be obtained by using a choke-input smoothing filter: see Fig 32, but this system is seldom used with silicon diode rectifiers.

In recent years silicon semiconductor power rectifiers are used in the place of valve or metal (selenium) rectifiers. Silicon junction diodes are extremely efficient rectifying devices having a low forward resistance and high reverse resistance, and maintain these characteristics much better than selenium rectifiers. They can be made almost as compact as small resistors and can similarly be supported in the wiring. Since they require no heater supplies, they are particularly useful for various bridge and voltage-doubling circuits which with valves would need several well-insulated heater supplies.

Silicon diodes are thus almost an ideal form of rectifier and can give excellent service over many years when correctly used; on the other hand they can easily be ruined—often causing further damage to mains transformers and other components—unless certain precautions are taken in the design of the power units. This is because silicon diodes are capable of withstanding only their rated peak inverse voltage (piv) and their rated forward current and these ratings must not be exceeded, even momentarily, whereas other rectifiers are usually much more tolerant in this respect.

When building power supplies making use of silicon diodes it is essential to ensure that no excessively high voltage or current surges, even of extremely short duration, can occur. Such surges or transients are normally found in power supplies for various reasons. There are, for instance, many short duration "overvoltages" on normal domestic mains supplies. Transients can also be caused by the inductive surges that build up when an inductive winding, such as a mains transformer or smoothing choke, is switched into or out of circuit. Strong current surges occur when a supply is switched on due to the rapid charging up of the high-value reservoir capacitor.

It is thus essential to ensure that not only is the rated piv of the silicon diodes well above that required in the normal functioning of the circuit but also, equally important, that there will be no excessively high voltage or current transient surges.

Fig 31 shows the normal piv conditions found in a number of typical rectifier circuits. This indicates that for the popular full-wave circuit the piv across each leg of the rectifiers is just over 2·8 times the ac voltage across each half of the transformer secondary windings. Thus with a 350-0-350V transformer, the minimum piv to each rectifier chain must be a least 980V. To provide a good margin of safety a piv rating of around 1,200V would be desirable in such circumstances. Diodes with a piv rating as high as this are rather expensive, and in practice one might replace each of the two diodes by three series-connected diodes each with a piv of 400V (or two of 600 or 800V) *provided that* steps are taken to equalize the inverse voltages across each diode. This can be done quite simply by connecting 100kΩ or 150kΩ resistors as shown in Fig 33, or alternatively by using equalizing capacitors.

In Fig 33 the use of the 100Ω resistor should also be noted. This is to limit the initial flow of current into the reservoir smoothing

## SAFETY FIRST

The voltages developed in many amateur stations are capable of causing injury or death. Reasonable precautions should always be taken.

All apparatus and wiring should be so placed and constructed that it is impossible to touch points of high dc, rf or ac mains potentials under normal operating conditions.

The antenna should never be directly connected to a point at high dc or mains potential (this is illegal and highly dangerous).

Use double-pole switches to ensure complete isolation of all mains transformers. These switches should be clearly marked with ON–OFF positions. Some other person in the house should know where to find the main switch for use in case of emergency.

High wattage bleeder resistors across power supply filter capacitors will prevent many shocks. If it is necessary to touch the transmitter while the power is ON, keep one hand behind your back or in your pocket: never wear headphones while working on the transmitter.

Remember domestic ac mains voltages can be fatal if contact is made over an appreciable area of skin. Avoid "live" chassis techniques for home-built equipment.

Think ahead and remember that isolating and protective components may fail. The greatest danger is exposed metalwork becoming "live" to ac mains or ht supplies.

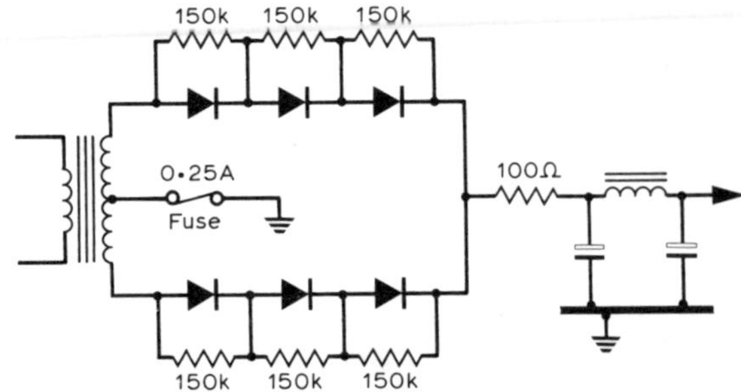


Fig 33. Typical 350V power supply using series-connected 400V piv silicon diodes

capacitor, and the value can safely be based on a figure of about 5 to 30Ω per diode in each leg. Thus in this particular design each rectifying leg has three diodes (six diodes altogether) so that 100Ω is more than adequate.

# BUILDING TRANSMITTERS

For many years, most amateurs built their own hf and vhf transmitters, often to published designs or from kits. For a.m./cw operation this presented relatively few problems, except perhaps those arising from the bandswitching of hf transmitters. Alignment and adjustment of the individual stages was quite simple since there was usually sufficient power to allow checking with pilot bulb and loop of wire, absorption meters or grid dip oscillators. The "distortion" of a Class C stage is automatically overcome in the tank circuits, although the circuit constants result in greater or lesser degrees of harmonic generation. While buffer stages and power amplifiers often required careful "de-bugging" to eliminate various forms of parasitic oscillation, such oscillation in itself seldom represented any danger to the valves or components, and the whole transmitter could be checked at leisure while running into a dummy load. Since the "duty cycle" of the output stages for a.m. operation meant that these had to be capable of running at full carrier power continuously (or at least to "intermittent" transmission periods), they were often capable of running for brief periods into highly mismatched loads. Generally the whole process of alignment and adjustment involved few risks of seriously damaging the equipment.

The swing to ssb and solid-state operation however has significantly changed the situation and it would now be unrealistic to suggest that many newcomers (unless possessing considerable experience in the other fields of electronic construction) are likely to find it easy to build an all-band hf ssb transmitter or transceiver, without first gaining considerable practical experience in tackling rather simpler projects. It is for instance difficult to align and adjust a home-built ssb transmitter without an oscilloscope and good knowledge of single-tone and two-tone testing.

## Components

A transmitter is a multi-stage system used as a means of producing electrical power at the required radio frequency, with the output varied by speech or keying in accordance with the required mode of operation. To do this we require active devices (valves or semiconductors) functioning as oscillators, frequency multipliers or frequency converters, isolating (buffer) stages, voltage or power amplifiers or intermediate power "driver" stages, modulators, speech amplifiers, etc. Diodes may be used for electronic tuning and/or switching, for frequency modulation, or as varactor frequency multipliers.

For many years it has been the practice (except for microwave transmitters) to generate the required frequency first at a low power level in the exciter section of the transmitter and then to raise the power to the required level in two or more power or intermediate-power stages; sometimes additional power may be achieved in a separate linear amplifier unit. A very large number of different circuit arrangements are to be found in hf, vhf and uhf transmitters, and this chapter can outline only a few of the fundamentals. Detailed information on transmitter design can be found in a number of amateur radio handbooks, such as the *Radio Communication Handbook* and the *VHF–UHF Manual*.

It should be appreciated that as the power levels are raised, involving either high voltages (for valves) or high currents (for semiconductors) or both (eg tuned circuits) there are increasing constraints on the choice of components. A high-power transmitting valve is different from those still found in receivers; an rf power transistor very different from a small-signal device. Variable capacitors used in the tank circuits of high-power transmitters need appreciable physical separation of the plates in order to avoid flashing over on high rf peaks; coils and chokes may need to be of heavy gauge wire; resistors rated for high-power dissipation and, like certain of the capacitors, able to maintain good stability over repeated heat cycles. In practice

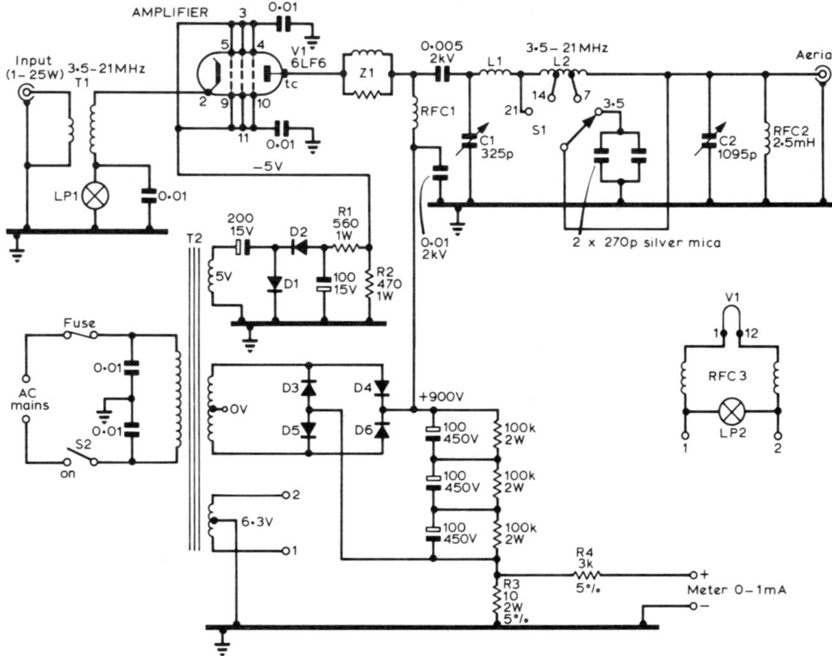


Fig 34. A 6LF6 linear amplifier with self-contained power supply. C2 formed from three-gang broadcast-type variable; D1, D2, 250V piv 1A; D3–D6, 1,000V piv 1A; L1, 5½t No 12, 1in dia, 1½in long; L2, 26t on T200 toroid, taps from C2 end, 13 for 7MHz, 22 for 14MHz; RFC, 3-bifilar wound filament choke, 50t No 20 enam on 4in length of ½in dia ferrite rod (or 75t No 20 on ¾in dia wooden dowel); T1 primary, 17t No 26 enam to cover two T-68-2 Amidon cores; secondary, 35t No 24 enam wound over primary winding; T2, 800V centre-tapped, 200mA; 6·3V at 5A and 5V at 3A. Z1, parasitic suppressor, 6t No 20 in parallel with 56Ω 2W carbon resistor; fuse, 1·5A for 240V supplies, 3A for 117V supplies

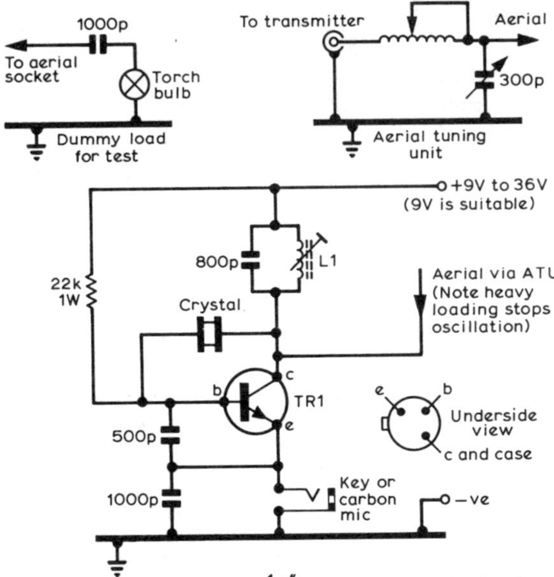


L1.... 40 turns of 32swg wire, 1/4"dia with slug, for top band
TR1..... BFY50, 51 or 52 with clip on heat sink

Printed circuit board, 3" x 2", viewed from copper track side

+ Power 9V to 36V
800p
Aerial
Crystal
22k
L1
Key or mic
500p 1000p
Crystal
–ve Power and earth
Key or mic –ve

**Fig 35. A simple 1·8MHz a.m./cw transmitter built by G3YUQ. With 9V supply it will produce about 0·25W a.m. and about 0·5W cw. About 0·75W a.m. with 27V. Usually intended only for contacts over a few miles**

considerably more electrical power has to be supplied to the transmitter than is available from it as rf output and the difference is dissipated within the units as heat. To prevent over-heating, particularly of output valves or transistors, it is necessary to remove this unwanted heat from within the enclosed units of the transmitter by means of ventilation (which may include fan-assisted air currents) or by heat exchangers such as large heat sinks. Power supplies have to be designed to cope with the "duty cycle" of the various operating modes and conditions under which the transmitter will be operated. For example, "on-off" keying for cw represents a duty cycle of about 0·5 whereas for nbfm or fsk the transmitter has to supply full output continuously: ssb normally has a very-low duty cycle (having to deliver maximum output only during the peaks of speech waveform which are of short duration and fairly widely spaced) and this is why relatively compact power transformers are often found in quite high-power ssb equipment; however if heavy speech processing, such as rf clipping, is used the duty cycle increases appreciably, and this needs to be taken into account (it is one reason why care should be taken when modifying an existing transmitter for rf clipping) both in the cooling of the power amplifier devices and the amount of "iron" needed in the power supply transformer etc. The efficiency (ie power output as a function of power input) of a power amplifier depends very much on the biasing conditions (Class A, B, C, D etc) and this in turn depends upon whether the amplification has to be linear, as for ssb, or can safely deliver pulses of output during only part of the conduction cycle as for cw, nbfm, fsk etc in Class C. Generally the efficiency of a linear stage (Class A, AB or B) is a good deal lower than with Class C, D or E.

Then again, parasitic oscillation or incorrect operation can destroy expensive rf power transistors in the twinkling of a few microseconds—and unfortunately even today rf power transistors are very prone to various forms of parasitic oscillation and any attempt to operate the device into a badly mismatched load can result in the destruction of the device.

For these reasons (and possibly also an element of change in attitudes within the hobby) the percentage of entirely home-built hf and vhf transmitters has dropped very sharply, to a level not very different from that found for communication receivers.

However this does not mean that there is no longer scope for home design and construction. The experienced amateur, with good test equipment, can confidently tackle ambitious projects. The newcomer can rapidly gain valuable experience from simpler designs: for example low-power 1·8/3·5MHz cw/a.m. transmitters will bring plenty of contacts on cw or from the amateurs who have stayed with a.m. (Figs 36, 37). For vhf and

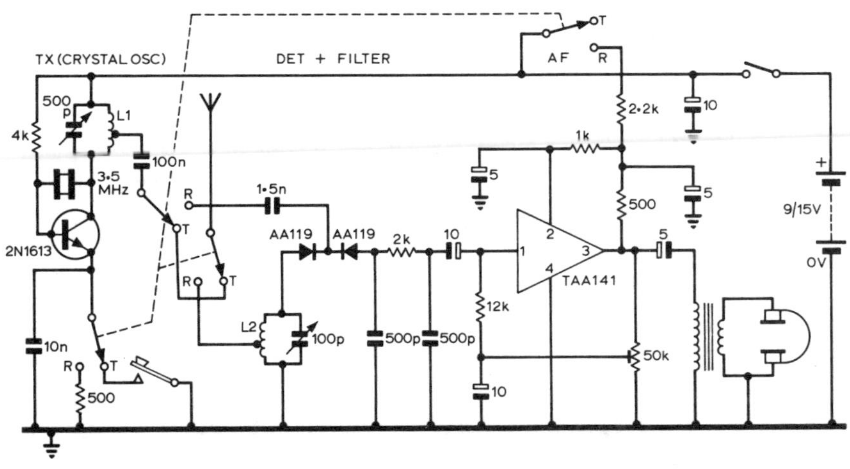


**Fig 36. Perhaps the simplest "all-solid-state hf trans-ceiver" ever designed. It provides a keyed oscillator providing about 1W output in the 3·5–3·6MHz cw band. For receive the same oscil-lator provides lower output for a direct-conversion fixed-channel receiver with balan-ced demodulator (two AA119 diodes) and a high-gain audio amplifier (TAA141). A BFY51 or BFY34 would be suitable alternatives to the 2N1613 and the TAA263 a suitable alternative to the TAA141**

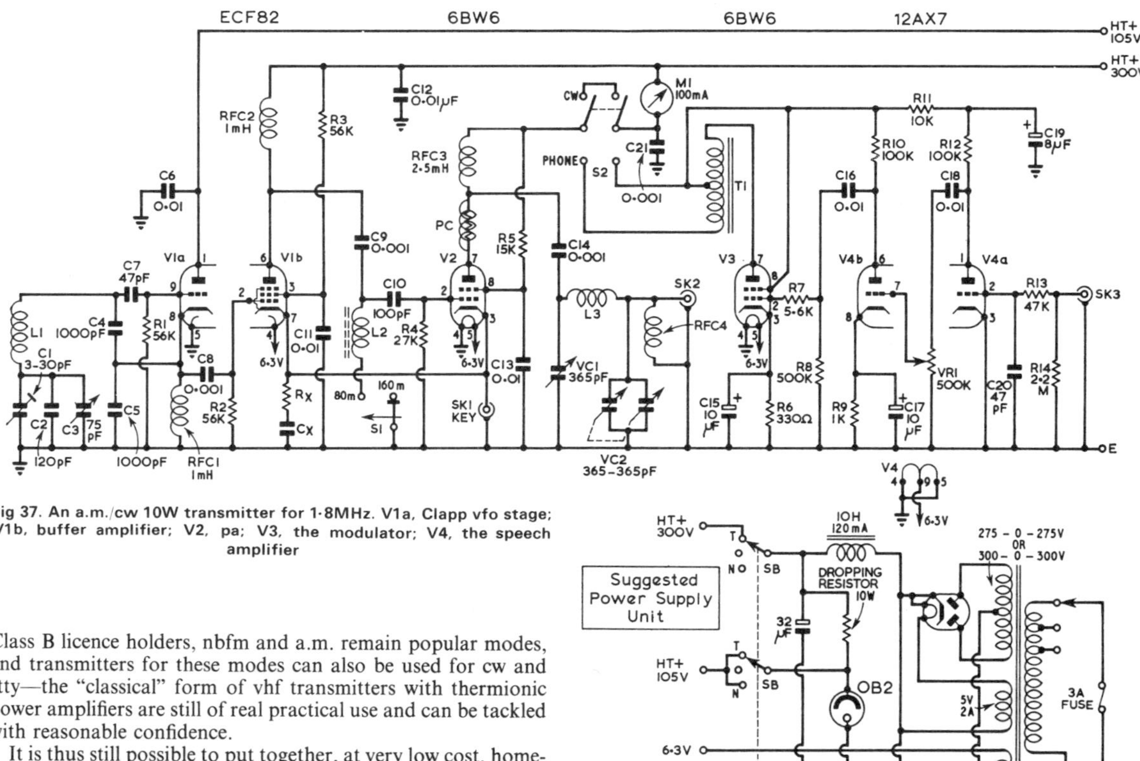


Fig 37. An a.m./cw 10W transmitter for 1·8MHz. V1a, Clapp vfo stage; V1b, buffer amplifier; V2, pa; V3, the modulator; V4, the speech amplifier

Class B licence holders, nbfm and a.m. remain popular modes, and transmitters for these modes can also be used for cw and rtty—the "classical" form of vhf transmitters with thermionic power amplifiers are still of real practical use and can be tackled with reasonable confidence.

It is thus still possible to put together, at very low cost, home-built equipment that will bring lots of interesting and enjoyable contacts, allow full study of propagation and antennas and the gradual build-up of test equipment.

As we move into the 'eighties, many once-common thermionic valves are becoming scarce and increasingly expensive. For most low-power transmitter applications, the valve is now being superseded by semiconductors, less for reasons of improved technology than from basic economics.

Fortunately, the initial problems of rf power amplification using bipolar transistors (in particular the ease with which such devices could be destroyed accidentally) have been much reduced. A very useful improvement has been the appearance of relatively low-cost vmos field-effect transistors such as the Siliconix VN10KM and the VN40AF, VN67AF and VN89AF series. Such devices work well in parallel and can be used for low-power transmitters providing, say, 5–10W rf output when operated directly from the 13·8V available from "12V" car batteries, or from nicad rechargeable cells, or from mains power supplies providing about 3–5A regulated output.

A very simple, experimental design for a 1·8–2·0MHz amplitude-modulated transmitter, using a crystal having a frequency in the range 3·6–4·0MHz, is shown in Fig 38. This uses cmos integrated circuits plus parallel vmos FETS to provide a simple transmitter of about 7W rating.

## Transmitter controls and adjustments

Most transmitters now use broadband tuning or coupling circuits in all stages except the pa and the vfo tuning control, all other stages being pre-tuned for the required bands. The pa stage will usually have a pi-network with an anode *tuning* control which is to set to maximum dip, with the *loading* capacitor then set to match the transmitter into the load impedance (note that these controls are interdependent). One or more current meters will usually be provided, with suitable shunts to allow them to be switched to measure the current or voltage at different parts of the transmitter (for example pa anode/collector current, screen current, grid current, ht voltage etc). A control will normally be fitted for adjusting the *drive* applied to the pa, although in ssb transmitters this may in effect be the gain control of the audio amplifier; it is most important to observe the correct range of drive to minimize the production of splatter due to driving the pa excessively and so producing intermodulation products and flat topping.

For hf transmitters there will be a *band* selection control and often a switch for selection of *upper sideband* (usb) or *lower sideband* (lsb) transmission. It is an accepted convention that ssb stations normally transmit the lower sideband below 10MHz and the upper sideband above 10MHz.

Mode switches for cw and a.m. (where incorporated) may often be combined with the sideband selection switch. Facilities such as vox (voice-operated transmit-receive switching), incremental tuning (on transceivers) to separate the transmit from the receive frequencies, and a switch to allow the use of either internal or external vfo facilities may also be found.

Most ssb transmitters incorporate a built-in tone generator at about 1,000Hz for routine and power adjustments. It is possible using a single tone and headphones to align ssb generators although for full checking of a linear pa a two-tone generator and oscilloscope are virtually a necessity. Generally the anode current reading of a linear amplifier on speech should not be more than one half the reading with a tone input and the audio gain or drive

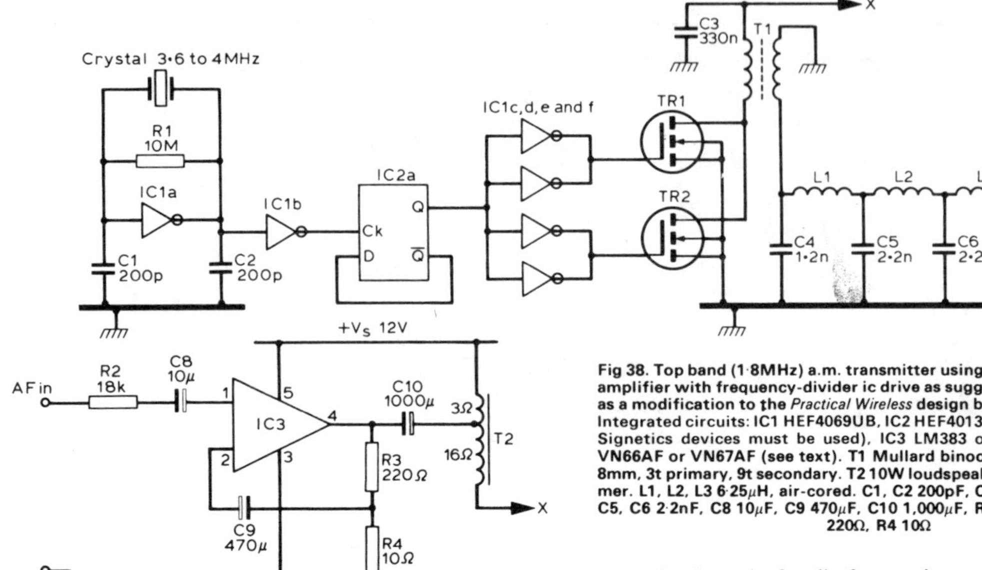


**Fig 38. Top band (1·8MHz) a.m. transmitter using low-cost vmos power amplifier with frequency-divider ic drive as suggested by J. W. Rimmer as a modification to the *Practical Wireless* design by J. R. Green, G3WVR. Integrated circuits: IC1 HEF4069UB, IC2 HEF4013B (these two specified Signetics devices must be used), IC3 LM383 or TDA2002. TR1, TR2 VN66AF or VN67AF (see text). T1 Mullard binocular balun 25 by 20 by 8mm, 3t primary, 9t secondary. T2 10W loudspeaker matching transformer. L1, L2, L3 6·25μH, air-cored. C1, C2 200pF, C3 330nF, C4, C7 1·2nF, C5, C6 2·2nF, C8 10μF, C9 470μF, C10 1,000μF, R1, 10MΩ, R2 18kΩ, R3 220Ω, R4 10Ω**

control should be adjusted in accordance with this recommendation.

There is a body of opinion which believes that every ssb operator should be able to check his transmissions by means of an oscilloscope and two-tone generator; it must be admitted however that large numbers of amateur stations based on factory-built ssb transmitters or transceivers do operate their stations without causing excessive interference without such aids, relying on the factory adjustments combined with intelligent appreciation of incoming reports. It should be noted that pa valves gradually deteriorate during their lifetimes, and as they age become less able to handle the very high peaks of current demanded by the voice waveform, resulting in a gradual increase of "splatter" unless compensated for by reducing the power output.

Speech processing, particularly the use of rf speech clipping, can result in a much more solid transmission, but such systems require careful engineering if they are not to result in distortion or rapid ageing of the pa valves due to the increased duty cycle. For this reason considerable care is needed when fitting an external speech processor to transmitters that may not have been designed to handle such waveforms.

It is important to note that when using an ssb crystal filter (for example a 9MHz filter with 2·7kHz bandwidth to the −6dB (voltage) points) optimum speech intelligibility can be achieved only when the filter response is very carefully matched to the appropriate carrier frequency used for the balanced modulator. Since this carrier frequency is largely suppressed in the modulator, before the signal passes through the crystal filter, it might be supposed that the relationship between filter and carrier frequencies would not be unduly critical. It is worth exploring why this is not the case.

Think of the original speech frequencies: for a male voice, these will include components up to about 7,000Hz, but for *communication* it is not necessary and not desirable to transmit more than about half this range. For this reason the speech amplifier of an ssb will normally include filtering which removes most energy above 3,500Hz. Many studies have shown that with a restricted range of audio frequencies maximum intelligibility with minimum frequency range is achieved in the range 300 to 3,000Hz, a bandpass of 2,700Hz. However any further restriction of frequency range or any significant shifting of the upper and lower limits will produce unnatural "toppy" or "boomy" speech which again reduces intelligibility.

Immediately following a balanced modulator the rf signal will consist of the two sidebands and partially suppressed carrier. For example, supposing the carrier frequency has been 9,001kHz, then 200–3,500Hz modulation will tend to result in rf sideband energy over the frequency range 9,001kHz ± 3·5kHz, ie 9,004·5 to 8,997·5kHz. If this signal is now passed through an ssb filter with 2·7kHz bandpass and a response characteristic of 9,000·5 to 8,997·8kHz, then it will allow through lower sideband energy corresponding to original audio frequencies represented by 9,001–9,000·5 to 9,001–8,997·8kHz, ie 500–3,200Hz. This is now some 200Hz higher than the optimum range of 300–3,000Hz and when received on a correctly-adjusted receiver it would sound rather deficient in low frequencies and rather "toppy". If the receiver was slightly off-tuned to make the speech sound more natural then if the receiver also had a 2,700Hz filter this could be done only by losing some of the sideband energy and by producing some of the undesirable effects of frequency-shifted speech.

In other words the audio range of 500–3,200Hz is some 200Hz higher than the optimum range. To hit the 300–3,000Hz range we would need to place the carrier frequency used in our balanced modulator some 200Hz lower, ie 9,000·8 not 9,001kHz. This example has shown that it is the relationship between the carrier frequency and the crystal filter that governs the transmitted audio range of a filter-type ssb transmitter.

Now in the above example a 200Hz error at 9MHz is relatively small and, even when using matched insertion-frequency crystals supplied by the manufacturer with a factory-made precision crystal filter, this could be caused by using a carrier oscillator having incorrect capacitance across the crystal or might even be caused by temperature variations. It is thus necessary to be able to adjust or trim the carrier crystal oscillator over a small frequency range. Ideally it should always be set 300–400Hz outside the appropriate −6dB voltage response point of the crystal filter. This will imply that the carrier is located around −20 to −30dB down a representative filter slope. It is yet one more reason why the adjustment and alignment of an ssb transmitter calls for understanding and accurate measuring equipment (though with experience it may well be possible to make adjustments with "makeshift" aids, including a good receiver, rather than high-cost test equipment).

## Transmitter power

There are various ways of defining the power of a transmitter and this has led to some confusion in the minds of many amateurs; a fact not helped by some advertisements which tend to take advantage of this confusion. It is necessary to understand the differences between power input and power output, dc input and *peak envelope power* (p.e.p.) and to appreciate how these relate to the power levels specified in the Amateur Licence.

Clearly what really matters from a communication viewpoint is the *useful* power radiated from the antenna (*effective radiated power* or erp) and this takes into account any directivity gain of the antenna along the particular path, the losses involved in coupling the transmitter to the radiating elements (eg feeder losses) as well as the actual rf power from the transmitter, noting that in A3 amplitude modulation the carrier power (which represents most of the power in an A3 transmission) actually contributes nothing to the signal/noise ratio of the speech information.

But erp is an extremely difficult characteristic to measure directly, and is normally only a calculated figure.

So the licensing authorities and amateurs take account only of the transmitter power and, whereas the professionals always rate a transmitter in terms of *power output*, the practice for a.m./cw/fsk/nbfm amateur transmitters is to rate the transmitter in terms of dc input power to the valve, valves or semiconductors delivering power to the antenna. This normally means the dc input to the anode(s)/collector(s) of the output stage, although strictly speaking in the grounded-grid (grounded-base) configuration some of the power output is provided by the driver stage.

So for a conventional pa the power input is found by measuring the anode (plate) current and multiplying this by the ht voltage between the cathode and anode (ie not counting any voltage dropped across a cathode bias resistor). For example with say 700V and 100mA anode current, the power input of the transmitter would be stated as 70W. The actual *output* power of such a transmitter would depend on the efficiency of the pa stage and this in turn will depend largely upon whether the correct load is presented by the tank output circuit or network and the bias conditions under which the stage is operating. For Class C (where the bias is significantly more than cut-off point), the efficiency may be as high as 75 per cent (and efficiencies up to 90 per cent or so are possible with the more specialized techniques of Classes D, E and F). So typically a 70W a.m./cw/nbfm/fsk transmitter might provide an rf carrier output power of about 50W on some bands. Because of the internal capacitances of valves and semiconductors the efficiency of many transmitters tends to fall with increasing frequency, so that output power at 28MHz might be significantly less than at 7MHz.

With an a.m. transmitter the output power at 100 per cent modulation will be increased by the power in the sidebands and the output will no longer be a sine-wave or have an "envelope" of constant amplitude; rather it will have an envelope comprising a series of peaks and troughs. The *instantaneous* power at the peak of a 100 per cent modulated transmitter will be *four* times carrier output power, but this would not show up on a typical power meter (although it can be confirmed by looking at the power envelope waveform on an oscilloscope). A conventional power meter would indicate a 50 per cent increase in power and an antenna current meter would show an increase of approximately 22·5 per cent when the transmitter is 100 per cent modulated by a sine-wave tone.

### POWER MEASUREMENT OF SSB TRANSMITTERS

For the UK amateur the maximum permitted dc power input to the anode circuit of the final amplifier stage of the transmitter is stated; for example in the 3·5MHz band the maximum permitted dc input is 150W. Assuming that the conversion efficiency of the final stage is 66·6 per cent the mean rf power delivered to the antenna with 150W dc input will be 100W, representing a peak envelope power of 400W under 100 per cent amplitude modulation conditions (Peak envelope power is the average power supplied to the antenna during one rf cycle *at the highest crest of the modulation envelope*).

In a.m. and cw transmitters it is a simple matter to measure the current and voltage to determine the dc power supplied to the anode circuit of the final stage. In single-sideband operation (A3A, A3H, A3J) the transmitter output stage operates in a linear condition where the 66·6 per cent efficiency no longer applies and dc power is supplied even when there is no rf output.

The most satisfactory method of assessing the power rating of a linear transmitter is based on the measurement of rf output power instead of dc input power. The official UK method of power assessment for amateur ssb transmitters states that the permitted p.e.p. output should not exceed that from an A1 or A3 transmitter using the maximum permitted dc power input.

The following is the approved method of measurement, which may be used during station inspections.

The transmitter output is terminated in a resistive load of appropriate impedance value, provided with an rf ammeter or rectifier/voltmeter to enable the power in the load to be calculated. The load should have low reactance at the frequency used and should be surrounded by an earthed screen. A cathode-ray oscilloscope is used to observe the output waveform in the load.

The first step in the measurement procedure is to adjust the ssb transmitter for suppressed-carrier operation and to connect two af tones of equal amplitude to the transmitter input. The transmitter is then adjusted to give a mean rf power output, as measured in the load, of the value appropriate to the band (eg 200W at 3·5MHz). If the timebase is correctly adjusted the oscilloscope will display the rf envelope waveform resulting from the combination of the two tones. Provided that the input tones are sinusoidal and that the transmitter is not overloaded the p.e.p. output will be twice the mean power measured in the load, and the maximum deflection on the cro display will represent this p.e.p. level. By this process the oscilloscope may be calibrated to indicate the maximum permitted p.e.p. for any frequency band.

The second step is to adjust the transmitter for A3A, A3H or A3J operation and to replace the two-tone input signal by speech signals, the level being adjusted so that the cro deflection on speech peaks does not exceed the calibration level. The transmitter is then providing the maximum permitted p.e.p. output and line-up conditions should not subsequently be altered in such a way as to increase this output.

In the case of vhf, uhf and shf measurements the use of an oscilloscope may not be practical, and the rf ammeter or voltmeter may be replaced by a crystal rectifier and calibrated meter; for shf measurements a bolometer may be used.

It should be noted that the maximum permitted *mean* rf output during a two-tone test is *half* the permitted p.e.p. output specified in the licence.

An a.m. transmitter with an rf carrier output of 100W (from say a dc input of 150W) will thus be providing 400W p.e.p. at the instantaneous peaks of the waveform, representing 200W p.e.p. carrier and 200W p.e.p. in the two sidebands. It should be appreciated that the extra power of a modulated a.m. transmission stems from the audio power in the modulator of a high-level anode and screen modulated power amplifier; similar efficiencies cannot be obtained with such techniques as grid and screen-grid modulation where the rf power amplifier requires to be adjusted for relatively low-efficiency operation.

A similar reduction in efficiency when compared to Class C operation is also a feature of any high-power linear amplifier. This will usually be biased to operate in Class A, AB or B, representing theoretical efficiencies in the range 20 to 65 per cent. In practice the p.e.p. dc input to a high-power linear amplifier will generally be of the order of twice the p.e.p. output, representing an efficiency of 50 per cent.

Where the linear amplifier is used for ssb operation, the measurement of power becomes considerably more complex than where there is a steady carrier. In the absence of speech there will be virtually no rf output and the dc input power under such conditions is no indication of the transmitter power, but only of the biasing conditions. During actual operation the power output reflects the "spiky" nature of the speech waveform. The rf p.e.p. output depends directly upon the peak input to the amplifier and it is meaningless to rate a linear amplifier unless a specified degree of maximum distortion is indicated.

By generating a steady sine-wave audio tone (a waveform not found in speech although approached by a steady whistle) the linear amplifier provides output on a single frequency (as for cw transmission) and this is often used for routine tuning up. It should be appreciated however that this gives no real indication of any lack of linearity which, when speech is used, would cause distortion and spurious products, as denoted by "splatter". For correct adjustment and power rating it is necessary to apply two different audio tones simultaneously. Because of the beating effect between the two af tones, this results in the production of a regular waveform pattern which, when observed on an oscilloscope, resembles that of a 100 per cent modulated carrier wave, and provides an indication of non-linearity and "flat topping" (indicating that the amplifier cannot deliver the peak of power demanded by the input waveform). With this *two-tone test* the power delivered by the transmitter to an external load represents *half* the p.e.p. output.

The UK authorities have provided guidance on acceptable methods of power measurement of ssb transmitters including the two-tone test.

Generally the p.e.p. output is determined under specified conditions of distortion. The rf output p.e.p., under linear conditions, is limited to 2·667 times the dc input power appropriate to the frequency band concerned: thus for 1·8MHz where dc input for A3 or cw operation is limited to 10W, the corresponding limit for A3A or A3J single sideband operation is 26·67W p.e.p.: for bands where 150W is the maximum dc input power the corresponding p.e.p. output figure is 400W.

It will be appreciated that during normal speech operation the current meter of a linear amplifier is unable to follow the rapid changes of the speech waveform and normally such an amplifier should not be "talked up" to more than *one-half* of the current indicated during the single-tone check.

# FREQUENCY MEASUREMENT

When operating a transmitter it is important to have a fairly accurate idea of the frequency of the radiated signal. In particular we should always be quite sure that the signal does not lie outside the amateur band.

With a crystal-controlled transmitter this presents no great difficulty because the frequency will be almost entirely dependent on the crystal.

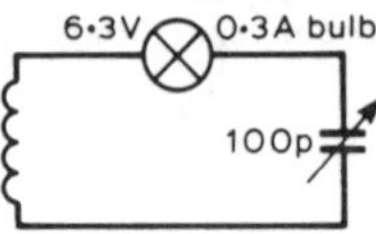


**Fig 39. Simple absorption wavemeter. This might have three plug-in-coils, on 1in dia formers, wound as follows, 90t 32swg enam (approx 1·5–4·5MHz); 32t 22swg enam (4–12MHz); 7t 22swg enam (11–33MHz)**

## The absorption wavemeter

Knowing the crystal frequency, however, is not in itself sufficient because there is the possibility that the final output from the transmitter may be on the wrong harmonic of that frequency. To ensure that this is not so, it is necessary to use an absorption wavemeter such as is shown in Fig 39. This would be built as a compact portable unit and, to test the output frequency of any stage, would be held close to the relevant tuned circuit. With the wavemeter resonated to the circuit to which it is coupled some energy is absorbed from that circuit. If the wavemeter is held close enough the rf current induced into it will be sufficient to light the bulb. Thus, provided that the wavemeter has been previously calibrated, it is only a matter of tuning for a resonance indication on the bulb and reading off the frequency.

A convenient method of calibrating the wavemeter is to use a communication receiver. An antenna is connected to the receiver, the lead-in wire being loosely wound once or twice around the wavemeter coil on its way to the antenna socket. If a suitable station is then tuned in it will be found that at a certain setting of the wavemeter capacitor the strength of the received signal will fall. This will indicate that the wavemeter is absorbing some of the incoming signal and that it must therefore be tuned to the frequency of that station. By repeating this operation with a selection of stations a number of calibration points can be obtained and it will therefore be possible to prepare a calibrated scale to fit under the knob of the variable capacitor.

A low power stage may not be capable of providing enough rf power to light the bulb. Under such circumstances resonance indication can be obtained from the anode current which will rise when the wavemeter absorbs energy.

## Crystal calibrator

When a vfo is used for transmission the absorption wavemeter is still invaluable for checking that circuits are tuned to the correct fundamental or harmonic frequencies but it is not capable of giving a sufficiently accurate reading to ensure that the transmitter is actually in the band, or to show the frequency other than very roughly. It is essential therefore to be able to check a vfo type transmitter against a more accurate standard, and this will usually include a crystal-controlled oscillator. The frequency meter may include a complete heterodyne unit with both stable variable oscillator and crystal reference, or alternatively a small crystal calibrator may be used to calibrate accurately the station receiver which is then used to check the transmitter.

It is unfortunate however that superhet receivers will generally respond to a very strong local signal at several spurious tuning points and is not always obvious which of the responses are spurious. The use of a crystal calibrator, in the absence of a heterodyne frequency meter, still requires a really good absorption meter or some other unambiguous means of checking the output frequency.

A versatile crystal calibrator based on a 1MHz crystal but capable of providing marker points at 100kHz, 10kHz or 12·5kHz points up to the 144MHz band, uses 7490 ICs as divide-by-10 or divide-by-eight devices. The 12·5kHz markers are particularly useful for checking 25kHz channels widely used for nbfm operation on 144MHz. The unit can be built on Veroboard and conveniently fits into a tobacco box together with three No 8

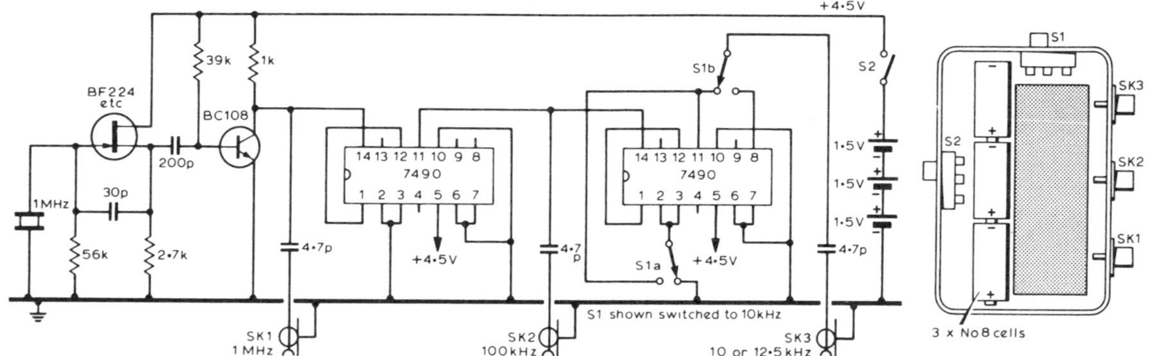


Fig 40. A versatile crystal calibrator providing output markers spaced at 1MHz, 100kHz, 12·5kHz or 10kHz. The second 7490 ic functions as either a divide-by-10 or divide-by-8 device

cells: Fig 40. The fet crystal oscillator can use a 2N3819 or almost any similar device.

## Checking 100kHz oscillators

The 100kHz harmonics can be checked against *standard frequency transmissions*, such as those radiated by stations MSF, WWV etc, on exactly 2·5, 5, 10, 15, 20 and 25MHz. In the UK there are fortunately also the accurate 200kHz BBC long-wave programme transmissions. These are most convenient for checking a 100kHz standard, since the BBC station can be tuned on almost any receiver and the difference between these transmissions and the local oscillator heard as beats. Although the calibrator frequency will be determined primarily by the crystal frequency, this can if inaccurate often be "pulled" into zero beat with the BBC station by means of a trimmer across the crystal.

## Heterodyne frequency meters

Many amateurs do not have a crystal calibrator built in as an integral part of their receivers but have what is usually called a heterodyne frequency meter as a separate instrument. This usually consists of an accurately calibrated vfo complete with a crystal calibrator similar to that which has been described here. With some instruments the calibrated oscillator must be compared with the transmitter vfo using a receiver. Others have facilities for accepting a small rf voltage from the transmitter, the comparison being effected by producing a beat note between the two frequencies within the instrument itself.

## Frequency-checking equipment

Applicants for Amateur Licence A, Amateur Licence B or the Amateur Radio Certificate have to sign the following undertaking:

"I undertake to maintain, at the station, frequency-checking equipment of sufficient accuracy to ensure that transmissions are within frequency bands allocated for amateur use, and equipment to enable me to confirm that harmonic and spurious emissions are suppressed. I understand that out-of-band working, whether intentional or not, will be regarded as a serious misdemeanour and could result in the withdrawal of the licence."

The Home Office does not endorse or recommend particular makes and types of frequency-checking equipment but does provide the following general guidance:

1. A licensee must:
(a) be able to verify that his transmissions are within the authorized frequency band, (ie that no appreciable energy is radiated outside the band).
(b) use a satisfactory method of frequency control.
(c) ensure that his transmissions do not contain unwanted frequencies (ie harmonics and spurious frequencies).

2. When his station is inspected by officers authorized by the Secretary of State, the licensee will be expected to demonstrate that he can conform with the requirements (a) to (c) above.

3. As a general rule, a station requires a crystal reference source to comply with 1(a) and (b) above so that:
(a) with a crystal-controlled transmitter an absorption device of suitable frequency range and accuracy is necessary to check that the desired harmonic of the crystal frequency is selected.
(b) with a transmitter that is not crystal-controlled a wave-meter based on a crystal oscillator is necessary.
Within these outline requirements the licensee is free to decide how he will meet the licence regulations.

4. The following comments may provide useful guidance:
(a) Frequency measuring equipment should be of sufficient accuracy to verify that emissions are within the authorized frequency bands. For example, operation in the centre of the 21·0–21·45MHz band would require frequency measurement to an accuracy of ±1·0 per cent to ensure that emissions were within band, whereas operation within, say, 10kHz of band-edge would require measurement to an accuracy of ±0·05 per cent. When determining the proximity of an emission to band-edge, the bandspread due to modulation, on the appropriate side of the carrier, needs to be added to the frequency tolerance of the carrier.
(b) Heterodyne wavemeters and crystal calibrators. When used in conjunction with a general-coverage receiver, a 100kHz crystal is usually adequate for checking frequencies up to 4MHz. For higher frequencies the spacing between 100kHz marker points is too small for accuracy, and a crystal of 500kHz or preferably 1MHz should be used in addition. If the receiver covers only the amateur frequency bands the bandspread scale will usually allow a 100kHz crystal to be used with sufficient accuracy throughout the hf bands.
(c) Absorption wavemeters and similar devices. The scale length and accuracy should be suitable for measurements of the required accuracy to be made, and the frequency coverage must extend up to the second, and preferably the third, harmonic of the radiated frequency so that the presence of unwanted frequencies may be detected. For vhf and uhf transmitters, probably the best technique is to measure the frequency of the fundamental oscillator as accurately as possible and to use an absorption device to confirm that the wanted harmonic has been selected. When a vhf or uhf converter is used in conjunction with an hf receiver and the calibration of the main receiver can be checked with sufficient accuracy, this will provide a means of frequency measurement but it is also advisable to use an absorption wavemeter to check the measurement and to confirm that no unwanted radiations are present.

CHAPTER 5

# The antenna

Whenever the intensity of an electric current passing through a wire changes, some amount of energy is radiated into space. The radiated energy is said to consist of radio waves. It is not necessary to be able to visualize these waves clearly, though it is useful to remember they travel outwards from the wire at 300,000,000 metres per second. Thus with a radio signal having a frequency of 30,000,000Hz (30MHz or 30 million cycles per second) the distance occupied by one complete wave (ie from one positive peak of current to the next) is 10m. We therefore say that a 30MHz signal has a wavelength (often abbreviated to $\lambda$) of 10m, which may be readily visualized.

It is important to note that *any* piece of wire or conductive metal tubing will radiate a signal if one can couple into it rf energy (which until it is radiated is in the form of an electrical alternating current in which the amplitude of the current and potential will usually be alternating between maximum positive and maximum negative values many millions of times per second). A short piece of wire will radiate energy less readily than a longer piece of wire of one half-wavelength ($\lambda/2$) or more long, but this is primarily because it is more difficult to couple energy into and out of a short length of wire and there will be increases in energy losses (IR or "current times resistance" losses), rather than because of any fundamental requirement that an antenna should always have a length of $\lambda/2$ or more.

Since an antenna radiates power when there is current flowing in it, it can be thought of as possessing *radiation resistance*. It should be noted that this is not the same as its characteristic impedance (or feed-point impedance) which will vary along its length.

The feed-point impedance comprises capacitive or inductive reactance and resistance. At the centre point of a half-wave ($\lambda/2$) dipole the impedance is low and purely resistive, with no capacitive or inductive reactance, and considerable use is made of this characteristic. The two ends of a resonant $\lambda/2$ dipole are also purely resistive, but with a feed-point impedance of several thousand ohms.

In practice an antenna system should normally aim at being resonant at the frequency of operation. This does not imply that the radiating wire or element itself need always be exactly $\lambda/2$ long, since the system may be "loaded" with lumped inductance (or its electrical length made less than its physical length by means of series capacitors) or by using the earth (or earth radials or counterpoise) as part of the resonant antenna element. Where the element is itself $\lambda/2$ long, or a multiple of this, then the operation does not rely upon the presence of the earth, although in practice an earth will normally be connected to the transmitter for purposes of safety.

A radiating element should be well removed from any energy-absorbing materials, such as buildings, trees, metal fences, metal drainpipes, house electrical wiring etc. Its height above earth (which acts as both an absorber and reflector of radio waves) has an important bearing on its operation, although for most practical purposes, the "higher the better". The electrical conductivity of the earth below and surrounding the antenna also affects its operation, and is especially important at the lower amateur frequencies where the "earth" may form the most important part of the antenna system, or where a vertical type of antenna is used. Since the rf energy may give rise to very high maximum rf voltages and currents, good insulation and sufficiently thick wire or tubing for the elements will reduce IR losses.

An antenna will almost always work better when erected outside a building, due to the attenuation of signals by the building or because of the close proximity of electric wiring, hot-water cisterns and systems etc. If there is no alternative to having the antenna within the building, the roof space is usually the preferred place to put it. For vhf and uhf, quite effective antennas can be taped to an outside window and even on hf it is surprising what can be done from the most difficult of situations. A short sloping radiator on the balcony of a high building, or even a thin "invisible" antenna lowered from such a balcony will usually produce hf contacts quite readily with stations up to hundreds of miles away, although will put the user at some disadvantage when seeking the longest-distance contacts.

The secret to using such antennas lies in the art of coupling into them the energy produced by the transmitter, in such a manner that the operating bandwidth is not unduly restricted and so that there is no risk of damaging the transmitter due to incorrect adjustments of the output coupling network. It is also important to ensure that the "hot points" (ie points of maximum rf voltage) do not represent a hazard to others; a medium-power transmitter can cause rf burns which would be extremely serious if unexpectedly incurred by someone standing on a ladder, for example. Do not be misled by the fact that birds may perch quite happily on the wire!

## Radiation

To radiate a strong signal we must persuade a large rf current to flow into the antenna wire, and in order to obtain some idea of how this may be done let us first consider the horizontal antenna shown in Fig 1. If the switch S is suddenly closed, current will flow into the antenna to charge up the capacitance formed by the wire and the ground. Now it is normally assumed that the electric current flow through a wire is instantaneous but in reality it has a finite speed which is approximately equal to that of radio waves in space. Thus the current will not flow immediately into all the capacitance. If we consider the antenna to be made up of a series of small elements we can think of each element as starting to charge up in sequence. There will in effect be a "front" of current

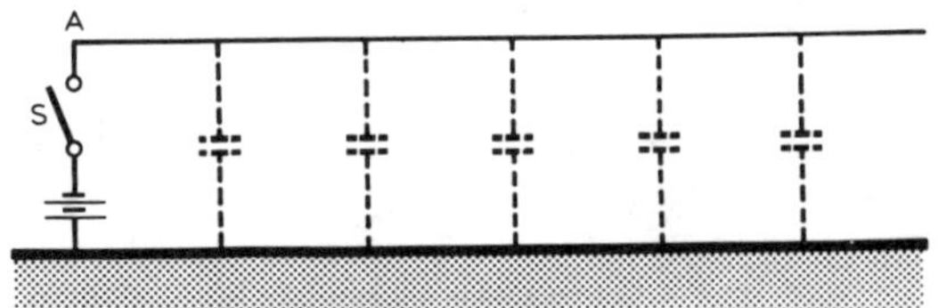

Fig 1. A horizontal antenna

speeding along the wire at 300,000,000 metres per second. Behind this front, current will be flowing, while ahead of it there will be no current. With an infinitely long wire this current front would be continually moving on and at an instant in time the current from the battery would be charging up some new section of the wire. At point A there would consequently be a steady flow of current and, as far as the battery was concerned, it would be feeding into a resistance.

The wire would still behave as a resistance if the battery was replaced by a source of rf energy. Peaks and troughs of rf current would then pass along the wire in the same way as did the current front. As the current passed along the wire some energy would be lost due to the resistance of the wire but, with a reasonably thick wire, this energy would be far less than that radiated as radio waves, and we would have an efficient transmitting antenna.

## The $\lambda/2$ antenna

A practical antenna cannot be infinitely long and its behaviour is not so simple. When the current reaches the end of the practical antenna it is reflected back.

A particularly important effect occurs when the antenna is exactly $\lambda/2$ long as in Fig 2(a). Suppose that a positive peak of current is moving from A to C. When it reaches C it will be reflected back again. The reflected current will have the same amplitude and the opposite polarity, and thus at C itself the net amplitude of the current will be zero. From B to C and back again is exactly $\lambda/2$. Thus by the time the reflected current peak, now negative, reaches B, it will meet the following negative peak which is on its outward journey from A to C. The two current peaks will therefore add together producing a very high peak of current at B.

When the reflected current peak reaches A it will be reflected back once more, this time as a positive peak. It will then have travelled exactly $1\lambda$ since first leaving A and will therefore be reinforced by the next positive peak leaving the generator.

The same sort of thing also happens to the negative peaks. If we used an ammeter to measure the force of the alternating current at various points along the wire we would find it to be as shown in Fig 2(b). At B each peak of current would be reinforced by a reflection from C and the amplitude of the alternating current there would be high. At C the incident current at any instant is always cancelled by its reflection and the net value of the current here is always zero. The net current at A would also be zero if the rf source were not connected at this point; in practice the ammeter would measure the current being supplied by the generator.

If we were to measure the rf voltages on the antenna we should find that each end of the antenna had a high voltage to earth and that the voltage fell as we approached the centre, at which point it would be zero. The voltages at the two ends would be equal in magnitude but opposite in phase.

The fact that the magnitudes of the rf current and voltages vary along the length of the wire is often described by saying that there are standing waves present. We have the interesting condition that the ratio of voltage to current, that is to say the impedance of the antenna, is high at either end but low in the centre. This is in contrast to the infinitely long antenna where the amplitudes of the current and voltage were always in proportion, gradually diminishing along the length of the wire as energy became lost.

We have assumed so far that the capacitance to earth is essential for the operation of the antenna. There is, however, always capacitance between the individual points on the antenna and, although the operation is then not quite so easy to understand, the electrical characteristics of a $\lambda/2$ antenna are in fact very similar with the wire perpendicular to the ground, or even without the earth being present at all.

If a $\lambda/2$ antenna terminated at the transmitter, it could be

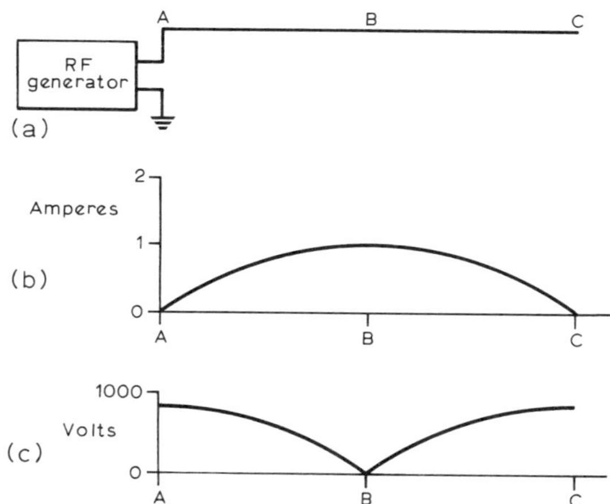


Fig 2. (a) Horizontal half-wave ($\lambda/2$) antenna. (b) Typical rf current distribution (100W rf). (c) Typical rf voltage distribution (100W rf)

connected in the manner shown in Fig 3. At resonance the antenna would appear as though it were a resistance of a few thousand ohms to earth. Assuming that the transmitter has been designed to give its greatest power output when feeding into about 72Ω† it would be necessary to interpose the rf transformer L1, L2, C. Then, although the antenna may be taking, for example, 0·1A at 720V, the input to L2 can be 1A at 72V so that the transmitter "thinks" it has a load of 72Ω.

Now if the antenna is not exactly $\lambda/2$ long the reflected current at A will not be exactly in phase with the applied voltage. This will give the effect of the antenna having some inductance or capacitance and will modify the tuning of L2,C. But as long as C is used to tune the whole system to resonance the load applied to the transmitter will appear to be a pure resistance.

It must always be borne in mind that for maximum power output the transmitter must feed into a resistive load. Various combinations of reactance (either inductive or capacitive) and resistance can give an impedance of 72Ω and any of these combinations will take 1A at 72V. The power taken, however, will only be 72W if the voltage and current are exactly in phase, ie if the load is a pure resistance.

An antenna of the type shown might have its length chosen so that its natural resonance lay in the centre of an amateur band. The rf transformer, more commonly called an antenna tuner or antenna coupler, could then be used to tune the antenna to any frequency in the band, at the same time ensuring that the transmitter "saw" a resistance of about 72Ω at each frequency.

## Directivity

A $\lambda/2$ antenna, or dipole as it is often called, which is well clear of the ground radiates quite well in all directions except those making an angle of less than about 30° to the wire. Thus a horizontal dipole lying north-south will radiate eastwards and westwards at all angles of elevation from horizontal to vertical. But northwards and southwards there will be little radiation at angles less than about 30° to the horizontal.

When the antenna is close to the ground, as must be the case in practice, the waves reflected by the earth interact with those radiated directly from the wire and modify the radiation pattern. This effect is most pronounced when the height of the antenna is $\lambda/4$ or lower, and under these conditions all radiation at low

†Feeder cables generally have characteristic impedances between 50 and 100 Ω, 72 Ω being a popular value.

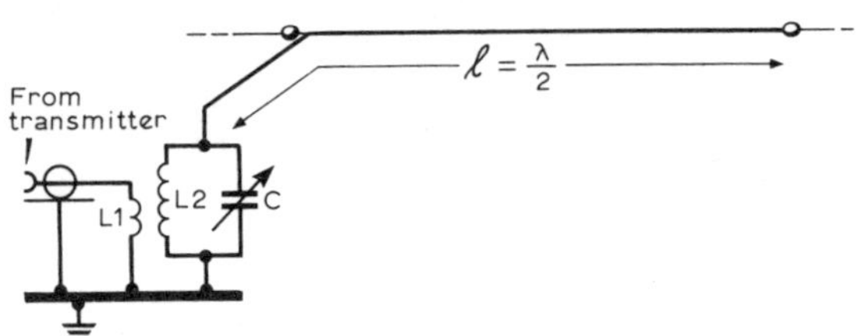


Fig 3. Half-wave antenna connected direct to antenna tuner. L ≏ 470/f ft, where f is the frequency in megahertz

angles to the earth tends to be suppressed and the high-angle radiation is accentuated.

For almost all hf communication we use those waves which are reflected back to earth by the ionized layers in the atmosphere. When transmitting over a long distance it is the low-angle radiation which is used and consequently an antenna height of $\lambda/2$ or more is to be preferred. On the other hand, for communication over distances of only a few hundred miles we normally use the high-angle radiation and a lower antenna can give excellent results.

If the dipole is mounted vertically there will be low-angle radiation in all compass directions but little high-angle radiation.

If we are to obtain the best results from an antenna some care must be exercised in siting it, not only to ensure that it radiates in the required directions but also to ensure that it is as clear as possible from buildings and trees which might absorb some of the radiated energy. Consequently some method must often be devised whereby energy may be fed to an antenna which is remote from the transmitter. To understand how this can be done we must first give a little thought to the theory of transmission lines.

## Transmission lines

A typical rf transmission line or feeder is shown in Fig 4. If an infinite length of this feeder were coupled to an rf generator it would behave in a similar manner to the infinitely long antenna. With the dimensions given the feeder would appear to have a resistance of about 600Ω and this value would be said to be its characteristic impedance. The current in the two wires at any instant would be equal in amplitude and opposite in polarity. Thus, because the feeder is symmetrical and the wires are close together, the net radiation from the feeder will be quite small.

Now suppose that we take a finite length of the feeder and connect a 600Ω resistor across the far end. As far as the rf generator and the feeder are concerned it is just the same as if an infinite length of feeder had been joined on the end. This means that any length of the feeder, when terminated by a load having a resistance of 600Ω, will present a resistance of that value to the generator. Moreover, nearly all the rf energy will be guided by the feeder to the load and not radiated en route.

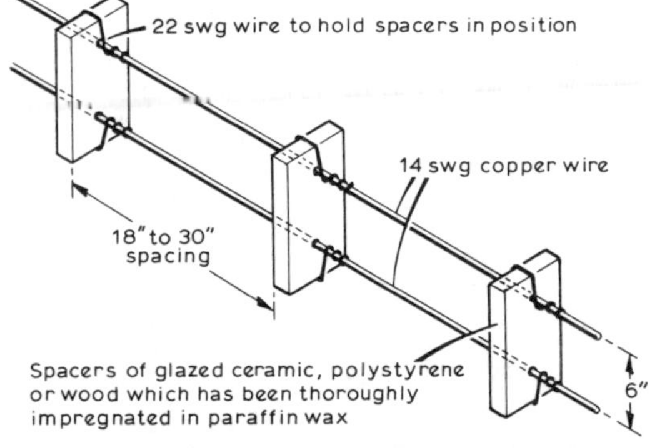


Fig 4. Typical amateur-built transmission line

If the feeder is not terminated by its characteristic impedance, reflections will take place at the end and standing waves will occur. An extreme case is that of a $\lambda/4$ line terminated by a short-circuit as in Fig 5(a). We can think of this as rather like a $\lambda/2$ antenna doubled back on itself. Thus although the load resistance is very low the resistance at A will appear to be high. A converse effect occurs if the end B is open circuit as in Fig 5(b). The apparent resistance at A is then very low.

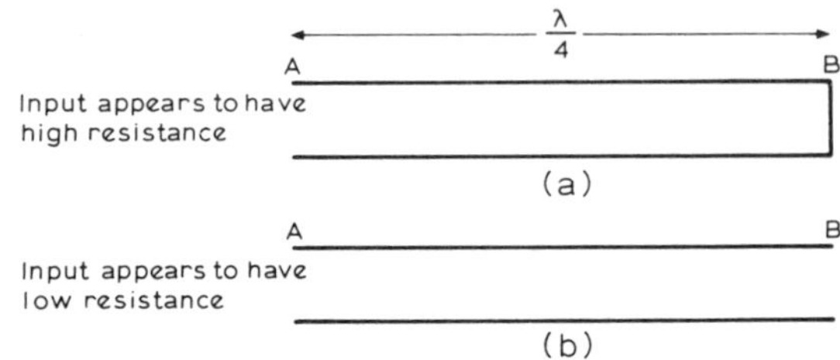


Fig 5. Quarter-wave line. (a) Short-circuited end. (b) Open-circuited end

The nearer the load resistance becomes to the characteristic impedance the nearer the input resistance becomes to it also. Mathematically the relationship is expressed $Z_i/Z_c = Z_c/Z_L$, where $Z_i$ is the impedance "seen" at the input to the feeder, $Z_c$ the characteristic impedance and $Z_L$ the load impedance. Thus with a 600Ω feeder and a resistive load of 300Ω the feeder input will appear to be resistance of 1,200Ω; with a 1,200Ω load the input would appear to be 300Ω. If any reactance is present in the load the input impedance will also contain some reactance.

The $\lambda/4$ line acts in effect like a transformer. If the line is $\lambda/2$ long we can think of it as being two of these transformers connected back to back. The net effect will consequently be as if there were no transformer at all and the input impedance will equal the load impedance. This condition will in fact hold for any even multiple of $\lambda/4$. With an odd multiple of $\lambda/4$ the net effect is the same as if there were only one.

Commercial feeders are obtainable with characteristic impedances of about 70 to 300Ω. There are also coaxial cables available which behave similarly and which usually have impedances around 50 to 100Ω. The coaxial cables have one conductor completely surrounding the other; the outer one is usually earthed so that it forms a screen. The commercial products are perfectly satisfactory provided that the feeder is operated "flat", that is to say, provided that the ratio of maximum and minimum rf currents along the wires (the standing-wave ratio) is less than about 3:1. With higher standing-wave ratios there are liable to be appreciable losses at the high-current points due to the resistance of the conductors and at the high-voltage points due to leakage in the insulating material. In such cases the use of a home-made feeder such as shown in Fig 4 is to be preferred.

## Centre-fed antennas

There are often advantages in feeding an antenna at the centre. We have already seen that the impedance here is low and if, in fact, we break open the centre of a dipole the two ends thus formed appear to have a resistance of about 72Ω between them. This means that we can take a 72Ω coaxial cable direct from the antenna to the transmitter as shown in Fig 6. Ideally a balanced feeder should be used. Since the impedance is low the unbalance caused by using coaxial cable is low but can cause television

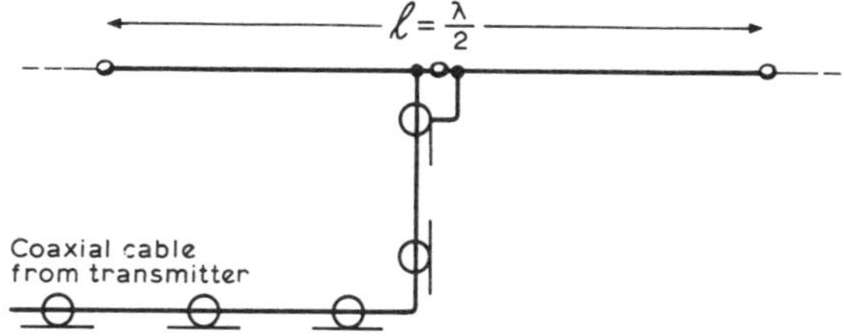


Fig 6. Half-wave dipole fed by coaxial cable. The length ≃ 470/f ft

interference: this problem can be overcome by using a *balun* (*balanced-to-unbalanced*) transformer.

It should be noted that 72Ω is the theoretical figure for a λ/2 antenna of thin wire which is well clear of the ground. With a practical antenna the value will often be more like 50Ω and some standing waves will occur. Nevertheless, provided that the antenna is not more than a few per cent off tune, the standing wave ratio will normally be less than 2:1. Any inductive or capacitive components appearing at the transmitter end of the coaxial cable, due to the latter not being an exact multiple of λ/4 long or due to the antenna being slightly off resonance, will be fairly small and will be compensated for in the tank circuit tuning.

Fig 7 shows the dipole centre-fed by a tuned line. Since the feed-point impedance of the antenna is low the impedance "seen" at the lower end of a given length of feeder will be the opposite to that with the end-fed version. Unlike the latter the whole antenna system is symmetrical and consequently the length of the radiating section is not particularly important since it can be compensated for by the antenna tuner without affecting the balance of the feeder.

## The folded dipole

If a dipole is composed of two parallel wires joined together at their extremities the current will be divided between them. If we break open only one of these wires at the centre we find that the resistance there is not 72Ω but more like 300Ω. The reason for this is not difficult to understand. We know that the power in any circuit is given by $I^2R$, where I is the current and R the resistance. If, therefore, for the same antenna power we are only going to cause half the current to flow, the resistance must be four times as large.

An antenna utilizing this fact, and usually known as a folded dipole, is shown in Fig 8. Advantages of this type of antenna are that it is broadly tuned and that the standing wave ratio on the feeder is normally very low. The radiating portion of the antenna can be composed of two parallel wires spaced a few inches apart, or can be constructed from a length of the 300Ω feeder itself.

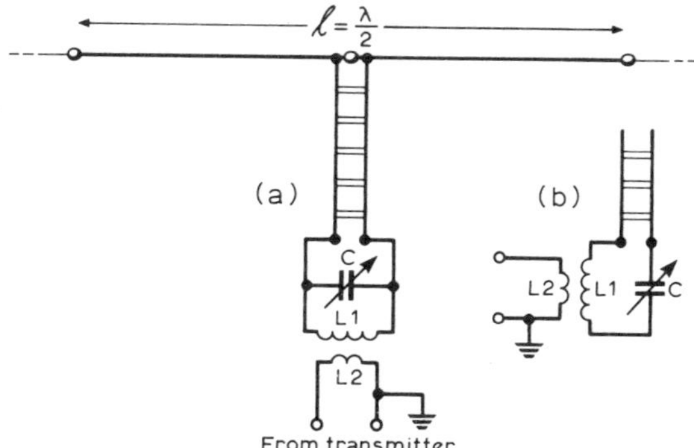


Fig 7. Half-wave dipole fed by tuned line. (a) Feeder length an odd multiple of λ/4 long. (b) Suitable antenna tuner for feeder whose length is an even multiple of λ/4. The length ≃ 470/f ft

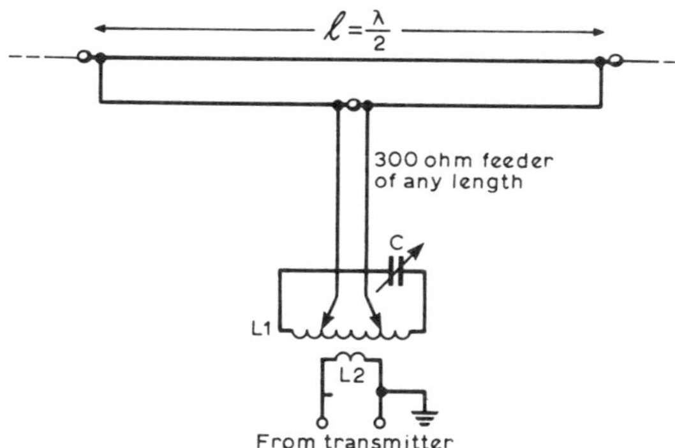


Fig 8. Folded dipole. The length ≃ 460 f where f is the frequency in megahertz. The taps on L1 should be about 2 5 of the way in from either end

## Multi-band operation

One advantage of the antenna in Fig 3 is that it can be operated on harmonic frequencies. At twice the design frequency, for example, the top is effectively two λ/2 sections joined together. The centre is no longer a high-current point but a high-voltage point. Nevertheless there are still high-voltage points at the ends and the antenna can be fed just as for the fundamental frequency. All that will be required is a suitable modification to the antenna tuner. When a horizontal antenna is several multiples of λ/2 long the low-angle radiation tends to be fairly evenly distributed in all compass directions.

The antenna in Fig 7 can also be operated on harmonic frequencies. On the second harmonic each section will be λ/2 and the feed point of the antenna will therefore have a high impedance. When operated on its second harmonic frequency the low-angle radiation tends to be concentrated in directions making angles of 45° to 135° to the antenna wire, but on higher harmonics the low-angle radiation tends to be omnidirectional.

The simplest and most versatile multi-band antenna of all is a single wire about 132ft long and terminated at the antenna tuner. This can be operated as λ/4 on 1·8MHz, as end-fed λ/2 on 3·5MHz and as end-fed multiples of λ/2 on, 7, 14, 21 and 28MHz: see Fig 9. Other techniques enable centre-fed antennas to be used on adjacent hf bands: see Figs 10, 11, 12.

While a λ/2 antenna radiates most of its energy in a broadside direction, longer elements will have radiation patterns that tend

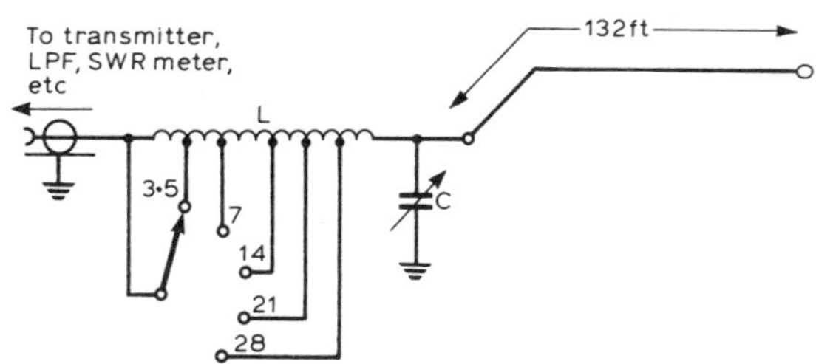


Fig 9. 132ft end-fed antenna used in conjunction with switched L-network matching unit. C is high-voltage 100pF capacitor. Diameter of L is uncertain but, typically, turns are 32 for 3·5MHz; 15 for 7MHz; 9 for 14MHz; 4 for 21MHz and 3 for 28MHz

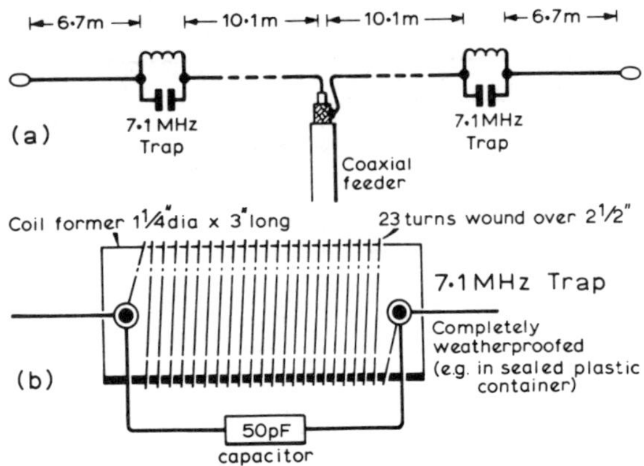


Fig 10. The W3DZZ multi-band trap dipole can be used on all bands from 3·5 to 28MHz. The 50pF capacitor must be a high-voltage type (eg 2kV)

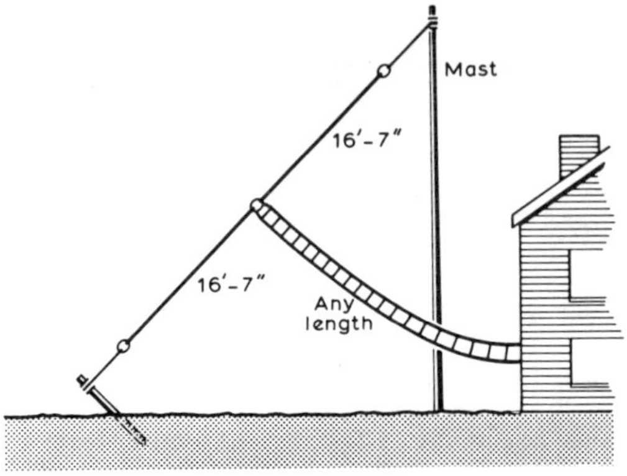

Fig 12. The simple "guy wire doublet" which can make an effective antenna for 14, 21 and 28MHz. The balanced transmission line can either be a home-made spaced line or 300Ω feeder

to break up into a number of major *lobes* and to radiate little energy in the intervening *nulls*. This will be indicated by the *horizontal radiation pattern* of the antenna, while equally important is the *vertical radiation pattern* showing the vertical angles in which the power will be concentrated. Fig 13 shows a number of horizontal patterns and clearly indicates how these are affected by the feed point of the antenna.

## Verticals

It has been noted already that a $\lambda/2$ dipole may be mounted vertically rather than horizontally. However, in practice, this results in the need for very high masts, particularly on the lower-frequency hf bands. It may also be found that the proximity of the earth to a high-voltage point of a dipole tends to unbalance the system and to result in high earth currents; results of vertical dipoles thus depend to a considerable extent upon the nature of the soil and the presence or absence of nearby metallic or conductive objects (including trees). Thus while some amateurs find vertical dipoles give excellent results, others may be disappointed and receive rather low signal reports.

However a $\lambda/2$ element can be thought of as a $\lambda/4 + \lambda/4$ element, each section of which is a mirror image of the other. The centre point has an impedance of about 70Ω at a point of maximum current and zero rf voltage; it is thus effectively at earth potential. If we remove the lower $\lambda/4$ section and replace this by earth, which may be true earth, or an artificial earth plane made from a number of $\lambda/4$ wires, or a single wire *counterpoise* (a

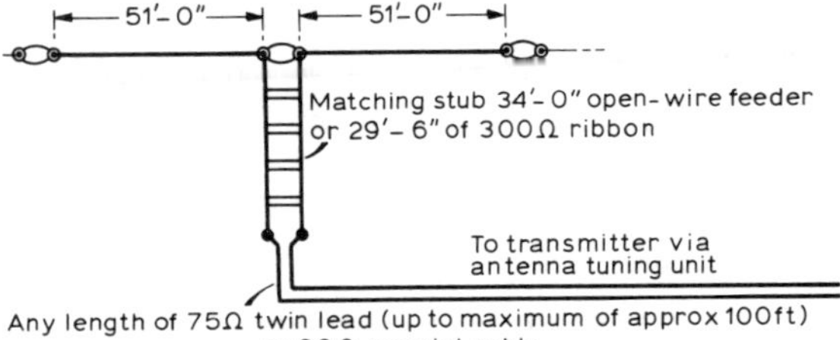


Fig 11. The popular G5RV "multi-band dipole" is a simple antenna which can be used effectively on all bands from 1·8 to 28MHz. If space is insufficient the final 10ft at each end of the "top" can be dropped downwards. Alternatively, a half-size version, with dimensions of the "top" and "matching stub" each scaled down to one-half, can be used on 7 to 28MHz bands

low wire more or less directly under the antenna insulated at the far end) or an *earth mat* made of a large number of wires on the surface of the earth or just below the surface, we still have the equivalent of a resonant $\lambda/2$ element, but now it is only $\lambda/4$ high and is much more manageable. Such an antenna is called a *vertical* $\lambda/4$ *monopole*, and has a feed point impedance of about 35Ω (or rather more if the *ground plane* slopes downwards); however most amateurs simply call it a "vertical" or, where mounted with its base at some height above ground and using a number of $\lambda/4$ *radials*, it may be called a *ground plane* antenna: see Fig 14. It is interesting to note that though theoretically there is no advantage in raising the base above ground, there may be practical advantages in doing so to keep the element well clear of metal drainpipes, fences, buildings etc. With the usual three or four radials such an antenna will be far from 100 per cent efficient, but nevertheless in practice it does form a useful and reasonably effective antenna which can conveniently be installed on the roof of a house or garage, without requiring any mast (the element being often a self-supporting tube of aluminium alloy) or any significant garden space.

If a vertical rod is correctly matched to a transmitter it need not necessarily be a resonant $\lambda/4$ but can be of any reasonable length. Vertical whip antennas of 25–30ft length are often used throughout the hf band, particularly in the maritime radio services. To do this effectively, however, requires the use of a matching network or switched networks located at the base of the antenna, whereas the resonant $\lambda/4$ rod can be connected directly to coaxial feeder. Often a *trap* may be inserted in a vertical element to provide in effect the W3DZZ form of multi-band dipole. Another interesting, though little used, form of vertical monopole is to have an elevated feed point to enhance low-angle radiation.

It is in fact often claimed that all vertical antennas have very good low-angle radiation compared with a horizontal dipole. While theoretically this is true, in practice it very much depends upon the conductivity of the soil and the absence of nearby objects, and on 14MHz and above there is usually little to choose between vertical and horizontal antennas in terms of low-angle radiation. In fact the usual ground-plane puts out considerable power at fairly high angles of radiation making it a very effective medium-distance antenna. However for 1·8, 3·5 and 7MHz a vertical antenna often provides more radiation at low angles than a horizontal dipole unless this is very high, and is recommended for long-distance working.

On 1·8 and 3·5MHz, $\lambda/4$ antennas tuned against earth (often

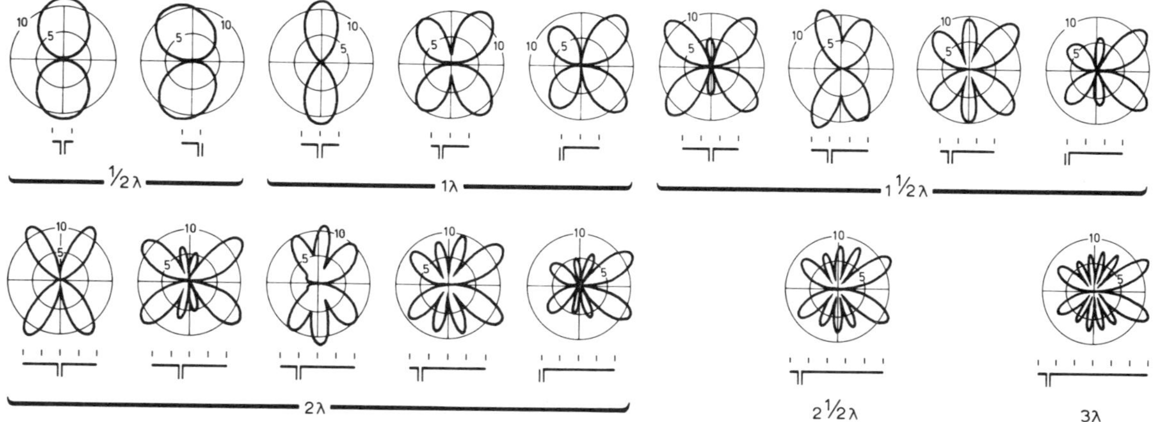


Fig 13. Horizontal radiation patterns of antennas indicating the effect of feed-point. Based on model antenna range measurements by D. C. Cleckner (*Electronics Manual for Radio Engineers*)

termed Marconi antennas) for 1·8 and 3·5MHz (Fig 15) again become difficult to erect as purely vertical elements, and the wire is often bent over to form an "inverted-L" or a "T", or "loaded" by means of a *capacity-hat* or other form of loading. The end of such an antenna is often brought directly into the shack and coupled directly to a low-impedance transmitter output, or preferably though an antenna tuning unit or *transmatch* (see later).

For all antennas of this general form, where an earth rod, ground mat, counterpoise or radials are used to form an essential part of a resonant system, it should be recognized that power will be absorbed in the ground. It is thus very important that the loss of power is minimized by using the most efficient earth possible. If care is not taken the earth connection may have a resistance comparable to the feed-point resistance of the antenna. It is common to find that less than half the rf power delivered to such antennas is being radiated, the rest merely heating up the garden! A simple and efficient earth for 1·8 and 3·5MHz can often be obtained by connecting the earth lead directly to a rising water pipe; alternatively a connection can be made to several plates of copper buried several feet below the surface. The closest to an ideal earth would be an enormous copper plate extending out many wavelengths; in practice this is impossible and the nearest we can hope to get to this is to use the good conductivity of salt water when operating maritime mobile. In a domestic environment much depends on the local conductivity of the soil, the depth and number of earth stakes we can sink, or the amount of wire laid on or buried just under the soil. Nevertheless, to improve the performance of a vertical or Marconi antenna the first thing to do is to make sure the best possible earth connection is being used.

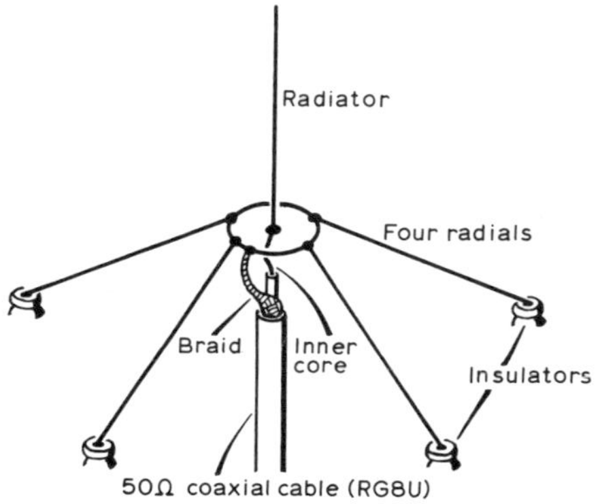

Fig 14. The ground plane antenna is one of the simplest yet most effective dx antennas for installation in restricted spaces. For operation on 14MHz the rod is 16ft 11in long and the radials can be made from hard-drawn copper wire, each 17ft 2½in long. The radiator is insulated from the fixing mount with the inner core of the coaxial wire soldered or bolted to the end. The radials are soldered to a copper ring together with the braid of the coaxial cable and kept well apart from the vertical rod and the inner core of the coaxial cable. For good match to the 50Ω feeder, the radials should slope downwards at roughly 120° to the radiator. Dimensions can be scaled up or down for other bands

## Resonant lengths

Table 1 indicates resonant lengths at selected frequencies in the amateur bands for wire antennas erected well clear of the ground and other large objects.

The second column shows the physical length of $\lambda/2$ converted from metres into feet. The next column shows a five per cent end correction, and the final column shows the correction deducted, ie 95 per cent of a physical $\lambda/2$ which, for dipole type wire antennas, is approximately equal to an electrical $\lambda/2$.

The end column is thus the resonant length for any type of $\lambda/2$ dipole clear of surrounding objects. Should the radiating section be bent a small additional length may be needed.

When using an antenna more than $\lambda/2$ long only one end

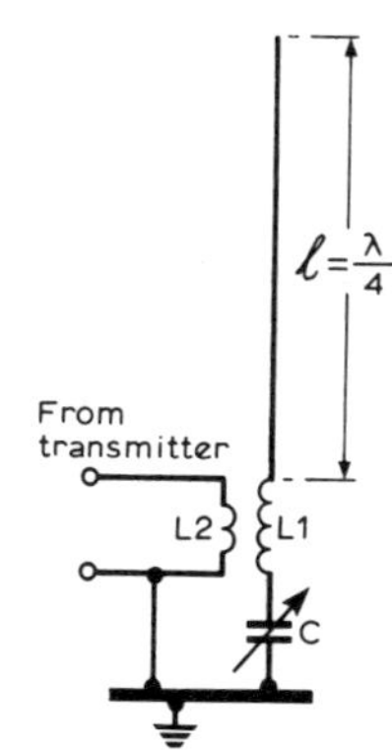


Fig 15. Quarter-wave (or Marconi) antenna. The length ≏ 230/f ft where f is the frequency in megahertz

**TABLE 1**

| Frequency (kHz) | λ/2 (ft in) | 0·05 × λ/2 (ft in) | 0·95 × λ/2 (ft in) |
|---|---|---|---|
| 3,500 | 140 7 | 7 0 | 133 7 |
| 3,600 | 136 9 | 6 10 | 129 11 |
| 7,000 | 70 3 | 3 6 | 66 9 |
| 7,100 | 69 3 | 3 5 | 65 10 |
| 14,000 | 35 2 | 1 9 | 33 5 |
| 14,100 | 34 11 | 1 9 | 33 2 |
| 14,200 | 34 8 | 1 9 | 32 11 |
| 21,000 | 23 5 | 1 2 | 22 3 |
| 21,200 | 23 3 | 1 2 | 22 1 |
| 21,400 | 23 0 | 1 2 | 21 10 |
| 28,000 | 17 7 | – 10 | 16 9 |
| 28,400 | 17 4 | – 10 | 16 6 |
| 29,000 | 17 0 | – 10 | 16 2 |

correction need be deducted. For example a 1λ antenna for 14,100kHz would be 2 × 34ft 11in less 1ft 9in, ie 68ft 1in.

Under normal circumstances antenna lengths are not unduly critical and an antenna cut for about the centre of an amateur band will usually function satisfactorily over the entire band. However with low radiation resistance antennas (for example, where reflectors and directors are used) accuracy becomes increasingly important.

After erection the resonant frequency of an antenna should be checked, if possible, by means of a grid-dip oscillator.

## The atu or transmatch

All modern hf and vhf transmitters are designed to work into a low-impedance, unbalanced coaxial-cable feeder of 50–80Ω resistive impedance, with a low standing-wave ratio. Particularly for solid-state transmitters or those using television line-output valves it is important to achieve these conditions in practice; otherwise damage to the transmitter may soon result.

It might therefore seem that the only types of antenna capable of fulfilling these conditions would be those with a driven element based on λ/2 or 3λ/2 centre-fed dipoles, carefully trimmed for resonance around the operating frequency. Such antennas when connected directly to the transmitter will, as we have seen, fulfil the required conditions. However this is a limitation that is frequently unacceptable, and indeed it may easily (and advantageously) be overcome by imposing between the antenna and the transmitter a variable impedance rf transformer so adjusted that the transmitter "sees" (works into) a resistive 50–70Ω impedance, while the output from the matching transformer is made to match accurately the impedance of the antenna or its feeder. Such an rf transformer is known as an *antenna tuning unit*, *antenna matching unit* or *transmatch*. A transmatch, in effect, brings the whole antenna system into resonance by compensating for any capacitive or inductive reactance at the connection point of the element or its feeder. In its simplest practical form the transmatch may consist of just two components, an inductor and a capacitor, both preferably variable. However in practice some additional components may be used to make the unit more flexible in the range of reactances and impedances with which it can cope. An antenna system that has been correctly resonated or matched in this way is virtually as effective as one in which the elements have been very carefully trimmed, even though there may be high standing waves (swr) on the feeder line. In such circumstances an swr meter (if used) is located between the transmitter and the transmatch unit, and *not* in the antenna feeder. There should be no significant swr between transmitter and atu when the antenna system has been adjusted correctly.

A transmatch/atu thus allows a transmitter to be used with a variety of antennas of varying length and varying impedance; the element may be centre-fed, off-centre-fed (as in the Windom) or end-fed etc. Where there is a high swr on the feeder between the transmatch and the radiating element(s), it is an advantage to use open-wire transmission line rather than coaxial cable.

One effective form of transmatch is a pi-coupler, similar to the network so often used in the output stages of transmitters; indeed this classic network was originally known to amateurs as the "Collins universal coupler", able to match any wire to any transmitter.

It is vitally important with any form of transmatch that it be adjusted correctly since for the unwary it is possible to appear to be loading a transmitter when all that is happening is the generation of very high circulating currents in the transmatch unit. The use of an swr meter between the transmitter and transmatch will assist in such adjustment; however with care and some experience it is entirely possible to adjust such a system without an swr meter, for example by using torch bulbs (suitably shunted) as current indicators and neon bulbs as voltage indicators. For a given antenna system (whether voltage-fed or current-fed) maximum current into the antenna is a valid indication of optimum adjustment.

A number of simple transmatch units have been illustrated in connection with various antennas. Two other arrangements are shown in Fig 16.

## SWR facts and fallacies

In recent years, partly resulting from the general use of techniques that measure (or at least provide an indication of) the standing wave ratio on the transmission line between transmitter and the antenna proper, a number of misunderstandings and fallacies have become widespread among amateur operators. Too many amateurs have come to believe that an swr meter indicates the "goodness" of an antenna system, equating a low or unity vswr with an ideal and correctly functioning antenna and significant swr with a poor system. It is forgotten that many hf antenna systems function efficiently with a relatively high swr on the transmission line, even when this is coaxial cable; that with open wire feeders the actual swr is of little practical significance; that reflected power does not normally represent lost power and is not dissipated in the power amplifier. It is frequently misleading to attempt to evaluate antenna performance on the basis of "the lower the swr the better." While it is true that many modern transmitters are designed for a nominal load impedance of 50Ω at a maximum vswr of 2:1 this does *not* mean that such

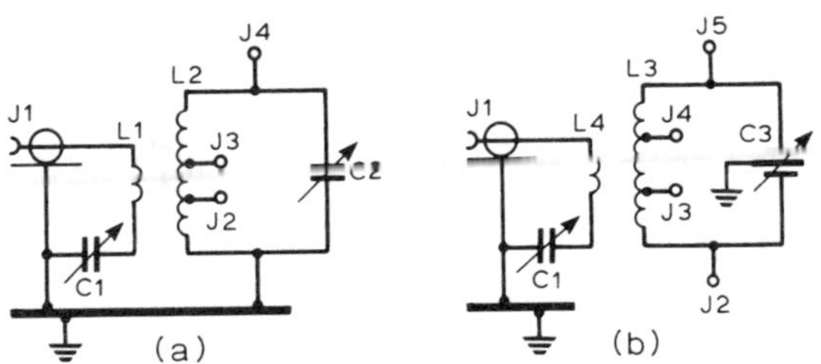


Fig 16. Typical atu or "transmatch" circuits, (a) for end-fed wires and (b) for either balanced feeders or end-fed wires, with balanced feed plugged into J2 and J5 or J3 and J4. The following values are those suggested by W1HDQ for 21 and 28MHz operation. C1, 140pF; C2 as C1 but higher voltage type; C3, 100pF per section split stator; J1, coaxial jack; J2-5, insulated-tip jacks; L2, 8 turns No 20 tinned 1¼in dia, 1½in long, tap J2 at 2 turns, J3 at 4 turns and J4 at top end; L1, 2t insulated hook-up wire, wound over bottom end of L2; L3, 8t No 20 tinned, 1½in dia; 1½in long, tap J2 and J5 at ends and J3 and J4 at 2 turns in from each end, L4 like L1 but wound over the middle of L3

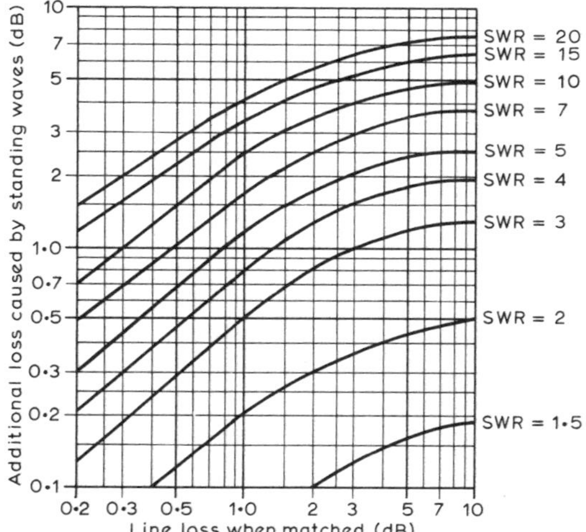


**Fig 17. This graph shows by how much (or at hf usually how little) additional feeder loss occurs with a moderate vswr on the line compared with the same line having an accurately matched 1:1 swr**

transmitters cannot be used with transmission lines of other impedance or having a vswr of more than 2:1 provided that a suitable simple matching network or atu is inserted between transmitter and feeder. It is particularly important to note that a low swr, particularly when achieved over a wide bandwidth, may often denote a *poor* rather than an effective antenna system!

(1) Reflected power does not represent lost power except for the (usually modest) increase in line attenuation over the matched line attenuation; in a lossless feeder line no power would be lost because of reflection *no matter how high the swr*. Note that both the attenuation of the line and the swr must be quite high to incur substantial additional loss. On all hf bands with low-loss cable, reflected power loss is usually insignificant; at vhf it may become significant; at uhf it may be extremely important.

(2) Reflected power does *not* flow back into a transmitter and cause excessive dissipation and other damage. The damage often blamed on a high swr is usually caused by improper output-coupling adjustments and not by the swr. Valve/semiconductor pa overheating is caused by either or both overcoupling and reactive (mistuned) loading. The transmitter does not "see" an swr but only the impedance that results from the swr: this means that the impedances can be correctly matched without concern for the swr on the feeder.

(3) *Efforts to reduce swr below 2:1 on any coaxial line generally represent wasted efforts from the viewpoint of increasing the radiation from the antenna.*

(4) A low swr is *not* evidence that an antenna system is a good one or that it is working efficiently. On the contrary *a lower than normal swr over a significant bandwidth is reason to suspect that a dipole antenna or a vertical antenna is being affected by resistance losses.* These may arise from poor connections, poor earthing systems, lossy cable or other causes.

(5) The radiator of an antenna system need not be of self-resonant length to achieve maximum resonant current flow, nor need the feed line be of any particular length. A substantial *mismatch* at the junction between feeder line and radiator does not prevent the radiator from absorbing all the real power that is available at the junction. Where suitable matching (atu, etc) cancels out the reactance presented by a non-resonant radiator and a random length of feeder, mismatched at the antenna junction, then the system is *matched* and all the real power may be radiated effectively. It is worth noting that relatively few mf broadcasting antennas are of resonant length—yet these radiate effectively.

(6) The swr on the feed line is not affected by any adjustment of an atu at the transmitter end, and a low swr achieved by this means is usually an indication of a mismatch between the transmitter and the input to the atu.

(7) With an effective atu and good open-wire feeder a 132ft centre-fed dipole does not (contrary to general belief) radiate significantly more power on 3·5MHz than an 80ft dipole fed with the same transmitter power. A dipole self-resonant on say 3,750kHz does not radiate significantly more power on 3,750kHz than on 3,500kHz or 4,000kHz with any normal length of feeder, although it is to be expected that the swr will rise to about 5:1 at these outer frequencies and that the coaxial will then, in effect, be working as a tuned feeder. Proper coupling between the transmitter and the coaxial feeder, however, in these conditions requires the use of a matching arrangement between transmitter and feeder. If the coaxial feeder of any antenna system requires to be a specific length to satisfy a specific matching condition, the same input impedance can be obtained regardless of the length of the coaxial feeder by the provision of a simple L network of only two components (either two capacitors, two inductors, or one of each).

(8) High swr in a coaxial feeder resulting from a severe mismatch at the antenna junction does not in itself produce common mode currents on the line or cause the line to radiate. At hf a high swr in any open-wire line caused by a severe mismatch will not produce antenna currents on the line, nor cause the line to radiate, if the feed currents in each wire are balanced, and if the spacing is small relative to the operating wavelength (this is equally true at vhf provided that sharp bends are avoided in the line).

(9) SWR meters do not provide a more accurate measurement of swr by being placed at the antenna/feeder junction.

(10) The swr in a feeder cannot be adjusted or controlled in any practical manner by varying the line length. If a meter provides significantly different swr readings when moved along the line, it may indicate "antenna" (common mode) current flowing on the outside of the coaxial cable, or an unreliable swr meter, or both: but *not* that the swr is varying along the line.

(11) Any reactance added to an already resonant (resistive) load of any value for the purpose of compensation to reduce the reflection on the line feeding the load will, instead, only increase or worsen the reflection. Lowest feeder swr occurs at the self-resonant frequency of the radiating element it feeds, completely independent of feeder length.

(12) Of the various types of dipoles (thin wire, folded, fan, sleeve, trap or coaxial) none will radiate more field than another, providing that each has insignificant ohmic losses and is fed the same amount of power.

Note that when an antenna element presents to the transmission line an impedance other than its characteristic impedance, the impedance offered to the transmitter at the input end of the line may be quite different from either the characteristic impedance of the line *or* (unless the line is an exact multiple of an electrical $\lambda/2$) the impedance at the antenna junction. The impedance represented by the line then depends on the length of the feeder (which acts as an impedance transformer). In such cases, *unless a suitable matching network is interposed* between transmitter and transmission line the impedance may be of a value (in the form $R + jX$) with which the transmitter output circuit cannot cope: in other words it may not be possible to *load* the transmitter properly without changing the length of the transmission line: it is this factor, rather than any *losses* associated with swr which leads to many misconceptions.

To sum up this information: it should always be possible to make any centre-fed antenna of any reasonable length, with any feeder, radiate quite effectively provided one has a good matching atu between a transmitter that is intended to work into a low impedance and the feeder. This is why the centre-fed dipole is such a dependable antenna.

## VHF/UHF antennas

Fundamentally there is no difference between an antenna for the hf bands and one for the vhf or uhf bands. However because the physical sizes of resonant elements are so much smaller and because it is unusual to design a vhf antenna for multi-band operation, in practice a rather different approach is usually adopted. Much more emphasis, for example, is given to achieving high power gains by using Yagi arrays with a dipole radiator, reflector and many director elements: see Fig 18. Such arrays will be familiar to most readers since the same approach is adopted for radio and television broadcast reception, and vhf and uhf Yagi arrays can be seen on literally millions of rooftops. However the amateur usually seeks to provide means of rotating his array to enable him to work stations in different directions.

Because vhf/uhf signals are not usually reflected from ionized layers it is much more important than on hf to use the same polarization for the antennas at both ends of the contact. The usual convention, at present, is to use horizontally polarized antennas for long-distance operation from fixed or portable locations, but to use vertically polarized ones for mobile and repeater operation. Many of the mobile antennas are based on the ground-plane antenna already described, the $\lambda/4$ radiating element being only 19in long for 144MHz.

Because the losses in transmission lines rise with frequency, it is important to use low-loss forms of coaxial or unbalanced cables and to ensure that the impedance is maintained throughout by using well-designed plugs and sockets. A high swr can result in appreciable loss of rf power at vhf and uhf.

The use of multi-element Yagi arrays results in very large power gains, so that the effective radiated power from an amateur vhf or uhf station may be many kilowatts. For example 10–15dB gain is quite usual on 144MHz and 15–20dB no means unusual at 432MHz. Very few wire antennas are used, most arrays being constructed from aluminium alloy tubing.

## Gain

A power gain of about 5–6dB (reference dipole) is usually the maximum that can conveniently be achieved from an hf rotary beam. 6dB is equivalent to increasing the power of the transmitter by four times. The theoretical maximum gain of a three-element Yagi is of the order of 7·5–8·5dB but this is most unlikely to be achieved in practice.

It is possible with care to achieve a gain of about 5dB with a two-element Yagi or two-element driven array, and the third element is then unlikely to add more than about 1dB additional gain. With care it should be possible to obtain a 3–4dB power gain with a lightweight two-element "mini-beam", although the effective bandwidth may be less than with a full-size array.

*Better results may often be achieved by increasing the height of a two-element array rather than using additional elements.* There is no optimum height for a horizontally polarized antenna: *the more height the better.* Field strengths at any given low angle when compared to the relative voltage of an antenna at a height of $0{\cdot}2\lambda$ are roughly $\times 2$ at $0{\cdot}4\lambda$, $\times 3$ at $0{\cdot}6\lambda$, $\times 4$ at $0{\cdot}8\lambda$ and $\times 5$ at $1{\cdot}0\lambda$.

Full-wave loop elements (eg quads) provide roughly an advantage of about 1dB over dipole elements, so that, for example, a two-element quad can be expected to provide about 6dB gain or about the equivalent of a three-element Yagi.

If one antenna appears to have markedly more gain than another, or to have significantly more than 6dB power gain over a dipole, suspect measurement errors, different polarization,

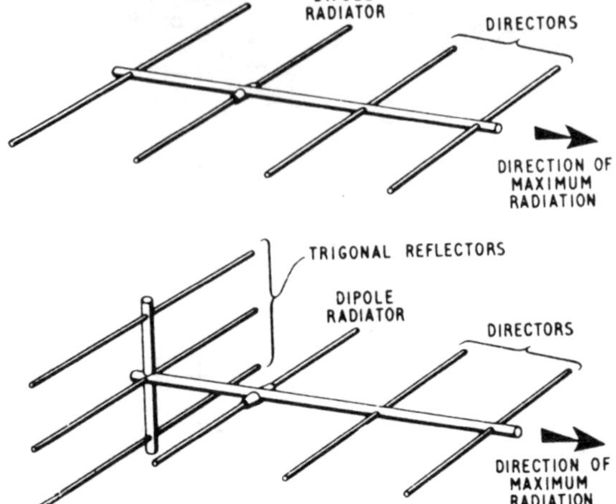


Fig 18. Two common forms of 4-element Yagi arrays. The use of trigonal reflectors improves the front-to-back ratio but does not increase the forward power gain which would be about 7dB

unnecessary losses in the reference antenna, or the effects of height or ground slope.

Do not adjust antennas for maximum front-to-back ratio but wherever possible for maximum forward gain.

In the case of loop (quad) elements the parasitic element should be tuned as a reflector, not as a director. The common practice of tuning a quad for maximum front-to-back ratio sacrifices just over 1dB of forward gain. Instead of tuning for maximum gain it may be convenient to tune for a null on signals 130° relative to the beam heading.

Interaction between different antennas at the same location is often important: trees have a great effect on vertically polarized antennas.

One method of breaking the 6dB gain barrier is to couple together the outputs from two separate beams: for example it is usually possible to obtain 2–3dB extra gain without critical phasing of outputs. Thus two two-element beams could provide up to 8dB gain.

With horizontally polarized arrays, the vertical-plane radiation pattern is determined by the array factor: there is thus no basis for the belief that one beam may be better than another because it lowers the angle of radiation. The array factor is the same for antennas at a given height and for a given angle of elevation.

The height below which vertically polarized antennas are markedly superior for dx operation at 14MHz varies from about 20ft with poor ground to about 60ft with good ground. With poor earth the difference in performance between vertical and horizontal polarization is normally less than 2dB over a range of heights from 20ft to 90ft.

A dipole has a 2·1dB gain over an isotropic antenna (an imaginary antenna that radiates equally in all directions) but gain should be defined with reference to a dipole and not an isotropic antenna.

*Long wire antenna*, power gain in the main lobes: $2\lambda$ 1·5dB; $4\lambda$ 3dB; $6\lambda$ 5dB.

*Bidirectional vee:* $2\lambda$ per leg 5·5dB; $3\lambda$ per leg 7dB; $4\lambda$ per leg 8dB.

*Rhombic:* $1\lambda$ per leg 3·5dB; $2\lambda$ per leg 6·5dB; $3\lambda$ per leg 8·5dB; $4\lambda$ per leg 10dB.

Misleading results can often result from attempting to

compare antenna power gain on the basis of ground-wave radiation.

**Power gain**

| | |
|---|---|
| +1dB = 1·3 times the power | + 6dB = 4·0 times the power |
| 2dB = 1·6 times the power | 7dB = 5·0 times the power |
| 3dB = 2·0 times the power | 8dB = 6·3 times the power |
| 4dB = 2·5 times the power | 9dB = 7·9 times the power |
| 5dB = 3·2 times the power | 10dB = 10 times the power |

## Erecting antennas

So far we have been mainly concerned with electrical details but there are a number of practical considerations which should be noted.

Due to various reasons the physical length of an antenna needed to produce resonance is usually a few per cent less than that which might be expected from simple theory. This discrepancy has been taken into account in the formulae which have been given here for antenna length. The physical length of a $\lambda/4$ open line is also a few per cent less than might be expected; with commercial twin feeder the physical length is often 20 per cent less and with coaxial cable 20–33 per cent less depending on the *velocity factor*.

Either wooden or metal poles are satisfactory for a horizontal antenna but a metal pole is not recommended for supporting a vertical antenna as it will couple to the radiator and distort the radiation pattern. Long guy wires are liable to interact with both horizontal and vertical antennas and should preferably be broken up electrically by inserting a compression type (egg-shaped) insulator about every $\lambda/10$. Trees and buildings can also have detrimental effects and the antenna should, therefore, be as much in the open as possible. The feeder should be kept at right angles to the antenna for as far as is practicable in order to minimize the intercoupling between them.

The quality of the insulators required will depend to some extent on their position. At the extremities of a $\lambda/2$ antenna the voltages will be high and the insulators should be of glass or glazed ceramic and be several inches long. But in the centre of a $\lambda/2$ antenna which is always used on its fundamental frequency the rf voltage will always be low and a poorer quality insulator will be acceptable. When high-voltage points of the feeder, or of the antenna itself, come into the house, good insulation must be provided here. On the other hand, if the feeder has a low impedance as it enters the house less care will be necessary.

The rope used for halyards should be of good quality and preferably waterproofed by impregnating it in heavy oil or wax. Some allowance should always be made for shrinkage in wet weather. Better still is nylon cord.

For long-distance work, consult a great circle map in conjunction with the radiation pattern of the antenna. Fig 19 shows a great circle map centred on the UK.

Antenna supports used by amateurs vary from making use of existing buildings and/or trees to many types of masts and towers. Typically the mast can be made from wooden or metal scaffold poles, or plastic (pvc) rainwater pipes. For example, Fig 20 shows the way in which an "A" mast of up to about 45ft can be constructed from lengths of 2in by 2in timber. Rather more elaborate towers may be needed to support rotary hf beams, and various forms of tilt-over and extendable masts are also used by a few amateurs.

Vertical antennas are usually self-supporting, with the elements made from aluminium-alloy tubing. A typical arrangement is the ground-plane antenna with a vertical rod element supported on a chimney with the associated radial wires sloping downwards in the form of guy wires.

For vhf operation, the most popular arrangement is the multi-element Yagi directional antenna (made familiar by its use by so many television viewers), preferably with a rotator to allow the array to be swung round to point in any direction. The much smaller element lengths involved on vhf compared with hf make it possible to provide highly directional antennas quite modestly.

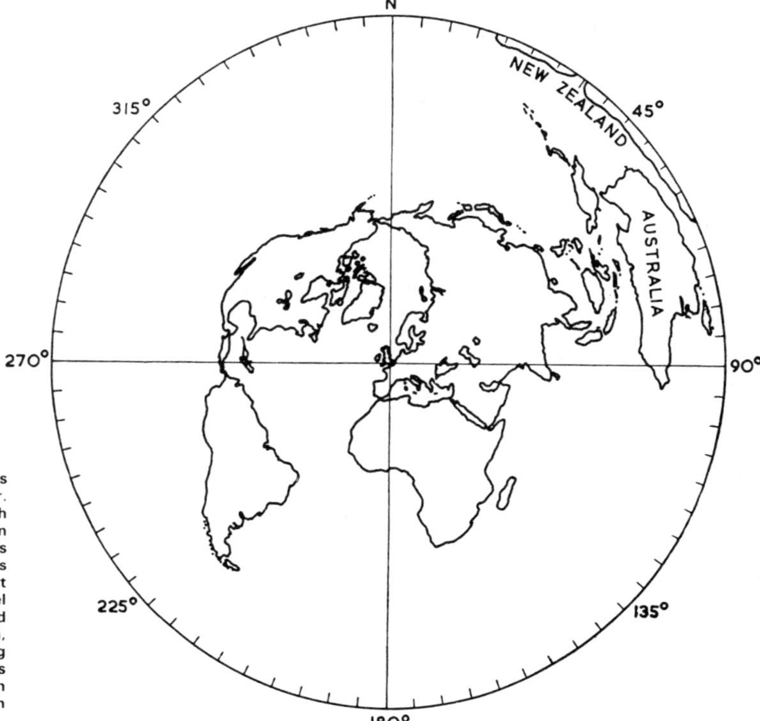


Fig 19. A great circle map, based on the UK. This is how the world appears to an amateur transmitter. Radio signals travel along great circle routes (which are the shortest routes on the globe and which on the map would be represented by straight lines radiating outwards from the centre). Such a map is essential when planning an antenna installation as it shows the directions along which signals will travel to particular countries. However, it should be noted that, in the mornings, signals to and from Australia, New Zealand and the Far East often travel the "long way round" across South America. These directions will be exactly 180° more than those indicated on the map. A special Great Circle Map centred on London is available from the RSGB

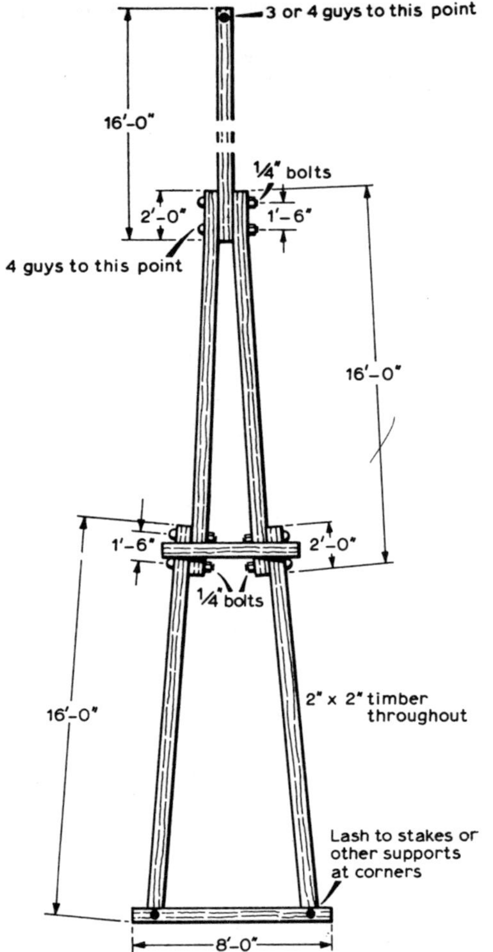


Fig 20. An "A" type timber mast. A minimum of three guys is required at the points indicated. With careful construction and good guying this form of construction is suitable for masts of from 30 to 45ft

While it must be emphasized that the antenna plays a tremendously important part in achieving good long-distance results, and that the ideal is a high, outdoor antenna, it should not be supposed that amateur operation is impossible without one. Quite a large number of amateurs, for example, use antennas entirely confined in loft spaces, or even vhf antennas taped to window panes. It is often possible to work long distances with indoor antennas, particularly on cw, though the operator will usually find himself at some disadvantage to those more fortunate amateurs with really good antennas.

Although there are now a very large number of different antennas in use by amateurs, these are mostly variants of a few fundamental types.

No matter how good the antenna and transmitter may be effective transmission will depend upon the transfer of the power from the transmitter to the antenna, usually along the *transmission line* which often consists of low-impedance coaxial line or 300Ω ribbon-type twin line feeder. For optimum transfer of power the feeder must be reasonably accurately matched in impedance at both the transmitter and antenna element ends, as discussed earlier. Where optimum conditions are not achieved, standing waves will exist upon the feeder and can be detected by a suitable *swr indicator*. Some very simple instruments for this purpose have been developed and are described in the handbooks. While an swr indicator is a very useful device to have available, it is perhaps worth emphasizing that quite considerable standing waves can often be tolerated on untuned transmission lines without seriously reducing radiation.

More than any other section of an amateur station, the antenna governs the final results. At long distance the difference between the signals from a good and a poor antenna of similar design may easily amount to the equivalent of some 100 or so times the power. However this does not mean that dx stations cannot be worked with simple or easily erected antennas—a simple wire antenna at no more than 30ft above the ground may easily out-perform an elaborate beam antenna that has not been properly adjusted or is wrongly matched to the transmitter. Until a sound understanding of the mysteries of antenna and feeder impedances has been achieved—and this will come best from a combination of practical experiment and a study of the theory—the newcomer will be well advised to concentrate on the simpler antennas, rather than immediately buying or erecting a more elaborate antenna, the finer details of which are not fully understood.

It is worth emphasizing that on hf it is possible to work all over the world using quite simple antennas, particularly on cw. Although when "rare" stations and unusual prefixes appear, there will often be many amateurs calling them, and then the contacts often go first to the amateur with the most effective beam antenna! But this should not deter the newcomer—if he waits a bit, or listens a little harder than the others he will usually find that he can contact many of the stations he calls. If the number who "come back" to him is very low then it is time to start checking that the transmitter really is delivering power to the antenna!

Generally the antenna should be erected as high as possible and well clear of all structures; this is of course the counsel of perfection, so do not despair if in your location there is room for only a short sloping span well below the building line or in an attic. Very often an apparently poorly sited antenna will give good results in a limited number of directions (often at variance with theoretical radiation patterns). A good tip in such circumstances is to have two or more antennas with provision for switching the transmitter and receiver to any one of them.

Because of the vital importance to the amateur of a really effective antenna, and also because of the severe physical limitations existing at many of the locations where an amateur station is to be installed, much attention has been paid in recent years to improving the performance of antennas of restricted size, and also in making then suitable for use on several different bands (*multi-band antennas*).

Band-switched hf transmitters and transceivers make it easy to hop from band to band—in order to follow changing propagation conditions or perhaps to work semi-local stations on 3·5MHz as a change from "chasing dx" on 14 or 21MHz. At one time, band changing involved quite an elaborate "tuning up" of the rig and its associated antenna tuning unit etc. But today many amateurs use multi-band antennas such as the W3DZZ trap dipole, the G5RV multi-band dipole or long-wire aerials with switched matching units, or one of the many multi-band vertical antennas. Often band-changing can be completed in a few seconds so that the amateur can at any time choose which band he wants to use.

## Summing-up

In this chapter, some of the more popular antenna systems have been described but it should be appreciated that there are numerous other arrangements in which the main differences are in the way in which the radiator is connected to the transmitter, or in the phase relationships of two or more radiating elements which form an array, or in the use of traps or phasing stubs to facilitate operation on more than one band; for example the Windom, the VS1AA (a multi-band form of Windom), the Zepp and the extended Zepp, the various "long-wires" and Beverage antennas, the inverted-vee, "bobtail" etc. Then there is a large number of two-element arrays in which, unlike the Yagi parasitically-excited systems, both elements are "driven" (that is to say are connected to the transmission line) including various forms of W8JK (Kraus), ZL-Special, HB9CV arrays, or phased $\lambda/4$ antennas in which vertical monopole type elements are driven in such a manner as to result in specific phase differences and where the earth is a fundamental part of the radiating system. Then again there has been increasing use of single-turn loop forms of antennas with a perimeter usually $1\lambda$ but occasionally only $\lambda/2$ or some intervening figure. Thus while the Yagi array, single-element verticals such as the "ground plane" and variations of the dipole remain the most popular systems, very many other systems may be encountered and can be very effective.

The total maximum power radiated by any antenna is the power output of the transmitter less any losses in the transmission line, coupling circuits and the IR losses in the element or (where this forms part of a resonant system) the earth. However the directivity of an antenna may considerably increase the *effective radiated power* (erp) in some directions while reducing it in others. Since we are dealing with three dimensions, it is possible to achieve gain with an omni-directional antenna (ie where the horizontal radiation pattern is circular) by concentrating radiation at specific angles of the vertical radiation pattern; this is usually achieved by *stacking* elements one above the other, to confine maximum radiation to the lower angles. An array such as a Yagi tends to concentrate radiation in both the horizontal and vertical planes.

The understanding and development of novel antennas is one of the most fascinating of all the technical aspects of amateur radio. Such work can be highly mathematical or purely practical, based on a few quite simple fundamental concepts, and the results of all this work account for a substantial proportion of the technical articles published in amateur radio journals and magazines. Many antennas are popular because they are convenient to use or install, rather than because they are particularly efficient. For example for those with plenty of space and several masts, there can be few better antennas than the 50-years-old "rhombic" long-wire beam. Because in residential areas such physically large systems are virtually impossible at hf (though they can be, but seldom are, used effectively at vhf/uhf), the amateur continues to try innumerable new antenna designs.

## CHARACTERISTICS OF TYPICAL RADIO FREQUENCY FEEDER CABLES

| Type of cable | Nominal impedance $Z_0 (\Omega)$ | Dimensions (in) Centre conductor | Over outer sheath | Over twin cores | Velocity factor | Approximate attenuation (dB per 100ft) 70MHz | 145MHz | 433MHz | Remarks |
|---|---|---|---|---|---|---|---|---|---|
| Standard tv feeder | 75 | 7/·0076 | 0·202 | — | 0·67 | 3·5 | 5·1 | 9·3 | — |
| Low-loss tv feeder (semi-air-spaced) Super Aeraxial | 75 | 0·048 | 0·290 | — | 0·86 approx. | 2·0 | 3·0 | 5·5 | Semi-air-spaced or cellular |
| Flat twin | 150 | 7/·012 | — | 0·18 × 0·09 | 0·71 | 2·1 | 3·1 | 5·7* | *Theoretical figures, likely to be considerably worsened by radiation. |
| Flat twin | 300 | 7/·012 | — | 0·405 × 0·09 | 0·85 | 1·2 | 1·8 | 3·4* | |
| Tubular twin | 300 | 7/·012 | — | 0·446 | 0·85 | 1·2 | 1·8 | 3·4* | |

| UR No | Nominal impedance $Z_0 (\Omega)$ | Overall diameter (in) | Inner conductor (in) | Capacitance (pF/ft) | Maximum operating voltage (rms) | Approximate attenuation (dB per 100ft) 10MHz | 100MHz | 300MHz | Approx RG equivalent |
|---|---|---|---|---|---|---|---|---|---|
| 43 | 52 | 0·195 | 0·032 | 29 | 2,750 | 1·3 | 4·3 | 8·7 | 58/U |
| 57 | 75 | 0·405 | 0·044 | 20·6 | 5,000 | 0·6 | 1·9 | 3·5 | 11A/U |
| 63* | 75 | 0·853 | 0·175 | 14 | 4,400 | 0·15 | 0·5 | 0·9 | |
| 67 | 50 | 0·405 | 7/0·029 | 30 | 4,800 | 0·6 | 2·0 | 3·7 | 213/U |
| 74 | 51 | 0·870 | 0·188 | 30·7 | 15,000 | 0·3 | 1·0 | 1·9 | 218/U |
| 76 | 51 | 0·195 | 19/0·0066 | 29 | 1,800 | 1·6 | 5·3 | 9·6 | 58C/U |
| 77 | 75 | 0·870 | 0·104 | 20·5 | 12,500 | 0·3 | 1·0 | 1·9 | 164/U |
| 79* | 50 | 0·855 | 0·265 | 21 | 6,000 | 0·16 | 0·5 | 0·9 | |
| 83* | 50 | 0·555 | 0·168 | 21 | 2,600 | 0·25 | 0·8 | 1·5 | |
| 85* | 75 | 0·555 | 0·109 | 14 | 2,600 | 0·2 | 0·7 | 1·3 | |
| 90 | 75 | 0·242 | 0·022 | — | 2,500 | 1·1 | 3·5 | 6·3 | 59B/U |

All the above cables have solid dielectric with a velocity factor of 0·66 with the exception of those marked with an asterisk which are helical membrane and have a velocity factor of 0·96. This table is compiled from information kindly supplied by Aerialite Ltd and BICC Ltd, and includes data extracted from Defence Specification DEF-14-A (HMSO).

CHAPTER 6

# Amateur radio equipment

During the past 40 years many hundreds of different equipments—receivers, transmitters and transceivers—have been marketed specifically for radio amateurs and SWLs. In addition a considerable number of receivers developed for professional and military communication have subsequently found widespread application in amateur radio.

Many of the better equipments have had operational lifetimes extending over 10, 25 or even 35 years, often passing from one amateur to another, with or without progressive modification. And today the oldest commercially produced equipments of the early and mid-thirties are often eagerly sought as "collectors' items". Good receivers it seems never die, they only drift away.

The newcomer thus finds that he is constantly coming across or hearing references to firms, brand names, model numbers and military designations, very often without any further explanation: KW2000, HRO, Super Pro, S-line, EA12, SB303, BC348, R1155, Hammarlund, Yaesu, Trio, Kenwood, Drake, AR88, National (both American and Japanese), Geloso ... and so on. It can be even more confusing than the Q-code!

It would be very difficult to explain each and all of these names and numbers. But the following notes on the evolution of this mass of amateur radio equipment may help the newcomer feel a little more at home when he hears equipment being discussed, even if he intends to "roll his own". They may also give at least a little guidance when scanning the classified adverts offering second-hand and reconditioned equipment.

## The early designs

Even in the 'twenties some amateurs bought ready-made receivers or had equipment made for them; but there was no large-scale production specifically for this market, and indeed the bulk of all equipment then in use was home-made, often including many of the actual components.

But quite early in the 'thirties, a few firms in the UK and USA began to find a growing market for receivers designed specifically for reception on the "short waves". These were often rather more expensive than domestic broadcast receivers of the period, but very much cheaper than those designed for commercial and military use. While some of the earliest sets of this type were "straight" receivers with plug-in coils (National SW3, Eddystone "All-world" series) the superhet soon came to dominate the scene, fitted with BFOs to allow reception of cw as well as a.m. (no amateurs normally used ssb before 1948 or so). Sets also had "send-receive" switching to make it easier to use them for two-way communication as well as for listening.

Gradually such receivers came to be called "communication receivers" and it is interesting that although this term is now used to cover many types of professional receivers, it originally came into use to describe the type of receivers designed for amateur operation.

Possibly the first to achieve wide recognition (mainly in the United States) was the National FB7 receiver of 1933, followed a year later by the FBXA which included a crystal filter (this type of filter was developed in the UK by Dr Robinson but was first applied to hf receivers by James Lamb of ARRL).

From 1934–1940 many low-cost and medium-cost communication receivers appeared: the HRO Senior, made by the National Company in the USA, in 1936; the Hammarlund Comet Pro and more expensive Super Pro and just before the second world war the Hammarlund HQ120, first of a long series of Hammarlund "HQ" type numbers that finally came to an end a few years ago with the HQ215. Another American firm, Radio Manufacturing Engineers Inc (RME for short), brought out the RME69, a nine-valve receiver covering from 550kHz–32MHz in six bands, a very popular design particularly when used with a DB20 pre-selector. There were also the early Hallicrafters receivers which ranged in the pre-war period from a £7 "Sky Buddy" up to the more advanced Super Skyrider. Many firms supplied receivers either with or without a crystal filter and thus models which included such a filter often denoted this by using an "X" in the model number. Bandpass filters, using several crystals, did not come into general use until many years later.

American communication receivers began to reach the UK around 1936–7 at a time when the only home product of this type was the £20 "Evrizone" single-signal superhet. A sign of the changing scene was the use by the leading British station in the BERU contest of 1936 of a Comet Pro, but the Australian winner used a "straight" O-v-2 receiver.

These early communication receivers were almost always single-conversion superhets with an i.f. of around 456kHz and the better models had two rf amplifier stages to reduce "image" and to overcome the high noise of multi-grid mixer stages. Most were general coverage from about 550kHz up to about 30MHz but a number covered only the amateur bands (which in those days did not include 21MHz).

The National HRO with its highly ingenious "PW" mechanical bandspread dial providing 500 clear calibration points and the Hammarlund Super Pro were both well established as effective and well designed receivers before the outbreak of war in Europe in 1939, and many thousands of these receivers were built for

**The HRO-60. A late model in a famous series of receivers that were in production for over 30 years from 1936**

The GEC BRT400 communication receivers were introduced in 1947 and continued in production for 20 years mainly for professional applications

military and surveillance purposes (eg HRO-5 while the Super Pro 200 became the BC779 etc). The HRO was even paid the compliment of being closely copied by both the Germans and Japanese as wartime general-purpose receivers.

Other famous wartime models were the RCA AR88D and AR88LF. These receivers were built to highly professional standards that would in normal circumstances have put them outside of the price range of most amateurs. But many thousands of these large (and heavy) receivers were brought into the UK for the Services and after the war were gradually released over a period of many years as "surplus". Even today, nearly 40 years after its introduction in 1940–1, the AR88 still deserves to be a highly respected hf receiver provided always that you have the space to accommodate it and the strength to lift it! It forms an excellent general-coverage receiver although the tuning rate is rather high for ssb reception; often available at around £35–45 or so it still represents excellent value for money.

In the UK the Eddystone company entered the superhet communication receiver field just prior to the war and an early wartime model was the 358X receiver. This, like the HRO, had a series of plug-in coil assemblies, rather than band-switching. Wartime receivers also included American military receivers such as the BC312, BC342 and BC348 series, the BC1147 and a large number of "Command" receivers, some with an i.f. of 85kHz that later made them popular as add-on units to provide improved selectivity for receivers having an i.f. of the order of 450–475kHz. In the UK many thousands of R1155 aircraft and CR100/2 (B28) receivers were made by The Marconi Company; the CR100 receiver is still widely used by amateurs, having two rf stages and crystal and af filters.

## The post-war period

In the immediate post-war period, American firms quickly turned over to the production of receivers for the amateur market, the first models showing strong likeness to their final pre-war models. But the stringent currency restrictions limited the number of such sets reaching the UK to a small trickle. Several British firms attempted to fill the vacuum: the Radiovision double-conversion "Commander" and the Denco DCR19 among them, but undoubtedly the most widely used were the Eddystone range; the 640 (often called the S640) was a popular general-purpose single-conversion (i.f. 1,600kHz) receiver with crystal filter introduced in the late 'forties. At this period the most widely used receivers in the UK were probably the AR88, HRO, Eddystone 640 and CR100.

Because of the relatively high cost of models designed to have two tuned rf stages (calling for four-gang tuning), most post-war models had only one rf stage and reduced "image" reception by using either single-conversion with an i.f. of the order of 1,600kHz or double-conversion with the second i.f. about 455kHz or 85kHz. Up to this time all amateur receivers having a "crystal filter" employed filters of the single-crystal type. Although crystal bandpass filters were developed in the 'thirties they did not really come into widespread use until the popularization of ssb in the 'fifties and 'sixties. The single-crystal filter, with phasing control and variable selectivity by changing the impedance into which it works, was and remains an extremely effective form of filter for cw operation, although (except in conjunction with stenode audio-response correction) rather too peaky for really satisfactory a.m. or ssb operation.

In order to provide high-performance at low cost many of the receivers of this period were designed for amateur-bands-only operation, although most firms also offered general-coverage models intended to appeal to SWLs interested in broadcast-station reception. The problems of accurate tracking of the local oscillator and signal frequency circuits are much reduced if reception is required only on amateur bands.

In 1947, the American firm of Collins, which had gained a high reputation as manufacturers of both professional and amateur communication equipment, introduced the 75A1, the first of a series of pace-setting receivers which relied on the firm's reputation rather than highly competitive prices to achieve wide acceptance. These receivers used many of the techniques found in the parallel range of high-cost professional receivers (Collins 51J, R390 etc); they exploited the then-novel technique of having a crystal-controlled hf oscillator in conjunction with a variable first i.f. (with a high-stability permeability-tuned variable oscillator), so providing a similar degree of bandspreading on all bands.

In the third of this series—the 75A3—Collins also introduced to amateurs the first high-performance mechanical bandpass filters (limited in practice to use on I.F.s below 500kHz). The 75A4, the final model in this particular series, probably remains one of the best hf receivers ever to have been marketed specifically for the amateur.

Collins Radio in the 'fifties undoubtedly set the pace in high-quality equipment for amateurs, and in 1959 introduced the conveniently-packaged ssb hf mobile or fixed-station transceiver (KWM-1), the forerunner of the many transceivers now available to the amateur. A complete KWM-1 incidentally cost about $1,500. Many of the transceivers which followed covered only one or a few bands, or segments of bands, but by about the mid-'sixties the era of five- or six-band hf ssb transceivers at moderate

More than 5,000 Collins 75A4 receivers were made from 1955–8 and it remains one of the most respected hf receivers ever made for amateurs

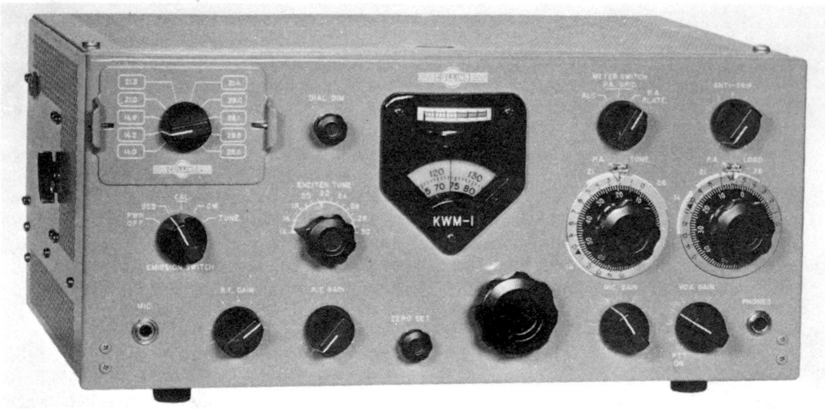


The Collins KWM-1 hf ssb transceiver introduced in the late 'fifties prepared the way for the present popularity of the transceiver approach

cost had arrived; although to reduce the basic cost they were not always fully equipped with crystals for all segments of such bands as 28MHz and often were fitted only with ssb (2·2kHz nose bandwidth) selective filters, and so were not always ideal for cw operation. One British transceiver to make a considerable impact has been the KW2000 series which provides operation on 1·8MHz as well as the usual facility of 3·5–30MHz bands.

The original compact KWM-1 was developed largely for mobile operation but most of the subsequent models, including the Collins KWM-2, have been widely used for normal fixed-station operation.

Until about the mid-'fifties very many amateurs used home-built transmitters but an increasing trend to more compact "table-top" models with built-in vfo control led to a growing dependance upon factory-built models. In the UK a popular transmitter of this type was the Labgear LG300 which has an 813 power amplifier, wide-band inter-stage coupling and separate units for the rf and modulator/power-supply sections. Labgear is a firm associated with Pye of Cambridge but no longer producing equipment for the amateur market. The basic transmitter section, which can be operated up to (and beyond) the UK legal power limit, still forms the basis of an excellent cw transmitter which can often be acquired very cheaply in view of the little use now made of a.m. on the hf bands. A lower-power table-top transmitter made by Labgear was the LG50.

Similarly the Heath DX100 and DX40 table-top transmitters (often bought initially as "kits") were widely used and are still capable of providing very satisfactory cw stations.

Professional receivers produced in the post-war period are also now quite widely available as second-hand models, particularly as many commercial users are now re-equipping with fully transistorized receivers. Among such receivers are the Racal RA17 with a highly ingenious Wadley-loop triple-mixing system that provides a drift-free system that can be readily set to within about 1kHz of a required frequency; the Hammarlund SP600JX general-coverage (540kHz–54MHz) receiver with double-conversion above 7MHz (i.f. 3,955kHz and 455kHz); the GEC BRT400 series of general-purpose single-conversion receivers (455kHz i.f.) which remained in production from the late 'forties to the mid-'sixties. Again Collins receivers deserve special mention since the military model R358 is considered to be one of the best general-purpose valved receivers ever to have been made in large numbers.

But many receivers that are still capable of giving good results on a.m. and cw can prove operationally unsatisfactory for ssb—either because of frequency drift or, even worse, frequency "jumps" which make it difficult to hold an ssb transmission, or because high-tuning rates (ie covering many kHz for each complete turn of the tuning knob) make it very difficult to tune a station without "safecrackers' fingers" to the accuracy of 25Hz or so required for ssb or because of insufficient bfo injection and the absence of the preferred "product" (synchronous) detection. This is not to say that all receivers built before the ssb era are unsatisfactory for ssb, but certainly this tends to be true of the vast majority of receivers designed for the amateur market, although these receivers often continue to be suitable for cw operation or to form the "tunable i.f." of a vhf receiver.

As a result the past decade has seen a complete swing towards ssb-orientated equipment (often well suited also to cw operation). At first the receivers were designed along established lines but with improved stability and fitted with product detectors, and gradually the crystal-controlled first mixer pioneered by Collins became almost standard practice.

The Eddystone EA12 receiver for amateurs continued in production after the firm became a subsidiary of The Marconi Co Ltd and found its primary market in professional communications. There was a series of popular Hammarlund models in the HQ100, HQ110, HQ145, HQ170, HQ180 range. Collins ended the 75A series and introduced the "S-line", again a new concept in the form of matched but separate transmitter-receiver combinations, eg 75S1 receiver, 32S1 transmitter with 175W p.e.p. ssb and an optional 30S1 1kW linear amplifier: an approach that has since been widely followed by other manufacturers. The National NC303 receiver, introduced in 1958, was a popular double-conversion receiver (i.f. 2,215kHz and 80kHz) having such features as both current and voltage regulation of the hf oscillator, and selectable sideband that provided good performance on ssb signals.

## The Japanese influence

The 'sixties saw increasing popularity of the products of relatively new British and American firms such as KW Electronics, Drake and Swan. But by the mid-'sixties a new challenge appeared that soon had a very significant impact on the range of products offered on the amateur market. This was the appearance in Europe and America of equipments made in Japan by such firms as Yaesu (initially marketed in Europe as Sommerkamp), Trio (now often known as Kenwood and some of whose models were widely sold under the American name of Lafayette), Star etc.

Initially the Japanese designs were basically very similar to the then-current American equipments but built to prices that compared very favourably with mass-produced consumer-type

**The TRIO JR310—popular Japanese receiver introduced about 1969**

domestic equipment. Increasingly however it must be recognized that the Japanese have become pace-setters in the adoption of semiconductor and frequency-synthesis techniques in amateur designs. Rather less successful have been the mechanical aspects of some of these models, deliberately built to sell at attractive prices; tuning mechanisms for example may be dependent upon cord drive and other low-cost approaches rather than precision mechanical gears. This is not necessarily a criticism of this equipment, which has proved very adequate for the amateur service, but it has widened the difference in constructional and other standards between the best professional communications receivers and those currently offered on the amateur market.

The new competition from the Far East soon had an influence on the long-established firms, a number of which have either disappeared, or entirely or partly withdrawn from the amateur scene. After over 30 years as one of the pioneers of good factory-built amateur equipment, National (the firm that had produced the HRO series and the many equipments with an NC prefix) ceased production and the brand name of "National" is now used by Matsushita of Japan. RME (of pre-war RME69 fame) had long become a subsidiary of Electro-Voice and had virtually withdrawn from the amateur receiver field even before the arrival of the main Japanese competition. Collins Radio has become part of the Rockwell International Group and although still producing amateur equipment had not until recently introduced any new designs for some years; Hammarlund no longer produce amateur receivers. The main American firms (some of them marketing equipment made or partly made in the Far East) include Drake, Heath (Heathkit), Swan, Sideband Engineers (part of the Raytheon group), Collins and Hallicrafters (now a brand name of Breaker Corp.). Atlas ceased trading in 1979.

In the UK, since the early 'sixties, Eddystone has been associated with The Marconi Company and now forms part of the GEC group, but continues to produce receivers for amateurs and SWLs; Heath (Gloucester) Ltd markets kits and assembled models as part of the Zenith group of companies; KW Electronics has become a subsidiary of Decca but is still active in the amateur market.

In both the USA and the UK (and indeed virtually every other country) the main streams of new designs are coming from Yaesu (Sommerkamp), Trio (Kenwood), Inoue (ICOM), FDK and other Japanese companies, although less favourable currency exchange rates and general inflation are tending to make the equipment subject to steadily rising prices. It is perhaps interesting to note that today there are more amateurs in Japan than in any other country, including the USA, although a high proportion of the Japanese licences are restricted to "novice" type operation, and many of these who qualify as "amateurs" do not operate their own stations.

## Current equipment trends

It has been indicated that many different categories of equipment for amateur-radio applications are currently being offered by Japanese, American and British firms, and the latest models are announced, advertised and reviewed in the amateur journals: *Radio Communication* for example regularly presents reviews based on very detailed measurements of the performance of tested models, as well as "on air" tests.

In general terms most new hf receivers are now all-solid-state although valved models are still available (where solid-state front-ends are used care is still necessary when choosing such receivers to ensure that they can cope adequately with a wide range of input-signal strengths and do not suffer too much from intermodulation, cross-modulation and other aspects of non-linearity). Most of the hf transceivers and transmitters are of "hybrid" form using semiconductors in the small-signal stages but incorporating valves for the driver and power-amplifier stages; some all-solid-state hf transmitters and transceivers are

**The Yaesu FTdX401 hf transceiver is representative of the Japanese equipments that became so popular in the 'seventies, providing complete ssb cw a.m. stations in a single unit at powers up to 500W p.e.p.**

The Heathkit SB104, introduced about 1974, was one of the first amateur transceivers to incorporate digital readout and all-solid-state design

available at ratings up to about 200W p.e.p., (eg Swan SS200, Trio TS120S, Yaesu FT301). However not even the most fervent solid-state enthusiast would claim that a watt output from a transistor is superior to one from a valve (although some of the solid-state units are built to broadband concepts and require less adjustment when changing frequency, and they are very attractive for mobile operation or for the traveller). For example the Atlas 180 for 1·8–14MHz measures only 3½in high by 9½in wide by 9½in deep.

The efficiency of the pa stages used in ssb transceivers and transmitters generally varies appreciably on different hf bands; for example, it may be only about 33 per cent on the higher-frequency bands, increasing to about 50–55 per cent on 3·5MHz. This difference may be further emphasized by variations of drive (often output on 21MHz may be less than on 28MHz) and it will usually be found that the output is less on 21 and 28MHz than on lower bands. This can mean, for example, that a typical compact transceiver of nominal 150W p.e.p. input rating may have an output which varies from perhaps 35W on 21MHz to over 100W on 3·5MHz. PA arrangements commonly used in current models include: two S2001 (6146A or B), two 6JS6A, two 6JM6, two 6KD6 etc. Where "sweep tubes" are used it is often necessary to ensure that tuning up the transmitter is carried out quite quickly as such valves are not intended to maintain continuous output for long periods; this point should also be considered when using any form of speech processing such as audio or rf clipping.

The use of hf crystal filters in ssb equipment reduces the number of frequency conversions and many filters are around 3,180kHz, 5·2MHz or 9MHz, but other designs continue to use high-performance mechanical filters at about 455kHz. The use of pre-mixer systems reduces the number of frequency conversions in the signal path of a receiver, although careful design is necessary to reduce the number of spurious responses.

For vhf enthusiasts, including the Class B licence-holders, a large number of low-power transceivers for fixed, mobile and hand-held operation are marketed, most of them manufactured in Japan. These include 144MHz and 432MHz units for nbfm, ssb, a.m., cw and fsk (rtty and sstv) operation, often fitted with toneburst audio generators for access to fm repeater stations. These units may be intended for use on a number of fixed crystal-controlled channels, or include vxo or vfo facilities for tuning over all or part of a band.

As the number of crystal-controlled channels is increased it becomes rather expensive to load the units fully with crystals, and often not all possible channels are so equipped.

A number of recent units make use of frequency synthesis to provide a large number of channels with the frequency stability of crystal oscillators but using only a small number of crystals. Various forms of frequency synthesis have been used, including systems using a limited number of crystals arranged to provide a large number of outputs by means of heterodyning systems; the present trend however is the use of digital frequency synthesis based on phase-locked loop techniques. The digital pll synthesizer can generate many channel frequencies at closely spaced increments from a single crystal-controlled reference oscillator by means of variable ratio dividers, and can now be constructed in very compact form by means of both special-purpose and general ttl logic devices. Although basically the pll synthesizer is a very complex system, the use of integrated circuits has reduced the cost to the stage where a pll synthesizer adds less to the cost of a vhf transceiver than the fitting of more than a few crystal-controlled channels, using individual crystals.

Most frequency-synthesizers have been for vhf operation, but hf systems have also been built by some amateurs, and indeed the Wadley-loop form of drift-free triple-mix system used in some receivers may be considered a form of frequency synthesis. It should be noted that although synthesis is an extremely effective system when well-engineered, it can present problems in achieving a spectral purity as good as that of a free running oscillator.

The same synthesizer may be arranged so that it can be used in transceivers on receive as well as on transmit. The receiver sections of such units are usually single- or double-conversion superhets, mostly with an i.f. of 10·7MHz, either demodulating directly at this frequency, or with a second i.f. of 455kHz. Single-conversion results in fewer spurious responses but it is difficult to provide a sensitive discriminator for nbfm at 10·7MHz unless a crystal discriminator is used.

## Prices and inflation

In a period of rapid inflation, it is difficult to give guidance on prices of equipment and this is best obtained from studying current advertisements. But very roughly one might expect an hf transceiver with ac power supply for domestic operation to cost from about £400 upwards, with a considerable spread that reflects the provision or absence of such features as voice-operated switching, the ability to tune receiver and transmitter to slightly different frequencies, whether or not crystals are provided that enable the equipment to operate over the complete amateur bands or only segments of them, the characteristics of the ssb filter and whether a narrow-band cw filter is also

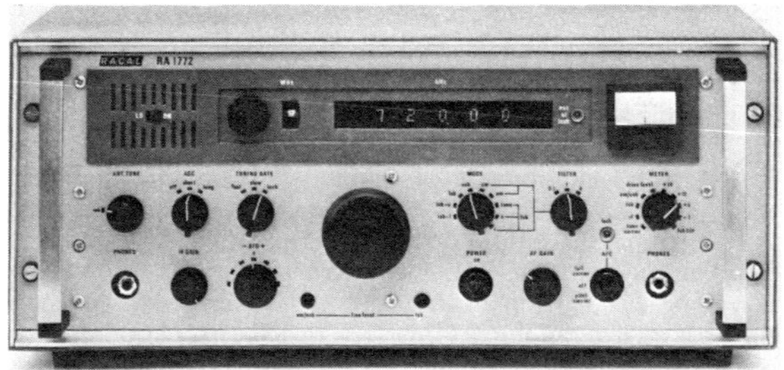

The Racal RA1772 high-grade general-purpose communication receiver (15kHz to 30MHz) with free-tuning synthesizer for search purposes. A feature of the design is the high dynamic range achieved by the use of a highly linear front-end which allows wide-band filters in the rf and mixer stages. A professional solid-state receiver of this grade can cost thousands of pounds

incorporated, the type of valves fitted and of course the power rating of the transmitter section. One or two models have had price tags in the region of £1,000 but much more typical would be £450 to about £700.

Prices of receivers intended specifically for amateur operation also vary considerably and have tended to rise rather rapidly of late as higher standards of stability are being offered. At the lower end of the market there are some direct-conversion receivers (say £25 to £35) and some conventional single-conversion receivers and also some with crystal-controlled hf oscillator and variable i.f. in the range £50 to about £150. The more advanced receivers, if bought new, may cost from about £250 to £350 or more, with the most advanced professional models now ranging up to several thousands of pounds. An ssb hf transmitter is likely to be from about £350 upwards; high-power linear amplifiers from about £150; rf type speech clippers about £50; vertical rod antennas from about £30.

For vhf an ssb transceiver for 144MHz might be around £250, roughly the same as for nbfm equipments, but it is easy to spend up to £500 or so.

Kits for home assembly will often reflect a significant saving on manufactured equipments.

This list of rather formidable prices does not mean that it will cost a newcomer a minimum of several hundreds of pounds to "get on the air". Considerable numbers of stations are based on second-hand equipment, wire antennas slung up to buildings and trees, converted surplus or reconditioned equipment or the like. Since many vhf enthusiasts use a.m. and crystal-controlled converters in conjunction with hf receivers, the a.m. operator can still make very effective use of the older hf receivers stemming from the pre-ssb era. Many of the older table-top transmitters can be picked up very cheaply and form effective cw stations. It is still possible to put a station on the air for around £50, although it must be admitted that an increasing number of stations today represent several hundreds of pounds expenditure.

## To build or buy?

It should also be appreciated that most modern amateur radio equipment is designed specifically to meet the budgetary restrictions imposed by the market; this does not mean that such equipment is badly made, indeed much of it represents very good value for money and a high standard of electronic design. But it must be recognized that, for example, a receiver with a retail price of £200 is likely to have been built with a cabinet and components that will have cost the manufacturer a maximum of £50 to £60, and to have labour costs, including assembly, alignment, testing and trouble-shooting of perhaps £25 to £35. Admittedly a manufacturer buying in large quantities will get much more for his money than an amateur buying new components and his production line will have been designed to minimize labour costs. On the other hand the amateur who builds his own equipment can often incorporate high-quality surplus or salvaged components and mechanical parts of a standard that no manufacturer could afford to put into a low-budget receiver.

Often it is in the mechanical details that modern factory-built receivers are less than ideal, using rather lightweight material for chassis and cabinet and merely adequate tuning mechanisms. The home builder can often use heavy-gauge aluminium panels, a rigid chassis and such tuning mechanisms as the Eddystone 898 or salvaged HRO-type tuning systems. A good general-purpose communication receiver represents a combination of mechanical and electronic engineering and both aspects deserve consideration.

On electronic design, the individual amateur may not be able to beat the professional designer but can often build a receiver that satisfies his own specialized needs without having the problem of meeting the requirement of producing a mass-market, multi-band, multi-mode, decorator-styled receiver that can sell at a popular price.

The future of home-built receivers, except those undertaken in order to obtain constructional experience, is probably in those designed for a specific, limited task—but with no compromise. There is at present a very considerable difference in the standards of performance found in high-cost military or professional receivers, in such matters as linearity, cross-modulation, intermodulation products, image rejection and spurii and those that can be expected from equipment built to a restricted budget. The home constructor can often aim at higher standards in these respects even if he cannot produce such a convenient multi-purpose design.

## Think about maintenance

One advantage of building or modifying equipment is that you can then be reasonably confident of maintaining it in good trim. When purchasing amateur equipment, thought should be given to the conditions under which it is likely to be operated and what problems may arise when the inevitable fault occurs. For example, equipment that is to be used in tropical countries must be able to withstand the effects of heat, humidity, dust and insects. Poorly tropicalized PCBS, for example, still give trouble in such conditions; and despite much progress some components remain vulnerable to humid conditions. Inductors and transformers at high voltage may suffer from "green spot" corrosion due to tiny imperfections in the coating of enamelled wire; variable-type carbon components such as potentiometers can be affected by moisture; push-button switches by dust; springs by rust, heat and humidity; rubber may deteriorate rapidly. Even in temperate countries, it should be recognized that complex equipment can involve maintenance problems and exact replacements may be difficult to obtain.

There are thus advantages for the amateur, particularly if he is likely to be a long distance from the suppliers or the original manufacturers, either to satisfy himself that servicing will continue to be available or to use equipment that he feels capable of keeping in good condition, or at least to use equipment that could be maintained locally.

Look for equipment that is accompanied by good instruction books and servicing instructions, and which is based on components, valves and semiconductor devices that are—and seem likely to remain—readily available. Take careful thought of future maintenance problems when contemplating the purchase of what is sometimes termed "semi-professional" equipment—that is highly complex equipment built to consumer-type standards. And if you do use complex equipment give some thought to the environment under which it is operated and try to avoid extremes of temperature and humidity, if this is at all possible. For semiconductors, extreme low temperatures are as damaging as high temperatures.

The disappearance of thermionic valves from the line (horizontal) output stages of colour television receivers has removed the last mass market for low-cost, mass-produced valves suitable for rf power amplification. Most modern hf transmitters and transceivers are now either all-solid-state or fit valves such as the 6146B which are specifically intended for transmitters and which continue to be manufactured in small quantities.

Newcomers buying both secondhand and new equipment which use valves should recognize the possibility that replacement valves will over the next few years become progressively more scarce and expensive. They should consider the advisability of building up a small stock of spare valves sufficient to last the anticipated lifetime of the equipment, remembering that, while most valves will last many years, the pa valves in some hf ssb/cw transmitters and transceivers have to handle high peak currents under linear conditions in an application for which they were not originally developed: such valves, if used regularly, may well need replacing in a period of about two to four years. Furthermore, some equipments may not work satisfactorily with all valves of a given "type" and it is advisable, when laying in a stock of power amplifier and driver valves, to check that they work correctly when fitted in the equipment. Receiver valves often last many years, but here again a small stock of replacements may avoid problems later, particularly if the equipment is fitted with older types of valves no longer manufactured.

When choosing equipment look for the finer points of design; the minimizing of spurious responses and good dynamic range in the front-end of receivers; transmitters which have "clean" signals with a minimum of splatter on ssb or are able to operate effectively with the recommended low deviation of nbfm; that produce minimum output on harmonic or spurii from mixers or from intermodulation products. Unfortunately it is not always easy to be certain from a manufacturer's technical data just how good or bad his equipment is really likely to be in operational use. Some guidance can be obtained by listening to other amateurs or by carefully reading reviews of new equipment (provided that this really does go into the performance details and is not just a superficial "write-up"). But in the end it is a question of *caveat emptor* (let the buyer beware), although the better-known manufacturers are usually very careful when introducing new models that these will not damage a hard-won reputation. Much of the equipment offered to the amateur today represents good value for money—but just as it would be foolish to expect a Rolls Royce performance from a Mini, so in amateur radio you must expect to get a specification closely related to the cost of the equipment.

And never forget that even more important for good results than the output of the transmitter is the antenna—and that simple wire antennas still give extremely good results if carefully put up and adjusted.

---

### COLLINS

**75A1:** High-performance amateur-bands receiver using variable i.f. principle, 1947–50. 3·5–30MHz in 1MHz segments. I.F. 1·5–2·5MHz, 500kHz. RF amplifier 6AK5.

**75A2:** Development of 75A1 but including 1·8MHz reception. 1950–3. Final i.f. now 455kHz. RF amplifier 6AK5.

**75A3:** Development of 75A2 but with mechanical instead of crystal filter. 1953–5. RF amplifier 6CB6.

**75A4:** Development of 75A3 with wider dynamic range. More than 5,000 produced; still regarded as one of the best amateur receivers ever produced. RF amplifier 6DC6.

**75S1:** Compact high-performance receiver. 3·5–30MHz in 200kHz segments (14 band positions). Introduced 1958. IF 2,955–3,155kHz, 455kHz. Mechanical filters. RF amplifier 6DC6.

**75S2:** Development of 75S1.

**75S3:** Development of 75S2.

**KWM-1:** Mobile ssb transceiver (175W p.e.p. ssb, 160W cw) any 10 100kHz segments between 14–30MHz. Output two 6146. I.F. 3·9–4·0MHz and 455kHz (mechanical filter). RF amplifier 6DC6.

**KWM-2:** Development of KWM-1 but any fourteen 200kHz segments between 3·5–30MHz. I.F. 2·955–3·155MHz, 455kHz. From about 1968 onwards. RF amplifier 6DC6. Often used also as compact fixed station equipment.

**Promark KWM380:** High-grade hf ssb/cw/rtty (and a.m. receive) transceiver with digital readout and general coverage on receiver (2–30MHz and 0·5–2MHz at reduced sensitivity) and 1·8–28MHz transmitter. Microprocessor frequency control with memory providing split vfo facilities. Provision for six bandwidth filters; internal ac power and speaker. Passband tuning. 100W p.e.p. ssb (or 100W average with optional blower). Introduced 1979–80.

### DRAKE

**1-A:** Triple-conversion amateur-bands receiver with crystal-controlled first oscillator, providing 600kHz segments in 3·5MHz–28MHz bands plus 10MHz WWV. Product detector. I.F. 2,900–3,500kHz, 1,100kHz and 50kHz. About 1957–60. RF stage 6DC6.

**2-A:** Triple-conversion amateur-bands receiver providing up to 12 600kHz segments between 3·5MHz and 30MHz normally fitted with five crystals (28·5–29·1MHz for 28MHz). I.F. 3,500–4,100kHz, 455kHz and 50kHz; vfo 3,955–4,555kHz. About 1960–61. RF stage 6BZ6.

**2-B:** Triple-conversion amateur-bands receiver providing up to 12 600kHz segments between 3·5 and 30MHz (normally fitted with five crystals). Passband tuning on final i.f. of 50kHz, otherwise as 2-A. Introduced 1961.

**2-C:** Triple-conversion amateur-bands receiver providing 500kHz segments on 3·5, 7, 14, 21 and 28·5–29MHz plus any additional 500kHz segment between 3·5 and 30MHz. I.F. 3,500–4,000kHz, 455kHz, 50kHz, vfo 3,955–4,555kHz. Valves in signal path (rf stage 12BZ6) but considerable use of semiconductors. Introduced about 1968.

**R-4:** Dual-conversion with pre-mixer amateur-bands receiver providing 500kHz segments between 3·5 and 29MHz (normally crystals fitted for five segments with provision for additional five). I.F. 5,645kHz and 50kHz. RF stage (12BZ6). Later models R-4A and R-4B (about 1969) increasing use of semiconductors. 50kHz tunable lattice filter with 5,645kHz crystal filter.

**R-7 (Model 1240):** General-coverage synthesized hf receiver with up-conversion (48MHz 1st i.f.) and double-balanced mixer. For usb/lsb/cw/rtty/a.m. modes.

**SPR-4:** Solid-state receiver (dual-gate mosfet front-end) which can provide up to 23 500kHz segments between 0·5 and 30MHz, often fitted for short-wave broadcast bands.

**TR-4:** Transceiver providing full coverage 3·5–28MHz amateur bands in seven 600kHz segments. 300W p.e.p. ssb, 260W cw and 260W p.e.p. a.m. PA three 6JB6 valves. Fitted with two 9MHz filters for sideband selection. Solid-state vfo 4·9–5·5MHz. Introduced 1965 with later version TR-4B about 1968.

**TR-7:** Solid-state hf transceiver with general-coverage up-conversion receiver (1·5–30MHz or 0–30MHz with Aux-7 optional unit). Transmitter 1·8–28MHz bands for usb/lsb/cw/rtty/A3h (ssb 250W p.e.p.) cw 250W, A3h 80W (carrier) plus usb. Partial synthesis and passband tuning. 1st i.f. 48·05MHz, 2nd i.f. 5·645MHz. About 1978.

### EDDYSTONE

**640:** General coverage 1·7–31MHz in three bands. Introduced about 1947 (sometimes referred to as "S640" model). Single-conversion with i.f. 1,600kHz, crystal filter and one rf stage (EF39).

**680:** General coverage 490kHz–30·4MHz in five bands. Single-conversion with i.f. 450kHz and two rf stages (two 6BA6). Introduced 1949.

**740:** General coverage 485kHz–30·6MHz in four bands. Single-conversion with i.f. 450kHz and one rf stage (EAF42). Introduced about 1950.

**750:** Dual-conversion general-coverage receiver 480kHz–32MHz in four bands. I.F. 1,620kHz, 85kHz. RF amplifier 6BA6.

**840:** AC/DC version of 740. Introduced about 1953. RF amplifier UAF42. I.F. 450kHz.

**888:** Dual-conversion amateur-bands receiver 1·8MHz–30MHz (six bands). I.F. 1,620kHz, 85kHz. Introduced about 1956. RF amplifier 6BA6.

**EA12:** Amateur-bands receiver with 10 600kHz segments between 1·8 and 30MHz. Variable first i.f. 1·1–1·7MHz, 2nd i.f. 100kHz. RF amplifier ECC189 (cascode). Introduced 1963.

### FDK

**Multi-700E "Perfect":** 144MHz mobile fm transceiver with synthesized 12·5 or 25kHz steps and digital display. Output 25W.

**Multi-3000:** 144MHz multi-mode transceiver (fm/ssb/cw) with digital display and switched synthesizer (10kHz or 1kHz steps) with memory and dual vfo system.

**M800D:** 144MHz fm transceiver with 800 synthesized channels (144–148MHz) and digital display. Manual or automatic tuning. Output 1–25W continuously variable. Memory for two channels.

**Palm IV:** 432MHz hand-held fm transceiver with six-channel capability and ±1·6MHz shift.

### HALLICRAFTERS

The following is a selection from the many models introduced by this firm since about 1936. Early models had names that mostly included "Sky" and most receiver model numbers start with "S" or "SX" (indicating that the receiver includes a crystal filter).

**Sky Buddy:** A low-cost (typically £9) general-purpose receiver in the period 1936–40; several versions but typically covered 545kHz–18·5MHz in three ranges. No rf stage (6A7 mixer). I.F. 465 or 455kHz.

**Sky Champion:** A series of general-coverage receivers from about 1936–45 including the S20-R from 1939–45. Typically covered 550kHz–44MHz in four bands. One rf stage (6K7), i.f. 465 or 455kHz.

**Super Skyrider:** Several models including SX16 about 1938. General-coverage receiver typically 545kHz–62MHz in six bands. One rf stage (6K7), i.f. 465kHz.

**S27:** Wartime tunable vhf receiver 27·5–145MHz in three bands (S27B 38–165MHz); i.f. 5·2MHz. Superseded by S36 with coverage 27·8–143MHz; i.f. 5·25MHz.

**SX28:** Wartime general-coverage receiver; 550kHz–42MHz in six bands with electrical bandspread and two rf stages (SX28A was military version to higher specification). I.F. 455kHz.

**S38:** Post-war receiver of low-cost Sky Buddy type. Various versions, typically covering 550kHz–31MHz in four bands. No rf stage and i.f. 455kHz.

**S40:** This was post-war replacement of Sky Champion series. Several versions, typically 540kHz–43MHz in four bands. One rf (6SG7) and i.f. 455kHz.

**SX42:** This was successor to SX28 but included vhf/fm bands with dual i.f. system. 540kHz–110MHz in six bands. Two rf stages (two 6AG5) and i.f. 455kHz (fm 10·7MHz).

**SX88:** General coverage with calibrated bandspread. 535kHz–33·3MHz in six bands. Introduced in 1954. Dual-conversion with i.f. 2,075kHz and 50kHz.

**SX101:** Amateur-bands double-conversion receiver. 3·5MHz–30MHz bands plus WWV and 30·5–34·5MHz band for vhf converters. SX101 introduced in 1957 and SX101A in 1959. One rf (6CB6), selectable sideband and crystal calibrator. I.F. 1,650kHz and 50·5kHz (SX101A 50·75kHz).

**SX115:** Triple-conversion amateur-bands receiver with nine 500kHz segments from 3·5–29MHz (usually only 28·5–29MHz crystal fitted for 28MHz band). I.F. 6,005–6,505kHz, 1,005kHz, 50kHz. Introduced 1961. RF amplifier 6JD6.

**SX146:** Single-conversion amateur-bands receiver with premixer technique to provide 500kHz segments from 3·5MHz–30MHz (usually only 28·5–29MHz crystal fitted for 28MHz band). I.F. 9MHz with 5·0–5·5MHz vfo. One rf stage (6JD6). Optional cw and a.m. filters. Introduced about 1967.

### HAMMARLUND

**Super-Pro:** A series of receivers for professional and amateur applications manufactured for about 30 years from 1936. Pre-war and war-time models (SP10, SP110, SP110S, SP110X, SP110SX, SP210X). Coverage either 540kHz–20MHz in five bands, or (with S suffix) 1,160kHz–40MHz in five bands. Single-conversion (i.f. 465kHz) with two rf stages and crystal filters in models with X suffix. Military versions included BC779, BC794 and BC1004.

**SP400X:** Post-war Super-pro introduced in 1946 covering 540kHz–30MHz in five bands. Two rf stages (6K7). Single-conversion with 455kHz i.f.

**SP600JX:** Later version of Super-pro produced in 1950 covering 540kHz–54MHz in six bands. Dual-conversion 3,995kHz and 455kHz i.f. Single-conversion below 7·4MHz.

**HQ100:** General-coverage 540kHz–30MHz in four bands with calibrated electrical bandspread and Q-multiplier. Single-conversion i.f. 455kHz with one rf stage (6BZ6). Introduced in 1956.

**HQ110:** Several versions including HQ110A. Amateur-bands only for 1·8–50MHz bands. Dual-conversion above 7MHz, with i.f. 3,045kHz and 455kHz. Product detector. One rf stage (6BZ6). Introduced in 1958, HQ110A in 1962.

**HQ129X:** General-coverage receiver for 540kHz–31MHz in six bands with calibrated bandspread dial for amateur bands. Single-conversion 455kHz i.f. with one rf stage (6SS7). Introduced in 1946. Variable selectivity i.f. filter.

**HQ145:** Later version HQ145X had provision for one crystal-controlled channel. General-coverage dual-conversion receiver with calibrated electrical bandspread. 540kHz–30MHz in four bands (dual-conversion above 10MHz). I.F. 3,035kHz and 455kHz. RF stage (6BZ6). HQ145 introduced in 1958, HQ145X in 1961.

**HQ160:** General-coverage dual-conversion receiver with calibrated electrical bandspread dial, 540kHz–31MHz in six bands. Q-multiplier and product detector, one rf stage (6BA6). I.F. 3,035kHz and 455kHz. Introduced 1958.

**HQ170:** Amateur-bands receiver with double- and triple-conversion. Covers 1·8–50MHz bands, selectable sideband, product detector, slot filter etc. Triple-conversion above 7MHz, i.f. 3,035kHz, 455kHz and 60kHz. One rf stage (6BZ6). Introduced 1958.

**HQ180:** (Later versions include HQ180A and HQ180AX). General coverage with calibrated bandspread on amateur bands above 3·5MHz. Covers 540kHz–30MHz in six bands. Triple-conversion above 7·85MHz. I.F. 3,035kHz, 455kHz and 60kHz. One rf stage (6BZ6).

**HQ215:** Solid-state amateur-bands receiver providing also 13 optional 200kHz segments. Mechanical ssb filter I.F. 3,055 ± 100kHz, 455kHz. Bipolar transistors. Introduced 1968.

### HEATHKIT

Equipments of this company are normally supplied either in kit form or fully wired and assembled.

**GC-1U:** All-solid-state (bipolar transistors) general-coverage single-conversion receiver for 580–1,550kHz and 1·69–30MHz in five bands. I.F. 455kHz with four ceramic transfilters. One rf stage (AF115).

**GR54:** General-coverage single-conversion receiver for 180–420kHz, 550–1,550kHz and 2–30MHz in five bands. I.F. 1,682kHz. One rf stage (6BH6).

**GR64:** Low-cost general-coverage single-conversion receiver for 550kHz–30MHz in four bands. No rf stage (12BE6 mixer).

**GR78:** General-coverage solid-state six-waveband receiver for ssb/cw/a.m. reception. Coverage 190–410kHz and 550kHz–30MHz with separate bandspread tuning. I.F. 455kHz (double-conversion on 18–30MHz band with a first I.F. of 4,034kHz. Rechargeable nickel-cadmium battery.

**HR1680:** Amateur-bands solid-state receiver for ssb/cw with six 500kHz segments between 3·5 and 28MHz, 100kHz calibrator and active cw af filter. First i.f. 8,395–8,895kHz; second i.f. 3,395kHz. Diode bandswitching.

**HW7:** Low-power, all-solid-state cw transceiver for 7, 14 and 21MHz bands (200kHz segments) with input power of about 3W. Direct-conversion receiver.

**HW8:** Low-power, all-solid-state cw transceiver for 3·5/21MHz cw bands. Input power varies from about 3·5W on 3·5MHz to about 2·5W on 21MHz. Direct-conversion receiver. Introduced 1976.

**HW100:** Five-band "economy" ssb/cw transceiver for 3·5–28MHz bands in 500kHz segments. PA two 6146 with ssb 180W p.e.p. input (cw 170W). SSB filter 3,395kHz, bandpass filter 8,395–8,895kHz, vfo 5·0–5·5MHz.

**HW2021:** Hand-portable 144MHz fm transceiver with capability of 5 receive and 10 transmit channels. Output 1W. I.F. 10·7MHz.

**HW2036-2:** Mobile 144MHz fm transceiver with frequency synthesis providing 5kHz increments over 2MHz band. Power output 10W. 10·7MHz crystal filter. Power 13·8V (nominal).

**HX1681:** HF cw-only transmitter with break-in facilities. Five amateur bands: 3·5–28MHz. Power amplifier two 6146. About 1979.

**RG-1:** General-coverage single-conversion receiver for 600–1,500kHz and 1·7–32MHz in six bands. I.F. 1,621kHz (half-lattice crystal filter). One rf stage (EF183). Introduced 1963.

**SB101:** Amateur-bands transceiver for 3·5–30MHz in 500kHz segments. PA two 6146, ssb. 180W p.e.p. input, cw 170W. SSB filter 3,395kHz, vfo 5–5·5MHz, bandpass filter 8,395–8,895kHz. One rf stage (6AU6).

**SB102:** Amateur-bands transceiver for 3·5–29·7MHz in 500kHz segments. PA two 6146 and details as for SB101 except that rf stage is 6HS6.

**SB104:** All-solid-state amateur-bands ssb/cw transceiver for 3·5–28MHz with digital frequency readout and broadband operation. Transmitter output 100W p.e.p. (or 1W). Basic unit for 12V operation but fixed-station power supply available. Introduced in USA in late 1974.

**SB300:** Amateur-bands receiver for 3·5–30MHz in 500kHz segments. I.F. filter 3,395kHz, bandpass filter 8,395–8,895kHz RF amplifier 6BZ6. Introduced about 1963.

**SB301:** Amateur-bands receiver for 3·5–30MHz in 500kHz segments. I.F. filter 3,396kHz, bandpass filter 8,395–8,895kHz. RF amplifier 6BZ6. Introduced about 1969.

**SB303:** All-solid-state amateur-bands receiver for 3·5–28MHz bands in 500kHz segments (plus WWV 15MHz). I.F. 3,395kHz, bandpass filter 8,395–8,895kHz, vfo 5·0–5·5MHz. Dual-gate mosfet devices in front end including 40673 rf amplifier. Introduced about 1971.

**SB310:** Receiver for hf broadcast bands and some amateur bands: 3·5–4·0, 5·7–6·2, 7·0–7·5, 9·5–10·0, 11·5–12·0, 14·0–14·5MHz, 15·0–15·5, 17·5–18·0, 26·9–27·4MHz. I.F. 3,395kHz. RF amplifier 6BZ6.

**SB400:** Amateur-bands ssb/cw transmitter for 3·5–30MHz in 500kHz segments. PA two 6146 with 180W p.e.p. input ssb and 170W cw. SSB filter 3,396kHz, bandpass filter 8,395–8,895kHz, vfo 5·0–5·5MHz.

**SW717:** General-coverage solid-state four-waveband receiver covering 550kHz–30MHz, fitted with bfo.

## ICOM

(Inoue Communication Equipment Corporation)

**IC21:** 144MHz solid-state mobile/base fm transceiver with 24-channel capability. Output 10 or 1W. I.F. 10·7MHz and 455kHz. 13·5V power supply. Built-in swr meter. Introduced about 1971.

**IC201:** 144MHz solid-state ssb/fm/cw transceiver with vfo control. Output 10W (cw/fm) and 10W p.e.p. (ssb). I.F. 10·7MHz (single-conversion), vfo 11–12MHz. Mains/13·8V power supply.

**IC202:** 144MHz solid-state hand-portable ssb/cw transceiver with vxo facility. Output 3W. I.F. 10·7MHz (single-conversion).

**IC202S:** 144MHz portable ssb/cw solid-state transceiver (usb/lsb/cw) with vxo and four sub-bands 200kHz wide (including 144·0–144·2MHz and 144·2–144·4MHz). 3W output, cw sidetone, rit, noise-blanking etc. For 13·8V supply.

**IC211E:** 144MHz fixed/mobile solid-state transceiver for fm/usb/lsb/cw modes with digital display, double vfo (pll), programmable, simplex or duplex, 10W output, i.f. (ssb) 10·7MHz, (nbfm) 10·7MHz and 455kHz. For 240V or 13·8V supplies.

**IC215:** 144MHz solid-state hand-portable fm transceiver with 15 channel capability. Toneburst generator. Output 3 or 0·5W.

**IC240:** 144MHz mobile fm transceiver with 80 channels (22-channel memory), 15 channels programmable with diode matrix, 12W output. For 13·8V supply.

**IC245:** 144MHz solid-state mobile fm transceiver with frequency synthesizer and four-digit frequency readout. Output 10W.

**IC245E:** 144MHz mobile fm/usb/cw transceiver with digital display, double vfo (pll system), simplex or duplex, two-speed tuning. 12W output. 13·8V supply.

**IC255E:** 144MHz mobile fm transceiver with digital display, double vfo system, five memories and built-in channel scanner. 25W (1W low power) output. 13·8V supply.

**IC260E:** 144MHz mobile fm/usb/lsb/cw transceiver with dual vfo/memory and band scanner. Seven-figure frequency display. Output 10W.

**IC280E:** 144MHz mobile fm transceiver with digital display, 80-channel synthesizer with four switched channels (three memorized to choice). Can be used as two sub-assemblies (main unit with remote control unit). 1 or 12W output. 13·8V supply.

**IC701:** Solid-state hf transceiver (usb/lsb/cw/rtty) with double vfo for 1·8–28MHz, 100W output, i.f. 9·0115/10·7015MHz.

## KDK

**FM2016E:** 144MHz fm transceiver with 1,000-channel synthesizer and four memory channels. Digital display. Output 15W or 1W. Receiver covers 144–148MHz, transmitter 144–146MHz. Monolithic roofing filter plus 15-pole ceramic filter.

## KW ELECTRONICS

Since 1974 the firm has been a member of Decca Communications Ltd.

**KW77:** Amateur-bands triple-conversion receiver for 1·8–28MHz bands in 600kHz segments. I.F. bandpass 3,500–4,100kHz, 455kHz, 50kHz. RF amplifier EF183. Introduced 1962.

**KW201:** Amateur-bands receiver for 1·8–28·8MHz in 11 200kHz segments. I.F. bandpass 2,955–3,155kHz, 455kHz (mechanical ssb 3·1kHz filter), vfo 2·5–2·7MHz, optional Q-multiplier for cw. RF amplifier EF183.

**KW202:** Amateur-bands receiver for 1·8–30MHz in nine 500kHz segments. 2nd I.F. is 455kHz (mechanical 2·4kHz ssb filter). RF amplifier EF183. Matches KW204 transmitter. Introduced 1970.

**KW204:** Amateur-bands ssb/cw/a.m. transmitter for 1·8–28MHz in nine 500kHz segments. PA two 6146 with input powers of 180W p.e.p. (ssb) and 150W (cw). 455kHz mechanical ssb filter. Matches KW202 receiver. Introduced about 1970.

**KW500:** Also known as "Viscount". Amateur-bands ssb/cw/a.m. transmitter for 3·5–28MHz bands with phasing-type ssb generation at 9MHz. PA two 5B/254M. VFO 5·0–5·35MHz for 3·5–21MHz and 5·0–7·0MHz for 28MHz.

**KW707:** Amateur-bands receiver for 1·8–28MHz bands in 600kHz segments. I.F. variable 5,000–5,600kHz and 455kHz (2·1kHz mechanical filter and half-lattice crystal filter for cw). RF amplifier EF183. Introduced 1963.

**KW2000:** Amateur-bands ssb/cw transceiver for 1·8–28·8MHz in 200kHz segments. PA 6146 with 90W p.e.p. on ssb and 75W on cw. Bandpass filter 2,955–3,155kHz, ssb filter 455kHz (2·1kHz mechanical filter), vfo 2·5–2·7MHz. RF stage EF183. Introduced 1963. Later versions include KW2000A (1964), KW2000B (1969) and KW2000E. These later models are rated at 180W p.e.p. (ssb) and 150W (cw) using two 6146 in pa. All models permit reduced power operation on 1·8MHz.

**KW Atlanta:** Amateur-bands ssb/cw transceiver for 3·5–28MHz bands. PA two 6LQ6 500W p.e.p.(ssb), 350W (cw), 125W a.m., ssb filter 5·2MHz (crystal).

**KW "One-sixty":** Transmitter for 1·8–2·1MHz for 10W (maximum 15 W) cw/a.m. operation. PA PL81 with pi-output.

**KW "Vanguard":** Table-top 50W cw/a.m. transmitter for 3·5–28MHz amateur-bands incorporating Geloso Type 4/102 signal shifter as vfo-multiplier. PA 6146 (QV06-20). Modulator two 6L6. Introduced in 1958 and supplied either as "kit" or complete.

**KW "Vespa":** Three-band ssb/cw transmitter for 1·8, 3·5 and 14MHz bands. 455kHz mechanical filter. PA 6146 with ratings of 90W p.e.p. for ssb and 75W for cw. Introduced 1963. Vespa MkII has rating of 220W p.e.p.

**KW "Viscount":** see **KW500.**

## LAFAYETTE

This range includes receivers made by **Trio** in Japan.

**HA50:** Double-conversion amateur-bands receiver for 3·5MHz–50MHz, using ceramic mechanical filters and one rf stage (6BZ6). I.F. 2,608kHz and 455kHz.

**HA350:** Amateur bands (plus WWV) in 600kHz segments for 3·5MHz–29·7MHz bands. Variable I.F. 3·5–4·1MHz and fixed i.f. 455kHz (single conversion on 3·5MHz band). One rf stage (6BZ6). Introduced about 1965.

**HA600:** General-coverage solid-state receiver covering 150kHz–400kHz and 550kHz–30MHz in five bands. I.F. 455kHz with ceramic mechanical filters and two FETS.

**HA700:** General-coverage single-conversion receiver covering 150–400kHz and 550kHz–30MHz in five bands. I.F. 455kHz. Valves and ceramic mechanical filters. Introduced about 1967.

**HA800:** Amateur-bands solid-state dual-conversion receiver with FETS and bipolars. All amateur bands from 3·5MHz–54MHz. I.F. 2,608kHz and 455kHz.

**HE30:** General-coverage receiver for 550kHz–30MHz in four bands with calibrated amateur-bandspread dial. I.F. 455kHz. One rf stage (6BA6). Introduced about 1961.

**HE80:** General-coverage receiver for 540kHz–30MHz plus 50MHz band. I.F. 455kHz. One rf stage (455kHz).

## NATIONAL

**HRO:** The original 1936 HRO-Senior was a single-conversion receiver with plug-in coil assemblies (standard 1·7–30MHz in four ranges) having provision for full bandspread on 3·5, 7, 14 and 28MHz. Two rf stages (6D6), crystal filter, i.f. 456kHz. 2·5V and 6·3V ranges of valves. HRO-Junior had no crystal filter or provision for bandspread.

**HRO-MX, HRO-M:** Military models produced before 1944, with glass valves. No bandspread facilities on coils and crystal filter omitted in M model (US Navy models RAS, RBJ, RCE).

**HRO-5, HRO-W:** Military models using metal-type valves from 1944, with general-coverage coils.

**HRO-5TA-1, HRO-5RA-1:** Post-war amateur models, based on HRO-5 with noise limiting and bandspread coils. 5TA is table model; 5RA rack model.

**HRO-7-T:** Restyled model with stabilized oscillator. Released 1947. RF amplifier stages 6K7.

**HRO-50:** Redesigned model in 1950 with direct-reading linear scale plus HRO-type dial. Coil assemblies include bandspread for 21MHz. I.F. 455kHz. RF amplifiers stages 6BA6.

**HRO-50-1:** Development of HRO-50 with additional i.f. stage.

**HRO-60:** Redesigned model with dual.conversion above 7MHz. I.F. 2,010kHz, 455kHz. RF amplifier stages 6BA6.

**HRO-500:** Compact solid-state hf/vhf receiver with partial-frequency synthesizer developed for professional market about 1965.

**NC300:** Dual-conversion amateur-bands receiver for 1·7–30MHz bands plus 30–35MHz for vhf converters. 1955–8. I.F. 2,215kHz, 80kHz.

**NC303:** Dual-conversion amateur-bands receiver 1·8–29·7MHz. Introduced 1958. I.F. 2,215kHz, 80kHz. RF amplifier 6BZ6.

## NEC

**CQ110:** Hybrid hf transceiver for ssb/a.m./cw/fsk operation (300W p.e.p. input) on 1·8–30MHz bands with digital readout, 15MHz standard-frequency facility and wide dynamic-range receiver front-end (valved).

**CQ-R700:** General coverage receiver (170kHz–30MHz) in six bands. For a.m./ssb/cw reception with "auto-spread" vfo. About 1979.

## PHILIPS

**FM321:** 432MHz mobile fm transceiver with 40-channel synthesizer with internal and external (microphone) channel-changing facilities. Transmitter output minimum 5W.

## RCA

A considerable number of RCA communications receivers came into the UK during the second world war and many have subsequently been acquired by amateurs.

**AR77:** This is a general-coverage receiver, 540kHz–31MHz in six bands developed for amateur and similar applications about 1940. Model AR77E has extended mains-voltage range for use on 200–250V etc, whereas the original AR77 is intended for 105–125V mains. Electrical bandspread, single-conversion, crystal filter. I.F. 455kHz with one rf stage (6SK7).

**AR88:** This was developed as a high-grade general-coverage receiver primarily for commercial applications. Covers 535kHz–32MHz in six bands. Introduced about 1940. Single-conversion receiver with i.f. 455kHz and two rf stages (6SG7). Variable-selectivity crystal filter.

**AR88D:** This is the basic wartime model made by several other firms as well as RCA. Similar to AR88 but usually without S-meter.

**AR88LF:** This is an lf model covering 73–550kHz and 1,480kHz–30·5MHz in six bands. Often fitted with 6V6 audio output valve instead of the 6K6. I.F. 735kHz.

## RME

**RME-45:** General-coverage receiver 540kHz–33MHz in six bands with mechanical bandspread. I.F. 455kHz, one rf stage (7B7). Released 1946.

**RME-69:** General-coverage receiver 500kHz–32MHz in six bands with electrical bandspread and crystal filter. I.F. 465kHz. Released 1937. One rf stage (6D6).

**RME-70:** General-coverage receiver 540kHz–33MHz in six bands. Electrical bandspread and crystal filter. Released 1940. One rf stage (6K7).

**RME-84:** General-coverage receiver 540kHz–44MHz in four bands with mechanical bandspread. I.F. 455kHz. Released 1946. One rf stage (7B7).

**RME-99:** General-coverage receiver 550kHz–33MHz in six bands with electrical bandspread and crystal filter. Released 1940. One rf stage (7A7).

**RME-4350:** Amateur-bands dual-conversion receiver for 1·8–28MHz bands. I.F. 2,195kHz and 455kHz. Released 1956 and Model 4350A released 1957. One rf stage (6BZ6).

**RME-6900:** Amateur-bands (plus 10MHz WWV) dual-conversion receiver for 3·5–28MHz bands. Product detector and notch filter. I.F. 2,195kHz and 57kHz. Released 1960.

## SWAN

**100MX:** Mobile solid-state hf usb/lsb/cw transceiver for 3·5–28MHz. Broadband power amplifier with 100W p.e.p. and cw, i.f. 9MHz, cw sidetone.

**Astro 102BX:** HF ssb/cw transceiver (1·8–28MHz) with pll synthesizer and digital display. Speech processing, i.f. passband tuning and dual vfo (pto) with 300Hz cw filter and ssb filter. 235W p.e.p.

**Astro 150:** HF ssb/cw transceiver with microprocessor control (variable-rate tuning and digital frequency readout. Model 150 covers 3·5–28MHz bands; Model 151 covers 1·8–21MHz bands. 235W p.e.p. and cw.

**SS200:** All-solid-state transceiver with 200W p.e.p. input for operation on 3·5–28MHz. Receiver has single conversion using pre-mixer technique.

**350:** Amateur ssb/cw transceiver for 3·5–28MHz using transistor bandswitched vfo (8,673–9,173kHz for 3·5 and 14MHz). PA two 6HF5 and 400W p.e.p. input. Receiver with 12BZ6 rf stage is single conversion with 5,174·5kHz ssb filter (7360 balanced modulator). Introduced about 1965.

**350C:** Later version of the 350 with two 6LQ6 in pa for 520W p.e.p. on ssb and 360W on cw. Filter 5,500kHz.

**500C:** Basically similar to 350C but with additional operational features such as sideband selection, sidetone, calibrator etc.

## TRIO

Also marketed under the brand names of **Kenwood** and **Lafayette.**

**9R-59DE:** General-coverage receiver for 550kHz–30MHz in four bands with calibrated bandspread dial for amateur bands. Single conversion with i.f. of 455kHz and one rf stage (6BA6). Introduced about 1967.

**JR60:** General-coverage receiver kit for 540kHz–30MHz plus 48–54MHz. Double conversion on 50MHz band.

**JR200:** Low-cost general-coverage receiver for 540kHz–31MHz in four bands. One rf stage (6BA6). Introduced about 1966.

**JR300S:** Amateur-bands receiver for 3·5–29·7MHz in 500–600kHz segments plus WWV (single conversion on 3·5MHz band). Variable i.f. 3·5–4·0MHz, vfo 3,955–4,555kHz, 2nd i.f. 455kHz with mechanical filter. One rf stage (6BZ6).

**JR310:** Amateur-bands hybrid receiver with valves, FETS and bipolar transistors introduced about 1969. Covers 3·5MHz–28MHz bands in 600kHz segments, 15MHz WWV or various optional ranges. Variable i.f. 5,955–5,355kHz, 2nd i.f. 455kHz.

**JR500-SE:** Amateur-bands receiver with transistor vfo covering 3·5MHz–28MHz bands in 600kHz segments. Introduced about 1968. Variable i.f.

8·5–9·1MHz, vfo 8,446·5–9,046·5kHz, 2nd i.f. 455kHz with two ceramic mechanical filters.

**JR599:** Solid-state amateur-bands receiver matching TX599 transmitter, and using field-effect and bipolar transistors, ic and zener diodes. Amateur bands from 1·8MHz–28MHz plus 144MHz and 10MHz WWV in 500kHz segments.

**QR666:** General-coverage broadcast and amateur receiver with separate bandspread-tuning control. Covers 170–410kHz; 525–1,250kHz; and 3·5–28MHz. I.F. 455kHz (plus 4,034kHz 1st i.f. on high range). FM detector fitted.

**R599S:** Solid-state amateur-bands receiver for 1·8, 3·5/28MHz plus 10, 27 and 144MHz with ssb/cw/fm/a.m. capability. Also a R599D version matching T599D.

**R1000:** General-coverage hf receiver for 200kHz–30MHz in 30 bands each 1MHz wide. With a.m. and ssb filters and noise blanker. Up-conversion 1st i.f. 48MHz, rf attenuator. Operates from 12V dc or 110/240V ac.

**T599S:** HF ssb/cw/a.m. transmitter for 3·5/28MHz bands. PA two fan-cooled 6146B. Also T599D version with darker cabinet.

**TR2200GX:** Hand-portable 144MHz fm transceiver with 12-channel capability. Output 2W or 400mW.

**TR2300:** Compact mobile/"portable" 144MHz fm transceiver with 80 synthesized channels (25kHz steps) and 600kHz shift for repeater operation. Output 1W minimum.

**TR2400:** Hand-held 144MHz fm transceiver with keyboard control and lcd channel display (800 synthesized channels for 144–148MHz) with 10 memory channels. 1·5W output.

**TR3200G:** Hand-portable 432MHz fm transceiver. Output 2W or 400mW. I.F. 10·7MHz and 455kHz.

**TR7010:** Mobile 144MHz ssb/cw transceiver with 48-channel capability. 10W output. Single conversion 10·7MHz i.f.

**TR7200G:** Mobile 144MHz fm transceiver with 23-channel capability. Output 10W or 1W. I.F. 10·7MHz.

**TR7500:** Mobile 144MHz fm transceiver with 80-channel synthesizer (25kHz steps). 15W output.

**TS120V:** Solid-state compact hf transceiver for mobile/fixed operation. Covers 3·5–28MHz (plus 15MHz standard frequency reception) with pll vfo, single-conversion receiver with i.f. shift, noise blanker. Digital frequency readout. 20W input.

**TS120S:** Higher-power version of TS120V with up to 200W p.e.p. input on 3·5–21MHz, 160W p.e.p. input on 28MHz.

**TS180S:** Solid-state hf transceiver (1·8–28MHz) with digital readout and control. Split-frequency operation. 200W p.e.p. ssb and 160W cw.

**TS510:** Amateur-bands ssb/cw transceiver for 3·5–28MHz bands using valves, transistors and FETS. PA two S2001. VFO 4·9–5·5MHz, bandpass filter 8,295–8,895kHz, ssb filter 3,395kHz. Rated input 160W p.e.p. (120W on 28MHz). Introduced about 1969.

**TS515:** Amateur-bands ssb/cw transceiver for 3·5–28MHz bands with fet vfo 5–5·5MHz. 180W p.e.p. 3·5–21MHz; 120W p.e.p. on 28MHz.

**TS520:** Amateur-bands ssb/cw transceiver for 3·5–28MHz. Solid-state except for driver (12BY7A) and pa (two 6146A/S2001). The fet vfo covers 5·5–4·9MHz. Bandpass i.f. 8,895–8,295kHz, ssb filter 3,395kHz. Introduced about 1972.

**TS520SE:** Economy version of TS520S.

**TS700G:** 144MHz transceiver (vfo or crystal) for ssb/fm/a.m./cw. Output about 10–15W. I.F. 10·7MHz for ssb/a.m./cw plus 455kHz as 2nd i.f. on fm. VFO range 8·2–9·2MHz. There are some differences between the TS700 and the TS700G.

**TS700S:** As TS700 but including 22 crystal channels with digital readout.

**TS770:** Solid-state two-band vhf/uhf continuously tunable transceiver covering 144–146MHz and 430–440MHz for usb/lsb/cw/fm modes. Output 10W (also 1W for fm). 1st i.f. 21·6MHz, 2nd i.f. 8·83MHz (fm 455kHz) for 12V dc and ac mains. About 1979.

**TS820:** HF ssb/cw/fsk transceiver for 1·8–28MHz bands. Six-digit frequency display. Phase-locked-loop vfo/synthesizer. Power 200W p.e.p. (ssb), 160W cw, 100W fsk. Incorporates rf speech processor.

**TS820S:** Hybrid hf transceiver (1·8–28MHz plus standard frequency reception) with digital frequency readout. 200W p.e.p. ssb, 160W cw, 100W fsk.

**TS900:** HF ssb/cw/fsk transceiver for 3·5–28MHz bands. 6GK6 driver, two 6LQ6 pa. Power 300W p.e.p. input ssb, 200W cw, 100W fsk. Introduced about 1974.

### YAESU/SOMMERKAMP

**FL200B:** Amateur-bands ssb/cw/a.m. transmitter with 600kHz band segments. Power ssb 240W p.e.p. PA two 6JS6A. VFO 4·9–5·5MHz. Mechanical ssb 455kHz filter. Introduced about 1966.

**FLdx500:** Amateur-bands ssb/cw transmitter with 600kHz band segments. Power ssb 270W p.e.p. PA two 6JS6A. Hybrid semiconductor/valve including dual-gate mosfet vfo. 9MHz i.f. with sideband selection. 25/100kHz calibrator. Introduced about1971.

**FRG7:** General-coverage hf all-solid-state receiver using Wadley-loop technique in the tuning system with 1MHz bands (500kHz–30MHz). I.F. 54·5–55·5MHz, 3–2MHz, 455kHz (vfo 2·455–3·455MHz). With rf attenuator and suitable for mains or 12V operation. Introduced about 1976.

**FRG7000:** General-coverage ssb/a.m./cw receiver covering 0·25–29·9MHz using Wadley loop technique (i.f. 55·5–54·5MHz, 3–2MHz, 455kHz. Digital frequency display and built-in digital clock. About 1979.

**FR100B:** Amateur-bands dual-conversion receiver providing 600kHz segments in bands between 3·5 and 28MHz plus provision for 10MHz WWV and other segments between 1·8 and 30MHz. Crystal filter for cw and 2kHz mechanical filter for ssb. RF stage (6BZ6). I.F. 5,955–5,355kHz and 455kHz, vfo 4·9–5·5MHz. Introduced about 1965. Companion to FL100B transmitter.

**FR101:** Series of hf/vhf solid-state receivers (FR101S standard; FR101D de luxe; FR101SD standard with digital display; FR101DD de luxe with digital display). Covers 1·8–144MHz amateur bands plus 5–27MHz broadcast bands (ie up to 30 500kHz segments). I.F. 6·02–5·52MHz bandpass; 3,180kHz.

**FT7:** Compact solid-state hf mobile transceiver (3·5–28MHz) for usb/lsb/cw operation. Input ssb/cw 20W.

**FT7B:** Later higher-power version of FT7 solid-state hf transceiver (3·5–28MHz) for ssb/cw/a.m. Input ssb/cw 100W dc, a.m. 25W. Schottky diode ring mixer plus mosfet rf. Noise blanker, 100kHz calibrator. Optional digital display (YC-7B). Audio cw filter. I.F. 9MHz. About 1979.

**FT100:** Transceiver for ssb/cw/a.m. operation on 3·5–28MHz (normally 28·0–28·5MHz crystal only). All solid-state except 12BY7A driver and two 6JM6 pa. Power 150W p.e.p. on ssb. Receiver vfo 5,220–5,720kHz, transmitter vfo 8,400–8,900kHz, i.f. and ssb filter 3,180kHz. Introduced about 1966.

**FT101:** Transceiver for ssb/cw/a.m. operation on 3·5–28MHz bands. All solid-state (including field-effect devices) except for 12BY7A driver and two 6JS6A pa. Power 260W p.e.p. ssb, 180W cw and 80W a.m. VFO 8,700–9,200kHz, bandpass filter 5,520–6,020kHz, ssb filter 3,180kHz. Introduced about 1971. Later models include FT101B and FT101E series. The FT101E has 12BY7A driver and two 6JS6C pa stage, with built-in speech processor. Later models include 1·8MHz band.

**FT101Z:** Hybrid hf transceiver (1·8–28MHz plus standard frequency reception). Input 180W dc (power amplifier two 6146B) with rf processor, variable i.f. bandwidth using two crystal filters. About 1979.

**FT101ZD:** Similar to FT101Z but with digital frequency readout.

**FT107M:** Solid-state hf usb/lsb/cw/fsk/a.m. transceiver for six bands (1·8–28MHz) with broadband power amplifier. 240W p.i.p. 12 memory channels with digital memory shift. Remote (microphone) up/down band scanning. About 1979.

**FTdx150:** Transceiver for ssb/cw/a.m. operation. All solid-state except for 12BY7 driver and two 6JM6 pa. VFO 8,400–8,900kHz. Bandpass filter 5,700–5,200kHz, ssb filter 3,180kHz.

**FT200:** (also **FT250**): Transceiver for 3·5MHz–28MHz. Mostly valves with 7360 balanced modulator, driver 12BY7A and pa two 6JS6A. SSB filter 9MHz with single conversion receiver using pre-mixer. VFO 5·0–5·5MHz. 500kHz segments and crystal normally supplied for 28·5–29MHz operation.

**FT202R:** Hand-held 144MHz fm transceiver with six channels (usually three fitted). 1W output. For nicad or dry batteries (12V). Receiver i.f. 10·7MHz, 455kHz. About 1979.

**FT207R:** Hand-held 144MHz fm transceiver with synthesizer (12·5kHz steps) with keypad and four memories. About 1979.

**FT220:** A vhf transceiver for ssb and fm operation in four 500kHz bands covering 144–146MHz. Power about 10W. I.F. 10·7MHz and 455kHz. For operation from 13·5V dc supplies or ac mains. Introduced about 1974.

**FT221:** Multi-mode solid-state transceiver for 144MHz (ssb/fm/cw/a.m.) with phase-locked-loop frequency synthesizer. Output 12W p.e.p. on ssb; 14W fm/cw; 2·5W a.m. I.F. 10·7MHz. Introduced about 1976.

**FT227RB:** Mobile 144MHz fm transceiver with synthesizer providing up to 800 channels (144–148MHz) with band scanner, memory channels etc.

**FTV250:** Transmit/receive transverter for 144MHz ssb/cw/fm/a.m. providing 10W p.e.p. on ssb and 4W on a.m./fm output from 28–30MHz input.

**FT277:** European (Sommerkamp) version of FT101.

**FT301D:** Solid-state hf transceiver with digital readout. Covers 1·8–28MHz (plus 5MHz standard frequency and 27MHz cb reception). 9MHz i.f. With rf speech processor and tunable i.f. rejection filter. 200W p.e.p. input (ssb/cw), 50W a.m./fsk. About 1978.

**FTdx400:** Transceiver for 3·5–28MHz with provision for additional 500kHz receiver segments. Power 500W p.e.p. for ssb, 440W cw and 125W a.m. PA two 6KD6.

**FTdx401:** Basically as FTdx400 but fitted with noise blanker, fan and cw filter as standard equipment.

**FTdx560:** Basically as FTdx400.

**FT901DM:** Hybrid hf transceiver with digital memory and built-in 8044 microprocessor keyer, rf speech processor and "tune" timer, noise blanker and offset tuning. Covers 1·8–28MHz amateur bands plus standard frequency reception. Companion units include: FTV901R vhf/uhf/Oscar transverter; FV901DM external vfo unit; YR901 cw/rtty electronic reader for use with domestic tv receiver; YO901 multiscope; and FC901 antenna coupler.

## MISCELLANEOUS

Many of the following selection of hf equipment were not designed specifically for the amateur market but have come to be used by radio amateurs.

**Atlas 180:** An all-solid-state compact ssb/cw transceiver for the 1·8–14MHz amateur-bands. 180W p.e.p. with broad-band transmitter stages. I.F. 5,520kHz. Mobile and fixed-station operation.

**Atlas 210/215:** Compact mobile/fixed all-solid-state hf ssb transceiver. Model 210 3·5–28MHz bands. Model 215 1·8–21MHz bands. Power input 200W p.e.p. (120W p.e.p. on 28MHz). Output 80W p.e.p. minimum 1·8–21MHz, 50W p.e.p. on 28MHz. Ladder crystal filter at 5,520kHz.

**B2:** Transmitter-receiver forming wartime "spy" and "special forces" equipment. Superhet receiver covers 3·1–15·5MHz in three bands. I.F. 470kHz. No rf stage.

**B40:** Post-war Admiralty receiver made by Murphy. 650kHz–30MHz in five bands. I.F. 500kHz. Two rf stages. Weight 114lb.

**BC312:** USA services. General-coverage receiver for 1·5–18MHz in six bands designed for 12–14V dc supplies (but often converted by amateurs for mains operation). Two rf stages (6K7), crystal filter, i.f. 470kHz. About 1942–5.

**BC342:** USA services. Similar to BC312 but intended for operation from 110V ac mains.

**BC348:** USA services. General-coverage receiver for 200–500kHz and 1·5–18MHz. Designed for 28V dc supplies. Two rf stages (6K7), i.f. 915kHz. About 1942–5.

**BC453:** USA services as part of "Command" equipment (also designated CBY46129 and R23/ARC5). Compact receiver for 24V dc supplies covering 190–550kHz with i.f. of 85kHz. This equipment has been widely used by amateurs to form an add-on "Q5-er" for receivers having an i.f. of the order of 455kHz. RF amplifier 12SK7. About 1942–5.

**BC779, BC794, BC1004:** These are USA services versions of Hammarlund Super-Pro.

**BC1147A:** USA services. General-coverage receiver for 1·5–30MHz in four bands designed for 110V ac supplies. I.F. 455kHz, two rf stages (6SH7, 6SK7). About 1942–5.

**BRT400:** Series of general-coverage professional receivers made in various forms (including rack-mounting BRT402) by GEC for 20 years from roughly 1947–67. Covers 150kHz–31MHz (less i.f. gap) in six bands. Two rf stages (W81), i.f. 455kHz with variable-selectivity crystal filter.

**Commander:** Double-conversion receiver made for amateur market by Radiovision Ltd about 1948. Covers 1·7–32MHz. I.F. 1,600kHz, 100kHz.

**CR100:** A series of wartime general-coverage receivers made by the Marconi Co Ltd including CR100/2, CR100/4, CR100/5, CR100/7, CR100/8 etc (also designated B28). Frequency range 60kHz–420kHz and 500kHz–30MHz in six bands with logging scale. Two rf stages KTW62 (6K7G), i.f. 465kHz with crystal and audio filters.

**DCR19:** Single-conversion general-coverage receiver with electrical bandspread made by Denco about 1948. Covers 175kHz–36MHz (less i.f. gap). I.F. 1,600kHz (crystal filter). RF stage EF9.

**DR30:** Compact solid-state amateur-bands and broadcast-bands receiver covering 3·5–50MHz (3·5–4·05, 7–7·55, 9·5–10·05, 14–14·55, 21–21·55, 28–28·55, 28·5–29·05, 29–29·5, 29–29·55, 29·5–30·05, 50–50·55MHz) in 550kHz segments. Bandpass i.f. 2,405–2,955kHz, fixed i.f. 455kHz (Collins 2·1kHz ssb filter). Field-effect devices in front-end. Introduced about1966. Made by Davco in USA.

**G207DR:** Amateur-bands double-conversion receiver for 3·5–28MHz bands made by Geloso in Italy. I.F. 4·6MHz and 467kHz (crystal filter). RF amplifier 6CB6. Introduced in mid-fifties.

**G209R:** Amateur-bands double-conversion receiver for 3·5–28MHz bands made by Geloso in Italy. I.F. 4·6MHz and 467kHz (crystal filter). RF amplifier 6BA6. Introduced about 1959.

**MR37:** Amateur-bands double-conversion receiver for 3·5–28MHz bands made by Minimitter in the UK. I.F. 1·5MHz and 465kHz. RF amplifier 6BY7. Introduced about 1957.

**R107:** Wartime British services general-coverage receiver covering 1·2–17·5MHz in three bands and incorporating ac mains power supply. I.F. 465kHz. RF amplifier (EF39) with bandpass tuning between rf stage and mixer (EF39) with four-gang tuning capacitor. Audio filter.

**R210:** General-coverage British services receiver covering 2–16MHz in seven bands (52in film-scale dial) for 24V dc supplies. I.F. 460kHz. RF amplifier CV131. Introduced about 1957.

**R408:** General-coverage all-solid-state double-conversion receiver for 13kHz–28MHz in 14 ranges. Made by Redifon. First i.f. depending on range is 470kHz, 1·5MHz or 4·5MHz. Second i.f. 80kHz (single conversion on lower frequencies).

**R388/URR (R390, R390A, R391):** Series of high-grade receivers for military surveillance produced during 'fifties for USA services and based on Collins 51J series. Typically covers about 500kHz–30·5MHz in 30 1MHz bands with setting accuracy of about 300Hz. R388/URR about 1952 with 6AK5 rf amplifier and basically as Collins 51J3. Subsequent models incorporated mechanical filters, different rf amplifier valves and some with digital mechanical readout.

**R1155:** Wartime RAF general-coverage receiver. The standard models cover 75–200kHz, 200–500kHz, 600–1,500kHz and 3–18MHz in five bands; the R1155N and R1155L however cover 1·5–3MHz instead of 75kHz–200kHz. The i.f. is 560kHz with the bfo on 280kHz. One rf amplifier type KTW61.

**R7020:** General-purpose all-solid-state receiver by C&N Electrical Ltd for 24V dc supplies (including dry batteries). Provides 11 3MHz segments between 600kHz and 32MHz. 2nd i.f. 455kHz.

**RA17:** Professional receiver made by Racal and the first to use the Wadley Loop principle to eliminate hf oscillator drift, providing general coverage to 30MHz in 30 1MHz bands. Original model (with wideband EF180F untuned rf amplifier) introduced in mid-'fifties. Later models from 1958 (RA17L etc) has tuned cascode (ECC189) rf amplifier. I.F.s 40MHz (bandpass ±650kHz), 2–3MHz (tunable) and 100kHz. 23 valves.

**RA71:** General-coverage receiver (1–30MHz) using the Wadley Loop principle and basically similar to RA17 but at lower cost with a view to amateur and general applications.

**RA217:** An all-solid-state receiver by Racal following basically similar lines to the valved RA17 with bipolar transistors (two rf stages). Veeder Root counter dial. Introduced about 1965.

**RA1772:** A high-grade professional receiver by Racal featuring wide dynamic range with all-solid-state techniques and many other features. Original price over £2,000. I.F. 35·4MHz, 1·4MHz. Introduced about 1972–3.

**RC410R:** All-solid-state general-coverage receiver with full-frequency synthesizer providing simulation of continuous tuning. Made for professional applications by GEC (1968) but a number have subsequently been sold as surplus equipment. I.F. 1·6MHz, 100kHz. Two rf amplifiers each comprising cascode-type junction fet stage.

**Uniden 2020:** HF ssb/cw/a.m. transceiver for fixed or mobile operation on 3·5–28MHz bands. Input 200W p.e.p. ssb/cw and 100W cw. 12BY7 driver and two 6146B pa. Power 12V dc and 220V ac.

*The above selection of almost 250 equipments represents only a part of many hundreds of equipments developed for amateur radio operation. While the list includes many of the more popular units, non-inclusion of other equipments is no reflection on their performance but is dictated by space and availability of information.*

CHAPTER 7

# Workshop practice

This chapter is based on an article by T. Kirk, G3OMK, originally published in *Radio Communication*.

Home construction requires many skills, all of which make this facet of amateur radio a very interesting one. Most amateurs deciding to build some piece of equipment start collecting information on circuits, source of components, and other technical data, but often forget to enquire about the manufacturing techniques.

The choice of the right material for the job, the tools and how best to use them, combine to confuse or disappoint many a home constructor, and may permanently deter him from attempting to make even the simplest of equipment.

The first essential is an elementary knowledge of the types of materials available and useful to the amateur.

## Metals

### Aluminium and its alloys

Good electrical conductor. Medium to high cost. Available in sheet, rod, tube and other forms. Usually bends and machines easily, but does not readily solder. With the unknown quality of this metal that most amateurs use, soldering is best left alone. Adhesive bonding is very good, when the adhesive instructions are meticulously followed, but such joints are not good electrically so do not go sticking wires onto the chassis.

It is non-corroding in normal use, but try to avoid the use of brass or copper in direct contact with the metal as these react and encourage corrosion with its attendant troubles of poor or non-existent electrical contact. Cadmium or other plated screws, nuts and washers should be used, particularly where the finished job is intended for long term usage.

If sheets of unknown properties are used, it is advisable to test its bending ability on a small strip (4in by ½in). For reliable bending of sheet up to 0·048in thick, it should be possible to bend the sample strip back on itself, and hammer the fold flat, without breakage.

A metal that looks like aluminium, and may come into the hands of the unwitting, is magnesium alloy. Filings and chippings from this metal are highly inflammable. They can burst into a glaring flame with the heat generated by filing or drilling. Trying to put out such a fire with water only makes matters worse. (Factories where this alloy is machined use a chemical fire extinguisher, one type of which goes under the title of DX powder).

### Copper

Very good electrical conductor. Very expensive. Available in sheet, rod, tube and other forms. Before work of a forming or bending nature is attempted, the metal should be annealed by heating as uniformly as possible to a bright red heat, and air or water cooled. If considerable bending is required, especially on tube or thick sheet, this annealing should be repeated as soon as the metal begins to resist the bending action. In the annealed state copper bends very easily indeed.

Soldering is easy, but adhesive bonding can be troublesome. Non-corroding, but reacts with aluminium and zinc.

### Brass

Good electrical conductor. Very expensive. Available in sheet, rod, tube and other forms. Soldering is easy, but like copper adhesive bonding is difficult.

For work involving bending or forming, the most suitable grade is ductile brass. For panels and non-formed parts, the half-hard and engraving brasses are adequate.

Can be annealed like copper, though some care is necessary as brass is nearing its melting point if heated to bright red.

Non-corroding, but reacts with aluminium and zinc.

### Tin plate

Good electrical conductor. Cheap, and available in sheet form around 0·020in thick. Soldering, adhesive bonding, bending and machining are easy.

Non-corroding under normal conditions. Cut edges should be re-tinned with solder if the full benefits of non-corroding properties are required.

### Steel

An electrical conductor. Cheap, and available in numerous forms, and qualities. Most sheet forms commonly available will bend, solder and machine easily. Unless plated or well painted, corrosion is a problem. For outdoor use galvanizing is perhaps the best form of protection.

* * *

All the above metals work-harden. That is, if repeatedly bent and straightened at the same point they will break. In the case of copper, brass and steel, annealing removes the effects of work hardening. Aluminium and alloys can be annealed, but this is a specialized process, normally outside the range of the home workshop.

Where copper and brass have to come into good electrical contact with aluminium, they can be tinned with solder, along the contacting face of the copper or brass.

If further information on these or other types is required, a useful book is *Metals* published by Product Journals Ltd.

## Plastics

### Laminates

Various base fabrics such as paper, cotton, glass, asbestos and others are bonded together by selected resins. The combination of the resin and the base fabrics produce laminates for many applications. Readily available forms are sheet, rod and tube. They cannot be formed, though certain shapes can be laminated to order, when the quantities etc warrant. Normal machining is possible, particularly if attention is paid to the lay of the base material. Drilled and/or tapped holes should be arranged so that they go through at right angles to, and not in the same plane as the laminations.

Most laminates are water absorbent to some varying degree. Where components made from this material are exposed to wet conditions and are expected to insulate, the glass or nylon fabric-based laminates should be preferred. This also applies for applications where the dielectric properties are important (vhf converters etc). The normal heat generated by valves will not harm these laminates. Where higher temperatures are encountered (150–300 °C), the glass or asbestos fabric should be used.

Costs range from cheap for the paper and cotton bases, to expensive for the nylon and glass bases. (The nylon-based laminates are not readily available in this country to date.)

**Acrylics (Perspex)**
Commonly available in clear or coloured sheet, rod or tube and may be encountered as injection or vacuum mouldings.

May be formed by heating, but not with a flame as this plastic is combustible and gives off toxic fumes if burnt. If placed in a pan of water and simmered, or in the oven—xyl permitting—at around 95 °C (200 °F), the plastic softens and can be formed easily. Stop forming and reheat the moment hardening is felt.

Bonding can be done with ease and success. A suitable glue can be made by mixing acrylic filings or small chippings with trichloroethylene, to the required consistency. This form of glue is also useful for securing coils or waterproofing antennas, traps etc. *Warning*—trichloroethylene fumes are highly toxic. Use only in the open air. Do *NOT* inhale the fumes, nor smoke whilst using this chemical, or the glue made with it. Do not let vapour come into contact with hot copper as this produces deadly phosgene. Never confuse trichloroethylene with the more dangerous trichoroethane.

The polishing of cut edges or scratch marks can be done by sequentially filing, emerying (using first the medium grade and working down to the Crocus paper grade) and finishing off with a mixture of metal polish and Vim, the proportions of which are reduced until the final polish is achieved with metal polish alone. Commercial compounds are available.

When drilling, sawing or filing, ensure that the work is adequately supported, as these plastics are not flexible and can shatter or crack very easily. It is important also to use a slow-speed drill, perhaps with paraffin lubricant, to avoid heating and melting the perspex. Where holes are used in such a manner as to put the plastic under a tensile or pulling load, it is advisable to chamfer or radius the edges of the holes on both sides. Most sheet forms are supplied covered with protective paper and it should be cut, drilled and filed with this left in place.

Do not use in direct contact with valves and lamps.

Cost: medium to expensive.

**Epoxides**
This covers a group of medium to expensively priced heat setting resins, which can be used for bonding, surface coating, laminating or encapsulation. Whatever the application the makers' instructions should be meticulously followed, or failure is certain. Encapsulation can be carried out very easily, at home, by using what is termed the cold-setting or working types of resins. The biggest hazard in home encapsulation is air bubbles (the professionals do their encapsulating under vacuum), but this hazard can be minimized by warming the work and the resin to around 40 °C and providing a generous shrinkage allowance, with a large pouring area, which can be cut off from the dried encapsulation.

Surface coatings can be applied by dipping, spraying or brushing. The resins for this type of work require to be "flexibilized" and in this state they are ideal for protecting beams, traps etc.

Laminating kits are available, and do not really warrant detailed description. The same goes for bonding resins with the exception of three general rules.

1. Thoroughly degrease, for even finger marks impair bonding.
2. Roughen the joint faces with a file or scratch card.
3. No joints under a peeling type of load.

Most of the disappointed users of epoxy adhesives can attribute failure to the non-observance of the above rules.

**General**
All the above mentioned plastics are electrical and thermal insulators. The degree of insulation can be considered as very good for most amateur purposes. PVC, ptfe and other plastics may be encountered, but only those so far commonly available and easily worked have been listed here.

For further information a useful book is *Plastics*, published by Product Journals Ltd.

## Tools required by the radio amateur

Any tool purchased with reliable use in mind should be the best that the xyl and the bank manager can afford. Bazaar type tools, especially those used for measurement, are a last resort.

Most amateurs have a shack, room or some place with a bench and a vice of one type or another. Accepting these, the range and type of useful tools available is virtually limitless, unlike most pockets. Basic tools and a few useful extras will therefore be listed here.

*Soldering irons*—15W instrument
50W electrician's and a
200W heavy duty

*Electrician's pliers, side cutters, watchmaker's shears, long nosed pliers* with and without grooved ends, 4*oz ball pein hammer*, 4*oz or* 8*oz softfaced hammer*. *Twist drills* in high speed steel (not carbon steel) in at least fractional sizes from $\frac{1}{16}$–$\frac{3}{8}$in diameter. *Hand or electric drill*—a drilling stand of the diy drill kits is a valuable addition. The drill chuck capacity should at least match the range of drills on hand.

*Large and small hacksaws* with blades (24 or 32 teeth per inch hand blades preferred)—a recent hacksawing addition is the type known as the "automatic or endless hacksaw." This is a conbination of a pad and jig saw; the saw can be used for sawing large sheets as it has no frame to interfere. *Screwdrivers*—4*in insulated* plus 6*in* or 8*in electrician's*. An instrument maker's set is also very useful.

*Spanners*—*box and open ended types*—in at least the BA sizes.
*Hexagon socket screw wrenches*—at least BA range.
*Files and handles*—6*in second cut. Hand, half-round, round* and *three square*, also the same shapes in *round handle needle files*.
*Surform* or similar type of file or plane.
*Repairman's reamer*.

Next a few extras, which for the serious or struggling constructor can help considerably to reduce the time spent "chassis bashing" and improve the quality of the work.

*Bending bars* with or without radius bending attachment— Fig 1. These will have to be a home-made item, and the things to check when acquiring the steel angle are its straightness and squareness. If either of these are not up to much the corrective actions will require skills which may be beyond those so far acquired. A milling or surface grinding machinist is a good friend to have in these circumstances. Surprisingly enough, old bed frame angle, provided it has not rusted too badly, usually makes very good bending bars. The length and the distance between the clamping bolts of the bars governs two things. First is the maximum width of metal or other material bought. Anyone who has tried cutting a 6ft by 4ft sheet of aluminium will know the difficulties. The answer here is, if buying materials in these sort of quantities, have them guillotined or sawn to a width which will fit in between the bending bar clamp bolts. The length of sheet will not matter. Second is the maximum size of chassis or cabinet which can be produced.

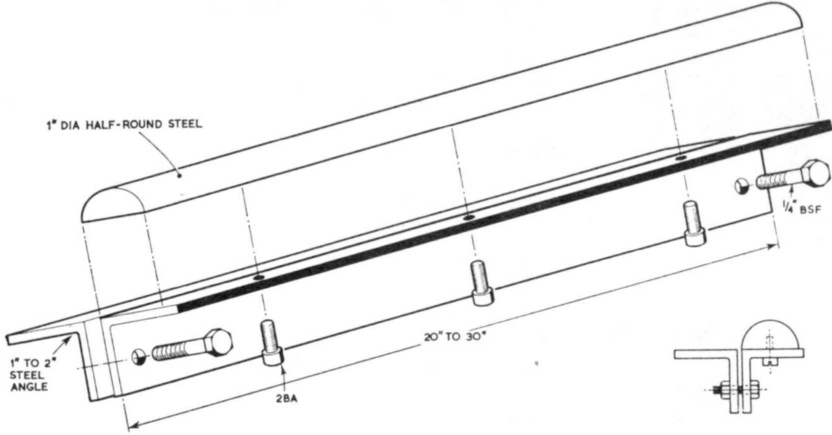


Fig 1. Bending bars with a radius bend attachment

*Tool-maker's clamps.* These are available in several sizes. Two 3in clamps are preferable for most amateur purposes. These clamps should not be confused with the Joiners' "C" or "G" types.

*Chassis screw-up punches or hole saws*—$\frac{1}{2}$–$1\frac{1}{4}$in diameter in $\frac{1}{4}$in steps. These can now be replaced by a stepped drill which is made especially for drilling sheet up to $\frac{1}{4}$in thick. These drills cover a range of sizes—one type will drill holes in $\frac{1}{16}$in steps from $\frac{1}{4}$–$1\frac{1}{4}$in diameter on one drill. They are expensive but good value when compared with cost of the same range of punches or saws.

Last, the measuring and marking out tools. It is very difficult to classify tools in order of usefulness, but these will, or should, be the most used.

*Engineer's combination square*

Various types and sizes are available. The type with the 12in rule and square head only is preferred from the usefulness to cost viewpoint. Ensure that the combination square purchased is *not* of the pattern makers' type, otherwise measurements will become very confusing indeed. To check, place a known standard 1in graduated rule alongside the combination square rule 1in graduations. If they match over the full 12in all is well. Most combination squares are supplied with a scriber, which fits into a hole on the square head, so it really is four tools in one. (Fig 2).

*4in or 6in spring dividers.*

Combination square used for squaring-up prior to bending or cutting

*4in or 6in Jenny calipers*—the type with the positive register shoulder is preferred.

*Automatic centre punch.*

This is a formidable list of tools, and if buying from scratch will be very costly. A retiring fitter or toolmaker often sells off his tool kit, and many good but used tools can be obtained at a reasonable price. Some tool shops run a "tool club" primarily intended for the young impecunious apprentices. Whichever way they are acquired, these tools are a must, if pleasurable and successful "chassis bashing" is the aim.

## How to use these tools

Hole drilling is straightforward, but often on thin sheet the holes produced are anything but round. In these circumstances drill the hole undersize and ream. This will produce a round hole, and provided the reamer has been allowed to cut without being forced, the edges of the hole will be burr free. The use of a large drill for hole deburring is not recommended on the softer metals, unless the "touch" for this method has been acquired. A countersunk or rose-bit is more suitable as they have more cutting edges than a drill and do not "dig in."

*Always* centre-pop the hole position before drilling.

Filing is an art which can only be acquired by practice. General rules are:

(i) Always use a handle with the file. This eliminates the possibility of running a file tang into the wrist, and enables the file to be properly guided.

(ii) Use a sharp file. (Use new files first on only the softer metals such as brass and aluminium and as their sharpness wears off use them for filing steel and harder metals).

(iii) Do not force the file to cut. Only a light and relaxed pressure is required which also aids the accuracy of filing.

(iv) Keep the file clean by brushing with a file card or by rubbing a piece of soft brass or copper along the teeth grooves.

When bending or forming sheet metal never use the hammer directly on the metal. Either use a soft faced hammer or a block of wood to act as a buffer for the hammer blows. This will stop all those humps and hollows along the bends.

When three sides are formed by bending, the point of intersection of the bend lines should be drilled, before bending, with a hole diameter of between two and four times the thickness of the metal. This prevents corner bulge. (Fig 2).

Tube bending for beams, tuned lines etc need not be difficult, and flattening or kinking can be avoided by observing the following:

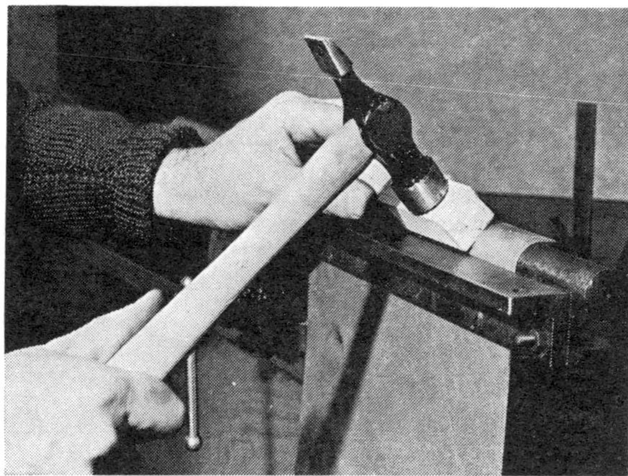

Buffer for hammering the bends

(i) Ensure the tube is suitable for bending. In the case of copper and brass this means annealing as previously mentioned. For aluminium/alloy carry out a sample bend on an unwanted piece of the tube.
(ii) Unless skilled or equipped with specialized tooling, do not attempt to bend to a radius of less than three times the tube outside diameter, eg ½in od tube—1½in minimum bend radius.
(iii) Always bend round a former shaped to the required radius.
(iv) Pack the tube with fine sand and cork both ends. This will minimize the risk of kinking during bending. The sand can be washed out afterwards.

Cutting long strips of metal with tin snips or shears is an expert's job. The cut edges usually produced by non-experts are anything but straight and they require flattening to remove the cutting curl. Tin snips are best used where a one-snip cut will remove the required amount of metal, such as 45° corners, or the trimming to length of narrow strips. Laminates and plastics

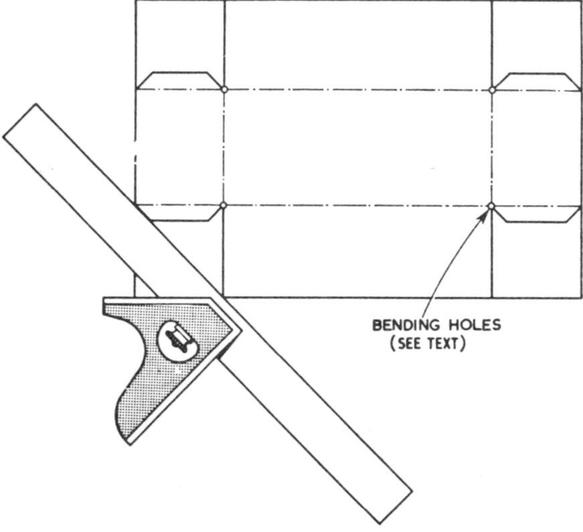


Fig 2. Marking out a box on sheet metal using a combination square

Bending bars used as a guide for sawing

should always be sawn. The use of bending bars as a guide for cutting and filing is shown in the photographs.

Large cut-outs for dials etc can be made in at least two ways, both simple and well within the range of the home workshop technician.

(i) Drill contiguous holes of around ¼in diameter on the waste side of the hole and about $\frac{1}{16}$–⅛in away from the finished size markings. Knock out the waste and file to size, using the bending bars as a guide for the straight portions. This method can be used for the large round meter holes, but the filing of round holes though not difficult requires patience, and a half-round file.
(ii) Drill holes at the corners and using a saw file (Abrafile) saw out the waste, leaving about $\frac{1}{32}$in all round for trimming by filing. Again the bending bars can be used as a guide, but care should be taken to ensure that the saw file does not cut them.

Large round holes can also be made using a tank or washer cutter. The biggest snag with this method is obtaining an even cut around the full circle. With hand drilling this is almost impossible, but by clamping the work onto a block of wood and drilling through into the wood with the centre pilot drill of the cutter, a guide is provided which improves things a little. Machine drilling requires the slowest speed possible (no more than 500rpm).

The de-burring of large holes can be awkward, particularly on existing panels where the finish needs to be left as intact as possible. The small half-round needle file enables this to be done easily. The file is used in the "draw" fashion. That is, both ends of

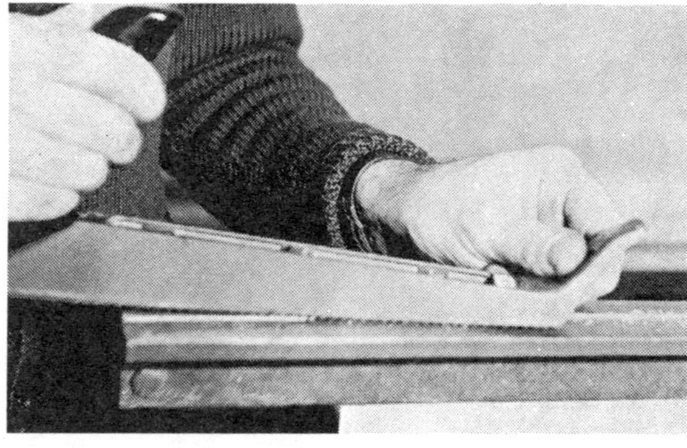

Bending bars used as a guide for filing. The Surform type file is ideal for most materials, as it will cut these and not the bending bars

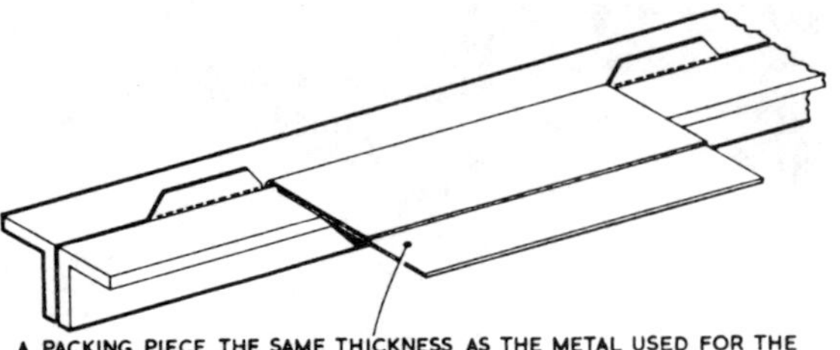


Fig 3.

the file are held and the file is drawn round the edges to be deburred, in a similar manner to a spokeshave, so producing a small chamfer.

Before any drilling, filing or bending can be attempted, the work must be marked out. The accuracy with which this is done governs the future results, so great care is necessary and worthwhile. Measurements and squareness checks should be made from one or two datum edges. Consider marking out the sheet for the box shown in Fig 2. The first operation is to obtain two edges straight and at right angles to each other. The method of doing this should be clear from the illustrations. These two edges are now the datum edges for vertical and horizontal measurements. The overall size of the sheet is marked out and cut to size. This results in a sheet with all corners square and all edges straight. In this case the box sides are to be the same depth; therefore adjust the Jenny calipers to the required measurement and locating from each edge in turn scribe round the sheet. Check with the combination square, from the datum edges, that these lines are parallel to their respective edges. If no Jenny calipers are available, the combination square can be used by setting the rule to the required measurement and scribing round the sheet with a scriber held against the end of the rule. If the box is now bent to these lines the sides will all be the same depth, and if the squareness of the sheet was originally right the bottom or width of the box will be true, with sides parallel and ends square. The illustrations for the remainder of the box marking out and bending are self-explanatory.

Existing chassis or panels should be similarly marked out using the datum edges. If a truly square corner is not available, then use only the longest and straightest side as a datum, and the end of the combination square rule to mark out the right angle lines. A line at right angles to this single datum edge can be used to mark off the measurements in this plane.

Hole centres should be scribed in and it is worthwhile to scribe each hole diameter around its centre. This helps in ensuring that the right size hole is drilled, and in checking that holes do not foul. When screw-up chassis punches are used these scribed diameters can be used to accurately locate the punch.

When a gleaming polished chassis is purchased, marking out directly on to it will spoil the finish, so obtain a piece of draughting film and mark out on this in pencil. Using draughting tape, stick the film on to the chassis and centre pop through the hole positions and around the outline of any cut outs. After removing the film scribe in the hole diameters and join up the cut out outline dots.

Fixing holes for meters, IFTs etc should be carefully measured and these measurements transferred to the work. In these cases, where holes or cut-outs are related to each other and not to some other edge of the chassis or panel, vertical and horizontal datums for these holes alone should be used. In the event of repeat sets of holes occurring it is often worthwhile to make up a metal drilling template on to which is clearly marked the location datum lines (eg the centre of the slug adjusting hole for IFTs, or the relay clamp screw for GPO type relays). These lines can then be matched up with those marked out on the chassis and the

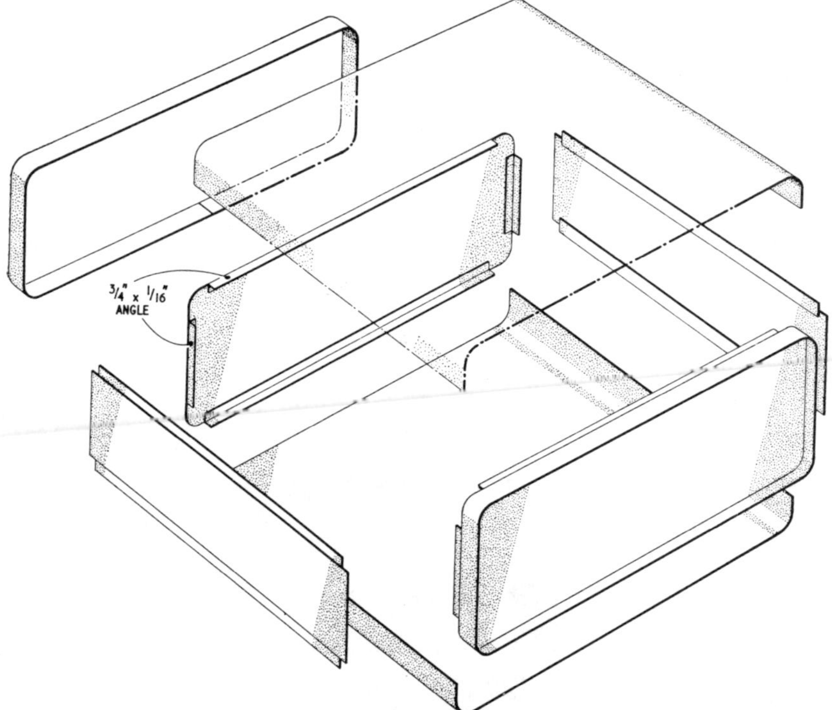


Fig 4. Cabinet design suitable for home construction

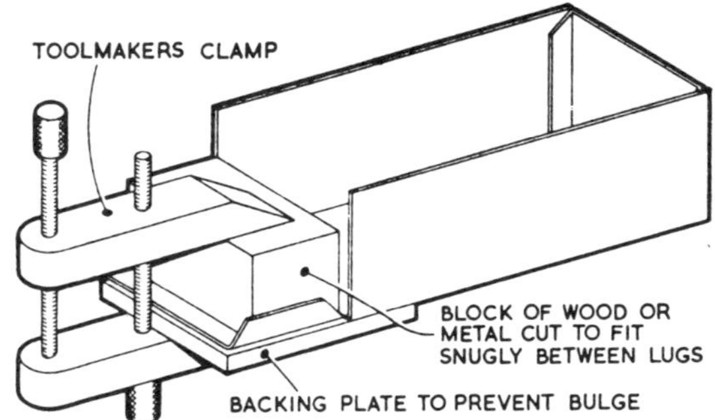


Fig 5. Using a toolmaker's clamp to position the bending block when forming the last side of a box

template clamped in place ready for drilling. Clearly mark the viewed side of the template. This will minimize the risk of using it the wrong way round. The centres of holes of the same diameter can be found by simply measuring from the inside edge of one hole to the outside edge of the other.

The adage "cut once measure twice" when applied saves a lot of wasted work.

The cabinet construction shown in Fig 4 is very simple to make, and looks very professional. The front and rear panels are made as a pair, and similarly for the sides. It is advisable to screw these panels to the angle pieces, which allows easy removal should a change of layout be required. Cadmium-plated self-tapping screws of the Pan-head type can be used with very satisfactory results. The bending bars with radius attachment are essential for the making of the top and bottom covers, which will usually be made from perforated metal.

The edging strips are best made from $\frac{3}{32}$in or $\frac{1}{8}$in thick metal, and the radius corners also require the radius attachment of the bending bars. The joint of each edging strip can be made simply by overlapping, marking off and sawing to give a butt joint. If this is too difficult the edging strips can be left off along the bottom edges.

Chassis panels are mounted on angle fastened to the front, back or side panels of the cabinet. All the angle pieces can be extruded section, or made from strip bent to suit. From the appearance viewpoint with this form of cabinet the front profile requires that the height be about half the width (6in high—12in or 13in wide). The top or bottom profile looks good in perforated metal with a depth to width ratio of one to one. One other feature is that of matching appearance. "Mother " receiver and "Father" transmitter look radiant when surrounded with obviously their "family" of swr meter, keyer, psu, etc!

## Finishing

Having built the equipment the urge to use it may be so great that there is no time to finish it off. This is a shame, because the finish affects the appearance, which in turn adds or subtracts from the pleasure of home-made equipment. The time spent on construction may be considerable but that spent on finishing is undoubtedly the most rewarding.

Painting is probably the most common form of finish used. The combination of two or more colours can enhance the equipment. The choice is limitless and most amateurs try to adopt some standard colour scheme. Brush painting very rarely gives a good surface and a considerable amount of elbow grease is required to burnish this form of painting into anything like a smooth finish. Brush applied hammer or crinkle finish paints are pleasing, but this type of finish is unsuitable for the larger front panels of, say, receivers or transmitters where a monotone finish looks far superior.

Spray painting is now within the realms of the home constructor, particularly with the aerosol paints. These are expensive in themselves but the cost of paint per square inch is reasonable. A few practice runs on an unwanted piece of metal enable the technique of spray painting to be quickly acquired. The spraying of corners sometimes gives trouble, and the paint may splodge. The technique here is to spray each corner in turn, masking off the other faces, which may or may not be freshly painted, by holding a piece of cardboard in the path of the unwanted spray.

For the painting of panels another useful and easily applied method is spin-coating. The panel is spun at about 50 to 100 rpm and the paint is poured onto the centre. This immediately flows outwards and a very thin and even film of paint is deposited on the panel. Paint flies off so ensure that this operation is carried out somewhere that cannot be spoilt by blobs of paint.

The face plate of the diy drill kits can be held in the chuck of a hand drill, which in turn is held upright in a vice, and the panel can then be stuck to this plate with double-sided adhesive tape, to provide one method of obtaining the spin. Another method is to use an old gramophone turntable and motor. The top speed of these is about right for good coating.

When cellulose lacquers are applied by this method the coating is usually dry by the time the spinning has stopped. This spin method of coating is also very good for the application of photo resists, when making printed circuits.

Before any painting can be done, the surfaces should be suitably prepared. The resultant ease of obtaining a smooth finish depends on the quality of the unpainted surface and the methods of preparation.

The first essential is to ensure the surface is burr free and smooth. Any scratch marks should be smoothed out or after priming filled in with one of the proprietary fillers, and this should be thoroughly blended in to leave a smooth flat surface. It is essential to degrease thoroughly the surface prior to painting. A very effective surface preparation is to wash and scrub the panels etc with hot water, cleaning powder and wire wool. This results in a very smooth surface which is also grease free.

Surfaces can also be prepared and sometimes finished by etching. As most of the substances used for this technique are dangerous to handle this method is not recommended, particularly in the home workshop over-run with young children. For copper, brass, steel and aluminium, a solution of ferric chloride will etch to produce a reasonable surface. The previously mentioned surface degreasing and scrubbing should be carried out prior to etching. The juice from stewed rhubarb cleans aluminium very well.

The next step in painting of metals is priming. For steels use the oxide primers to suit the type of paint that will be used for the finish. For aluminium alloys the etch type primers should be used, particularly on parts subject to flexing. It is a pity to paint brass or copper, but if unavoidable the priming can be done using the undercoat paints: do not use any of the zinc- or aluminium-based primers.

After priming and when the work has thoroughly dried, a light rub down with very fine wet and dry paper or pumice powder and water will smooth off ready for the under or finishing coats.

With modern paints the use of undercoats is not really necessary, as their function is primarily to ensure the colour coverage and most top or finishing paints will cover and colour anyway.

There are so many types of paint to choose from that a little personal experimenting is advisable. Most paint manufacturers will supply literature on their products on request. A few types are listed below.

*Acrylic and alkyd paints*—good gloss; flexible; durable;

suitable for metal protection and decoration; reasonably priced.

*Bituminous paints* —useful and cheap; suitable for protection of masts and other outdoor metalwork; usually black semi-gloss finish.

*Cellulose lacquers*—reasonably priced; quick drying with high gloss which can be improved by polishing; available in aerosol cans. Preferred because of its general ease of use and quality of finish. The fast drying properties considerably reduce the risk of dust marred work.

*Polyurethane paints*—high-gloss hard abrasion-resistant finish; reasonably priced; sometimes difficult to apply.

*Stoving paints*—usually of the acrylic or alkyd types which rely on stoving to complete the chemical changes of drying. The resultant finishes are hard, durable and can be very glossy.

The number of coats of paint and method of application is entirely a matter of choice, but all painting should be carried out in an atmosphere as warm, dry and dust free as possible. If the painted work can be popped in an oven at around 80 °C (180 °F) drying will be greatly assisted.

Decorative laminates and thin self-adhesive plastic sheets (ie Fablon, Contact etc) can be used to good effect for the front panels of receivers and transmitters. They should all be backed with a metal panel to give adequate support and good rf screening or electrical contact. The laminates can be glued or screwed to the metal backing panel. On some designs it is possible to utilize the dial, potentiometer or switch fasteners to secure the laminate facing. If subsequent modifications are likely this last method of fixing is probably the best.

The self-adhesive plastic sheets are relatively cheap and easy to use. Their one drawback is that of shrinkage with age. This can be minimized by heating the panel to which the plastic is to be applied to a temperature of around 60°C (just too hot to hold with comfort) and then in this condition apply the plastic sheet. A generous overlap should be left, $\frac{1}{2}$–$\frac{1}{4}$in all round, and the now-faced panel should be clamped between two sheets of blockboard or similar and allowed to cool before trimming the facing to size. The trimming is easily done using a sharp modelling knife and experience suggests that the folding over of the plastic sheet, in an attempt to cover the panel edges, is inadvisable.

Self-adhesive coloured and chrome tapes can be used to line-in features of the panel or cabinet. There is a strong tendency, due to the ease of application, to overdo this and finish up with something that looks like a cross between a juke box and a car radiator grill.

Lettering panel work used to be a problem. Machine engraving is undoubtedly the best way of doing this, but at today's prices it is prohibitive. The stamping or the hand-held vibro-engraving techniques usually leave much to be desired. Engraving by hand is fascinating to watch but the men who can do this today are very few.

The wet slide-on transfers are effective but are fast being replaced by the self-adhesive, rub-on type of lettering, such as Letraset, Letterpress and others. These are so easy to use and completely effective that the initial expense of buying the lettering sheets is not worth considering. The type faces available are extensive and it is advisable to choose one style and stick to this. Variations of signal importance can be made by using upper and lower case and different point size of the same style characters. Most of the type styles are available in black or white and this should cover most requirements.

Unfortunately these letters can be easily rubbed off, therefore a protective covering of some kind is required. The completely lettered panel can be coated with a clear lacquer. This lacquer may react with the surface of application, so a little test beforehand will save a lot of trouble.

Another method is to cover the panel with a self-adhesive plastic sheet. This has the same drawback as previously mentioned, shrinkage, and one other: being self-adhesive it must be applied right first time as any attempt to peel off and start again will remove all the lettering.

The easiest protection is obtained by facing-up with a thin ($\frac{1}{32}$–$\frac{1}{16}$in thick) sheet of transparent perspex, held to the front panel by screws or the existing component fasteners.

Etch engraving is very effective but requires techniques which are not within the scope of this book. Those familiar with the production of printed circuits can utilize the experience gained from this for the making of etch-engraved panels.

Self-adhesive labels made on the hand-operated lettering machines are a very useful way of marking controls etc but do not add much to the appearance of the equipment.

Dial calibrating is often a problem and can be tackled in many ways depending on the requirements and the dial used. Probably the best dial available today is the Eddystone 898. It is so expensive that good clear calibration is a must, and many a trembling hand has drawn back from attempting to mark the dial.

All preliminary marking should be made lightly in B or HB pencil, which for line work should be sharpened to a chisel point. The squareness of the lines can be obtained by using the combination square as a guide. When all is satisfactory inking-in can proceed. The now pencil-marked surface should be thoroughly cleaned using "Pounce" or french chalk applied lightly with a soft duster. This will not rub off the pencil marks, but will allow splodge free inking. Using a draughtman's lining pen, set to the correct width of line, and the combination square rule as a guide, ink in any horizontal lines. Allow this to dry thoroughly. Next proceed to ink in all vertical lines working from one side and doing every line as it occurs whether on different levels or not. To minimize the risk of line thickening at the finishing ends, work such that the line and the pen finish on one of the horizontal lines. If a mistake is made, dab the area dry with blotting paper and wipe off with a wet cloth; any slight smudges can be removed later with a rubber.

The ink for lining in should be good quality Indian suitable for paper. Colour inks can be used but greater care is necessary as these inks tend to run very easily. All figures and letters should be of the rub-on type, and applied as the manufacturer recommends. There is no need to protect these letters in this instance, as there is little likelihood of abrasion once the dial is in place.

Some dials are made of plastic and these are calibrated in the same manner, but using an ink of the acid etching type. This ink etches into the plastic and any errors should be wiped off immediately.

The lining pen should be filled very sparingly and a column of ink about $\frac{1}{8}$in long from the tip is about the maximum amount, if blots are to be avoided. The pen should be wiped clean before each refill as any whiskers or dried ink will cause uneven line thickness or splodging.

Dials for gdo, field strength meters etc can be made by sticking white smooth-surface paper on to a metal or plastic supporting plate (tobacco-tin lids are ideal). Calibrate and letter as required, spray with a clear lacquer and a lasting job is produced.

Other finishes, such as chrome plating, colouring, anodizing etc, which have to be carried out by an appropriate firm are well worthwhile but a few words of advice may be helpful.

Most finishing firms do not pre-polish or dress the work prior to finishing. With plating, the article will be plated as received, scratch marks and all. This is very disconcerting so ensure that all work for plating is highly polished and scratch free. A hard felt pad on the electric drill together with buffing compound cuts out the hard work. Fingers get a bit dirty but the work shines nicely.

Brass, copper and steel all polish easily. Aluminium is slightly more difficult as the metal tends to pick up on the polishing head and produce deep scratch marks. The risk of this happening is reduced by using oil with the buffing compound and ensuring that the metal being polished and the polishing wheel are kept cool. Aluminium is usually anodized and coloured, which does not necessitate a highly polished surface. To improve the finish on this material the contractor can be asked to carry out some form of etch treatment to improve the surface finish prior to anodizing or colouring.

All these finishes are expensive, but do provide corrosion protection as well as enhancing appearance.

## Design thoughts

Initial planning of the home-constructed project involves time, observation and experience. The more of each that can be devoted at this stage, the better results are likely to be.

Truly experimental work will not by its very nature require an attractive cabinet, and versatility of chassis or panel work are probably the most important hardware requirements. Modular type of construction is rapidly gaining favour and with it the chassis becomes nothing more than strip supports for boxes and PCBS. The cabinet illustrated in Fig 4 is an ideal box for much of this type of work. When the experimenting has proved satisfactory or when a proven circuit is being made, and a permanent job is the aim, some thought should be devoted to the outward appearance of the finished project.

A considerable amount of time has been and is being devoted to finding out how to think a design. It should be possible to glean a few ideas from this effort and utilize it for the amateur.

One piece of terminology is "value engineering" which aims at seeking lowest cost paths. Similar terms have been devised to describe systems and methods by which the design processes can be rationalized. All these systems and methods can be related to the use of common sense combined with personal and other people's experience.

Most amateurs employ some degree of value engineering or analysis: a field strength meter with an Eddystone 898 dial is very nice, but is a waste of money and a good dial. Alternatively a KWM2 with a crank-up and tilt-over tower supporting a TH6, which is fed by 5p a yard type of coaxial cable, is equally stupid. With today's high cost of components the value side of designing is very important even for one-off amateur jobs.

"Design optimization" is another of these expressions. Is it possible, at the design stage, to ensure that the design chosen is the right one? Are better ways available? If this valve base goes with pin 1 to the south what will happen to the resistor coming in from the ht line? Obviously the experienced constructor works out much of this sort of thing subconsciously.

For the novice this is the point where observation can be applied to great advantage. A notebook can be used to jot down the way so-and-so does this or that, and why, if this is possible. Eventually a store of useful knowledge is collected and combined experience can be utilized. It is not suggested that this should be the sole duty, but in the same way that knowledge is gained on the radio and operating sides so it should with design, and the idea is to gain as much experience as possible in the shortest time.

At the design stage a list of "fors" and "againsts" could be made of various layouts. Eventually the finalized layout will in all probability look nothing like the first one, but will be a combination of the best points of each. Anyone who has tried to make a pcb from a schematic circuit diagram of something involving six or more transistors and their associated bits and pieces will know this layout design technique.

The chance to study commercial equipment should never be passed by. It is amazing what the professionals seem to get away with, and the thinking out of how it is done adds greatly to the funds of knowledge.

Last is the artistic side of design. The professional types responsible for this are usually titled *industrial designers*, and they like to be in on the design at any stage but the last. One of their functions of design is to ensure that it has functional as well as aesthetic appeal.

The functional side can be evaluated very easily for home-made gear. Consider a receiver design and list, in order of preference, what functions (knobs, switches, dials etc) are required to appear on the front panels. The list may finish up like this:

| | | |
|---|---|---|
| (i) Dial | (iv) S-meter | (vii) AF gain |
| (ii) Tuning | (v) RF gain | (viii) Selectivity control(s) |
| (iii) Bandset | (vi) Phones socket | (ix) AGC on off |

and so on.

This forms the basis for the panel layout. In this instance if the design has fulfilled its intent, when looking at the receiver the first thing noticed should be the dial, the next thing the tuning, and so on. A rough sketch, to size if possible, can be made of this before work commences, and tried on a few friends (the xyl is usually a biased judge of these matters) to see if the desired effect has been achieved.

Most commercial firms producing amateur equipment employ such techniques and their "styling" can be scrutinized to decide what each considered important.

Many insignificant features of a front panel can be brought into prominence, if required, by the use of colour or by the size of lettering used to indicate its function. The techniques of doing this are very difficult to apply and great care has to be exercised if the "juke box" effect is to be avoided. For a trial run of this technique, limit the colour or tone changes to a maximum of three, and leave the letter size changes out altogether. Areas of functions can be outlined in chrome or coloured self-adhesive strip.

The dial is usually the significant feature and the use of multi-colours on this should be treated with very great care indeed, or avoided completely.

Wherever possible stick to one style of knob and switch. The functional importance of these can be brought out by using various sizes of the same general pattern. Communication equipment ought to have business-like controls, and the use of the domestic radio-type knobs just does not look right.

Electronic equipment styles change very much like dress fashions. This leaves plenty of scope for the amateur to try some "way out" designs, the results of which would be interesting to see. Maybe, like dress fashions, the old polished mahogany box with lift-up lid and a $\frac{3}{8}$in thick ebonite panel concealing a wonderful breadboard layout, will return. Perhaps someone will then write on joinery and wood polishing for the radio amateur!

CHAPTER 8

# The licence examinations

Unlike radio and television receiving licences, amateur transmitting licences are not issued just on payment of the appropriate fee, but are granted only to those applicants who have proved their competence to control and operate a transmitter, and an awareness of their responsibilities to the users of other radio services. This is a reasonable precaution when one considers the havoc that even a low-power transmitter could cause to radio and television reception, to world radio communication, or to important navigational services.

The danger of unskilled use of transmitters has been emphasized deliberately. Too often one hears expressed the erroneous belief that the conditions imposed by the authorities have been devised to discourage the experimenter. Such is very far from the truth: the newcomer can be confident that his desire to obtain a licence will meet with courtesy, assistance and every encouragement from the authorities—provided that he does not expect "special concessions" in his particular case.

Detailed information on how to become a radio amateur can be obtained, free of charge, from the Home Office, Radio Regulatory Division, Licensing Branch (Amateur), Waterloo Bridge House, Waterloo Road, London SE1 8UA. For the benefit of those who have not yet obtained these details here are a few notes on the more important points.

(1) Applicants must be over 14 years of age and provide evidence of British nationality (a birth certificate, a valid passport, or a nationalization certificate is sufficient). Applications from persons under 18 years of age have to be countersigned by parent or guardian.

(2) Applicants must show that they have passed the Radio Amateur's Examination, a 3h written examination which covers the elementary theory of radio communication, knowledge of transmitting techniques, amateur operating procedure and licensing regulations.

(3) Applicants must have passed the Post Office morse test not more than 12 months before applying for a licence. This test involves the receiving and sending of plain language text at an average speed of 12 words per minute (wpm). This does not apply to the Amateur Licence B.

(4) Before an Amateur Licence A or B is issued the fee must be paid, and a renewal fee becomes due each year on the anniversary of the date of issue of the licence.

Since the morse test must be passed within the year before applying for the licence, it is usual to obtain a pass in the Radio Amateurs' Examination (RAE) before attempting the morse test; otherwise there is always the possibility that the time limit will have expired before the written examination is passed, making it necessary to retake the code test.

Exemptions on the grounds of Service or civil qualifications are not granted from either of these examinations, and even if the applicant has previously held an amateur licence he must show that he has passed the RAE (which was first held in 1946) and must retake the code test unless he has passed this within the last twelve months.

## The Radio Amateurs' Examination

The City and Guilds of London Institute, Electrical and Telecommunications Branch, 76 Portland Place, London W1N 4AA, holds the Radio Amateurs' Examination No 765 twice yearly in spring and autumn (usually May and December). This can be taken, by arrangement with the local education authority, at centres throughout the country. Full details of the examination and sample questions may be obtained from the institute at the above address for a small charge. Applications to sit the examination usually have to be made at least three months in advance, so do not leave it too late.

The syllabus and examination format of the Radio Amateurs' Examination changed in May 1979. The current examination consists of two separate papers: *Licensing Conditions and Interference* (1 hour) and *Operating Practices, Procedures and Theory* ($1\frac{3}{4}$ hours) with a break of 15min between papers. The first paper contains 35 multiple-choice questions, the second 60 multiple-choice questions—all questions must be attempted. A candidate must take both papers at the first entry, but if he or she fails one paper then only that paper need be retaken.

A multiple-choice question consists of a sentence in which a question is either asked or implied, followed by four possible answers, only one of which is correct. The candidate is required to select the correct answer. Here is an example:

Poor frequency stability of an amateur transmitter can result in:
(a) The generation of parasitic oscillations.
(b) Operation outside the amateur bands.
(c) A reduction in power output.
(d) Difficult adjustment of the power output stage.

The syllabus of the examination is shown opposite. Further details of the examination are given in *How to Become a Radio Amateur*, and are also obtainable from the City and Guilds of London Institute.

## A summary of licence conditions

Full details of the conditions of the Amateur Licences are supplied by the Home Office with the leaflet *How to Become a Radio Amateur*. The following is an informal summary of those points which are most likely to be needed for the licensing condition questions of the Radio Amateurs' Examination, arranged according to their subject matter.

**Location:** An amateur station may be used at the main address shown in the licence or any temporary premises (with callsign suffix /A) or any temporary location (suffix /P) for separate periods, none of which exceed four consecutive weeks, or in any alternative premises provided that notice in writing has been given at least seven days in advance to the Post Office Telephone Manager of the area concerned. The Telephone Manager must be notified when the station is no longer at the alternative premises. In the Channel Islands, the Director of the Telecommunications Board of the appropriate Bailiwick has to be notified in advance and at the conclusion of the use of alternative premises.

Stations can be in any vehicle or vessel (suffix /M) but not on the sea or within any estuary, dock or harbour or in any vessel or aircraft or any public transport vehicle. A station may be operated as a pedestrian mobile station (suffix /P).

When not at the permanent address, particulars of the address or location in use must be sent at the beginning and end of each contact or at intervals of 15min, whichever is the more frequent. The station, licence and log must be available for inspection at all reasonable times by duly authorized officers of the ministry. The station must close down at any time

## SYLLABUS OF THE RADIO AMATEURS' EXAMINATION

(up to and including the December 1981 examination)

### LICENSING CONDITIONS AND INTERFERENCE

**1. Licensing conditions:** Types of licence available and qualifications required of holders. Conditions (terms, provisions and limitations) laid down by the Home Office in the Amateur Licence A, including the notes appended and the schedules of classes of emission and frequency bands.

**2. Transmitter interference:** Frequency stability; consequences of poor frequency stability: risks of interference, out-of-band radiation, difficulties in communication. Spurious emissions, causes and methods of prevention; harmonics of the radiated frequency, direct radiation from frequency determining and frequency changing stages of a transmitter, parasitic oscillations, key-clicks, excessive sidebands due to over-modulation. Mains-borne interference; causes and methods of suppression. Audio-bandwidth limitation: limitation and methods. Home Office requirements for frequency checking equipment: Appendix F to *How to become a Radio Amateur.*

### OPERATING PRACTICES, PROCEDURES AND THEORY

**1. Operating practices and procedures:** Calling procedures in telegraphy and telephony: general calls to all stations and calls to specific stations. Log-keeping: Clause 6 of the Amateur Licence A. Use of satellites and repeaters; accessing a repeater. Use of Q-codes and other abbreviations appropriate to the Amateur Service. The phonetic alphabet: reasons for its use; recommendations in *How to Become a Radio Amateur.* Safety in the amateur station; recommendations of the Radio Society of Great Britain. Discharging of capacitors and mains disconnection.

**2. Electrical theory:** Basic electrical terms, their meaning and use: emf, current, conductor, resistance, insulator, power, series circuit, parallel circuit. SI units, their use and relationship to each other: volt, coulomb, ampere, ohm, watt, hertz. Current, power and resistance; Ohm's Law. Total current and combined resistance in series and parallel circuits. Resistors, types and applications; resistors in electronic circuits. Power in a dc circuit. Magnetic and heating effects of currents: applications. Inductance and capacitance; appropriate units; effects in ac circuits. Total inductance and capacitance in circuits. Meaning of inductive and capacitive reactance. Factors affecting capacitance value. Sine wave. Definition of terms: amplitude, period and frequency; instantaneous, peak, peak to peak, rms and average values. Power, reactance, impedance and resonance in ac circuits; simple explanation of terms: phase angle, phase difference, phase lead and lag, reactance, impedance, series resonance, parallel resonance, resonant frequency and Q (magnification) factor. Transformers: function and operation. Tuned circuits: series and parallel ac circuits, resonant frequency data and calculations; voltage amplification and current amplification effects. Maintenance of oscillations in tuned circuits.

**3. Semiconductors:** Characteristics and principles of operation of npn and pnp semiconductor devices; principles of diode rectification; control of output current and voltage when transistors are used as audio frequency and radio frequency amplifiers. Use of semiconductor devices in radio equipment as (a) oscillators (crystal and variable frequency types); (b) amplifiers (audio frequency and radio frequency types); (c) frequency changers; (d) frequency multipliers; (e) demodulators; and (f) signal detectors. Typical power supply circuits; power rectification; smoothing and voltage stabilization systems.

**4. Radio receivers:** Principles of reception of continuous wave, double sideband and single sideband and frequency modulated signals in terms of radio frequency amplification, frequency changing (where appropriate), demodulation or detection and audio amplification. The superheterodyne principle of reception. Advantages and disadvantages of high and low intermediate frequencies; adjacent channel and image of frequency interference and its control. Frequency modulation and demodulation. Typical receivers; use of a beat frequency oscillator. Characteristics of a single sideband signal and the purpose of a carrier re-insertion oscillator.

**5. Transmitters:** Oscillators used in transmitters; stable variable frequency and crystal controlled oscillators; their construction and factors affecting their stability. Transmitter stages: operation of frequency changers, frequency multipliers, high and low power amplifiers and power output amplifiers (including linear types). Methods of keying transmitters for telegraphy; advantages and disadvantages. Methods of modulation and types of emission in current use including single sideband and frequency modulation; emissions in the A2, A3, A3J, F2 and F3 modes; relative advantages.

**6. Propagation and aerials:** Basic terms, their explanation: ionosphere, troposphere, atmosphere, field strength, polarization, maximum usable frequency, critical frequency, skip distance. Production of electromagnetic waves; relationship between electric and magnetic components. Structure of the ionosphere. Refracting and reflecting properties of the ionosphere and troposphere. Effect of sunspot cycle, winter and summer seasons and day and night on the ionization of the upper atmosphere; effect of varying degrees of ionization on the propagation of electromagnetic waves. Ground wave, ionospheric and tropospheric propagation. Fade-out and types of fading: selective, interference, polarization, absorption and skip. Velocity of radio waves in free space; relationship between velocity of propagation, frequency and wavelength; calculation of frequency and wavelength. Receiving and transmitting aerials; operation and construction of typical aerials including multiband and directional types; their directional properties. Coupling and matching. Transmission lines; balanced and unbalanced feeders; principles of propagation of radio waves along transmission lines.

**7. Measurement:** Purpose, operation and use of absorption wavemeters, heterodyne wavemeters and frequency counters; relative accuracies. Dummy loads, their construction and use in tuning transmitters. Use of standing-wave-ratio meters. Measurement of (a) dc power input to the final amplifier of a transmitter; (b) radio frequency power output of linear power amplifiers; and (c) current at radio frequencies. (Reference to *How to become a Radio Amateur.*) Setting up and use of a cathode ray oscilloscope to examine and measure waveform and to monitor the depth of modulation.

---

on the demand of a person acting under the authority of the Secretary of State.

Provision is made for revoking or varying the terms of the licence. A licence is not transferable and should be returned to the Home Office when it has expired or been revoked.

**Messages:** Amateurs may send messages in plain language which are remarks about matters of a personal nature in which the Licensee, or the person with whom he is in contact, has been directly concerned, and use signals (ie procedure signals etc) which are not in secret code or cypher in relation to such messages. This does not include messages about business affairs. The use of the station for business, advertisement or propaganda purposes or (with the exception of the "disaster" message, described below) for the sending of news or messages on behalf of, or for the benefit or information of, any social, political, religious or commercial organizations, or for anyone other than the Licensee or the person with whom he is in contact, is not authorized. Messages which are grossly offensive or of an indecent or obscene character must not be sent. Note that it is an offence to send certain misleading messages. As part of the self-training of the Licensee in wireless telegraphy messages may be sent in speech, morse, radio teleprinter signals, facsimile signals, visual images (see frequency table for classes of emissions).

Stations may be used during disaster relief operations—or exercises relating to such operations—to send messages on behalf of the British Red Cross Society, St John Ambulance Brigade, the Emergency County Planning Office or the Police.

Messages must not be "broadcast" to amateur stations in general but only to amateur station(s) with which contact has been established. The appropriate International Telecommunication regulations must be observed.

It is permitted to re-transmit recorded messages only when intended for reception by the original station concerned, and the callsign of that station must not be included in the re-transmission.

The station may not be modulated by recordings other than special constant or gliding test tones. Entertainment-type gramophone or tape recordings may not be transmitted for any purpose. The station can be used to receive standard frequency transmissions.

**Operation:** Stations may be operated only by (1) the amateur himself; (2) a person holding another current Amateur Licence issued by the Secretary of State or an Amateur Radio Certificate issued by the Secretary of State and such operation must be in the presence of and under the direct supervision of the Licensee.

**Logs:** (1) A record must be kept in a book (not loose leaf) showing: (a) date; (b) time of commencement of period of operation of the station; (c) callsigns of the stations from which messages addressed to the Licensee's station are received, or to which messages are sent, times of making and ending contact including CQ calls and the frequency band(s) and class or classes of emission; (d) time of closing the stations; (e) the address of the temporary premises or alternative premises or particulars of the temporary location when the station is not used at the main address. Times must be in gmt and no gaps should be left between entries. All entries should be made at the time (ie the log should not be "written up" afterwards). (2) Should the station be operated by anyone other than the Licensee, the log must be signed by that person with his full name and callsign, or number of his Amateur Radio Certificate. A separate log book may be kept for mobile or pedestrian use. Entries made in respect of calls when operating from a vehicle or vessel, or as a pedestrian should be made as soon as practicable after the end of the journey and must consist of date, geographical area of operation, frequency band(s) used and times of beginning and ending the journey.

**Callsign:** The allotted callsign must be used, except that where appropriate the correct country prefix (GM, GW etc) should be substituted and /A or /P used as already indicated. The callsign may be sent by telephony or by telegraphy at not more than 20wpm at the beginning and at the end of each period of sending, and whenever the frequency is changed. If the period of use exceeds 15min, the callsign must be repeated at the commencement of each 15min period. On telephony the callsign may be confirmed by the use of well-known words having the same initial letter as the letters of the callsign but the words used must not be of a facetious or objectionable character. When sending high-definition tv signals, the callsign sent for identification purposes must be adjusted to the centre of the video channel.

**Equipment:** A satisfactory method of frequency stabilization must be used in the transmitter; frequency measuring equipment capable of verifying that the transmitter is operating with emissions within the authorized frequency bands must be provided. The station equipment must be so designed, constructed, maintained and used as not to cause: (a) any undue interference with other "wireless telegraphy" (this includes any service using radio waves: tv, radio broadcasting, radio communication, radio navigation etc).

On telegraphy precautions must be taken to eliminate the risk of key clicks. At all times, every precaution must be taken to avoid over-modulation, and to keep the radiated energy within the narrowest possible frequency bands having regard to the class of emission in use. In particular, the radiation of harmonics and other spurious emissions must be suppressed to such a level that they cause no undue interference. Tests must be carried out from time to time to ensure that these conditions are fulfilled and details of the tests entered in the log.

A receiver must be available for the frequencies and types of emission in current use at the station for the purpose of sending.

Antennas and masts must not exceed 50ft above ground level within half a mile of the boundary of any aerodrome. Antennas crossing power cables etc must be guarded to the satisfaction of the owner of the cables concerned. Note that antennas should be sited as far as possible from any existing tv or other receiving antennas in the vicinity, particularly indoor antennas. In some circumstances it might not be possible to use an indoor antenna.

For Radio Teleprinter (rtty) operation the International Telegraph Code No 2 (5-unit start stop) at speeds of 45·5 or 50 bauds may be used.

**Frequencies and powers:** These and the types of emissions are given opposite.

## Preparing for the technical examination

It is difficult to suggest just how long it should take to prepare for the RAE. So much depends upon the radio knowledge already possessed. Obviously, a complete beginner who does not know how to read theoretical diagrams or who does not have at least a vague idea of how simple receivers work will need longer to reach the necessary standard than someone who has been interested in radio for some time but who has only just become keen on amateur transmission. But whatever the state of your present knowledge, it can soon be improved beyond measure by a little careful study and by taking note of what is read.

Many candidates (but by no means all) study for the exam at a local technical college or evening institute, or attend the special lectures often held by local radio clubs—so check to see if there is a suitable course in your locality. Many others follow correspondence courses or are entirely self-taught.

Just as most of us forget the finer details of a novel as soon as we have reached the end, so it is possible to read hundreds of pages of excellent technical books without assimilating more than a tiny fraction of the information contained in them. Haphazard reading is often worse than none at all: the mind becomes confused with half-forgotten and never fully understood technicalities. The answer lies in choosing one or two books which will help you most, and then reading slowly through them, jotting down in a study notebook each new idea, and all the useful *simple* circuits. After you have copied a circuit, close your book, and then try to reproduce it from memory, explaining to yourself the purpose of each component.

In this way you will soon learn a tremendous amount about the relatively few *basic* circuits around which almost all amateur apparatus is designed. Start with "straight" receivers, noting the rf and af amplifiers, detectors, regenerative circuits and power supplies. Once you know these elements you will be surprised how often the same basic circuits appear time and time again even in the most complex superhets and multi-stage transmitters. Never try to commit to memory a complex design as a whole but learn to split it up into its various "stages".

Although the full syllabus for the examination appears at first sight rather formidable, it is possible—where necessary—to compress the amount of preparatory work a little by "intelligent anticipation" of likely questions. This is not just a matter of hopefully guessing at what the test paper will contain. But when the aims and purposes of the examination are considered, it soon becomes clear what type of question is almost certain to occur in some form or other. As one of the main objects of holding the examination is to ensure that amateur operators will not cause interruption to other radio services, it follows that emphasis will always be placed on the causes of broadcast and tv interference, harmonic radiation, the suppression of key clicks and parasitic oscillation, overmodulation, the keeping of log books, and similar problems. Particular attention should therefore be given to items 1 and 2 of the syllabus.

Another tip to candidates is to make sure that their knowledge of simple radio mathematics is sound. The questions set require only comparatively elementary knowledge of how to use a few basic formulae; yet year after year the examiners report that the mathematical questions—which are almost certainly the most straight-forward ones on the paper—are the most poorly answered. So make sure by careful revision of any of the many textbooks on elementary radio theory that you are familiar with the formulae governing resonant frequency, the value of capacitors, resistors and inductors in series and parallel, Ohm's Law and the calculation of power, and that you are not unduly put off by such symbols as $\pi$ and the square root sign.

## Passing the examination

From the examiners' reports it can be seen that many of the candidates who fail do so from a failure to train themselves in examination technique. The older candidates, who may not have taken a written examination for many years, often find that they seem to have lost the knack of putting their knowledge on to

## AMATEUR BANDS IN THE UK (CLASS A LICENCE)

| Footnote No | Frequency bands (MHz) (See (A)) | Classes of emission (See (B)) | Power: Maximum dc input power (See (C) and (D)) | Power: RF output p.e.p. for A3A and A3J emissions only (See (D)) |
|---|---|---|---|---|
| 1 and 5 | 1·8–2 | | 10W | $26\frac{2}{3}$W |
| 2 and 10 | 3·5–3·8 | | | |
| 10 and 12 | 7–7·10<br>14–14·35<br>21–21·45<br>28–29·7 | A1, A2, A3, A3A, A3H, A3J, F1, F2 and F3 | 150W | 400W |
| 1 and 3 | 70·025–70·7 | | 50W | $133\frac{1}{3}$W |
| 4, 10 and 12 | 144–145 | | 150W | 400W |
| 10 and 12 | 145–146 | | | |
| 1, 7 and 8 | 430–432 | A1, A2, A3, F1, F2, and F3 | | — |
| 1 and 11 | 432–440 | | | |
| 1 | 1,215–1,225 | | | |
| 1 and 11 | 1,226–1,290 | A1, A2, A3, A3A, A3H, A3J, F1, F2, and F3 | | |
| 1 | 1,290–1,325 | | 150W | 400W |
| 1 and 11 | 2,300–2,450 | | | |
| 1 | 3,400–3,475 | | | |
| 1 and 11 | 5,650–5,850 | | | |
| 1 and 11 | 10,000–10,500 | | | |
| 9 and 11 | 24,000–24,050 | | — | — |
| 1, 9 and 11 | 24,050–24,250 | | | |
| 1 and 6<br>1 and 6<br>1 and 6 | 2,350–2,400<br>5,700–5,800<br>10,050–10,450 | P1D, P2D, P2E, P3D and P3E | 25W mean power and 2·5kW peak power | — |

**Footnotes**

**1.** This band is allocated to stations in the amateur service on a secondary basis on condition that they shall not cause interference to other services.

**2.** This band is shared with other services.

**3.** This band is available to amateurs until further notice provided that use by the licensee of any frequency in the band shall cease immediately on the demand of a Government official.

**4.** The following spot aeronautical frequencies must be avoided whenever this band is used: 144·0, 144·54MHz.

**5.** The type of transmission known as radio teleprinter (rtty) may not be used in this band.

**6.** Use by the licensee of any frequency in this band shall be only with the prior written consent of the Secretary of State.

**7.** This band is not available for use within the area bounded by 53°N 02°E, 55°N 02°E, 55°N 03°W and 53°N 03°W.

**8.** In this band the power must not exceed 10W erp (effective radiated power).

**9.** Use by the licensee of any frequency in this band shall only be with prior written consent of the Secretary of State and such consent shall indicate the power which may be used, taking into consideration the characteristics of the licensee's station.

**10.** Slow scan television may be used in this band.

**11.** High definition television (A5, F5) may be used in this band.

**12.** Facsimile transmission (A4, F4) may be used in this band.

**13.** Data transmission may be used within the frequency bands 144–145MHz and above provided (a) the station callsign is announced in morse or telephony at least once every 15 minutes and (b) emission is contained within the bandwidth normally used for telephony.

**(A)** Artificial satellites may not be used by stations in the amateur service except in the bands 7–7·10MHz, 14–14·25MHz, 21–21·45MHz, 28–29·7MHz, 144–146MHz, 435–438MHz, 24,000–24,050MHz.

**(B)** The symbols used to designate the classes of emission have the meanings assigned to them in the Telecommunication Convention. They are:

**Amplitude modulation**

- **A1** Telegraphy by on-off keying, without the use of a modulating audio frequency.
- **A2** Telegraphy by on-off keying of an amplitude-modulating audio frequency or frequencies or by on-off keying of the modulated emission.
- **A3** Telephony, double sideband.
- **A3A** Telephony, single sideband, reduced carrier.
- **A3H** Telephony, single sideband, full carrier.
- **A3J** Telephony, single sideband, suppressed carrier.
- **A4** Facsimile (with amplitude modulation of main carrier either directly or by a frequency modulated sub-carrier).
- **A5** Television (vision only).

**Frequency (or phase) modulation**

- **F1** Telegraphy by frequency shift keying without the use of modulating audio frequency, one of the two frequencies being emitted at any instant.
- **F2** Telegraphy by on-off keying of a frequency modulating audio frequency or on-off keying of a frequency modulated emission.
- **F3** Telephony.
- **F4** Facsimile (with frequency modulation of main carrier directly).
- **F5** Television (vision only).

**Pulse modulation**

- **P1D** Telegraphy by on-off keying of a pulsed carrier without the use of a modulating audio frequency.
- **P2D** Telegraphy by on-off keying of a modulating audio frequency or frequencies or by on-off keying of a modulated pulsed carrier—the audio frequency or frequencies modulating the amplitude of the pulses.
- **P2E** Telegraphy by on-off keying of a modulating audio frequency or frequencies or by on-off keying of a modulated pulsed carrier—the audio frequency or frequencies modulating the width (or duration) of the pulses.
- **P3D** Telephony, amplitude modulated pulses.
- **P3E** Telephony, width (or duration) modulated pulses.

**(C)** DC input power is the total direct current power input to (i) the anode circuit of the valve(s) or (ii) any other device energizing the antenna.

**(D)** As an alternative, for A3A and A3J single sideband types of emission, the power shall be determined by the peak envelope power (p.e.p.) under linear operation. The radio frequency output peak envelope power under linear operation shall be limited to 2·667 times the dc input power appropriate to the frequency band concerned. This column gives the maximum power determined by this method which may be used.

**(E)** Double sideband suppressed carrier emissions are permitted within the terms of this licence.

paper. Fortunately, general examination technique can soon be learned or regained. The main points to remember are:

(1) Always read the question paper, and particularly the introductory remarks, through very carefully—at least two or three times—until you are sure that you know exactly what the examiner requires. Many candidates read the paper too hurriedly and fail to grasp the real point of the questions. There is, for instance, no point in wasting time in drawing a full circuit diagram if the examiner has asked for a block diagram.

(2) Attempt to answer the exact number of questions asked for; neither more nor less. Allot yourself a certain time for each question and try to keep reasonably closely to your provisional schedule. The introduction in the UK of multiple-choice questions has changed the emphasis and style of the examination, but it is still important to check through carefully the choices you have made.

(3) There is no need to keep to the order in which the questions are set, provided you clearly indicate the number of the question you are answering. It may be far better to gain confidence by tackling what you consider to be the easy questions first rather than ploughing doggedly through the paper. But remember point (2) and do not spend too long on your favourite subjects.

(4) Before you begin to answer, note down in your answer-book a short list of the main points of the answer as they come to mind, a word or a phrase representing a paragraph. You can then select these points in a logical order. Afterwards draw a line through your original notes to show the examiner that they are not part of your answer.

(5) Pay reasonable heed to the neatness and style of your answers. The examiners are only human and cannot be expected to decipher illegible handwriting. As you complete each question, read carefully through what you have just written, correcting spelling and grammar. Carefully check all drawings—it is all too easy to make an elementary mistake when in a hurry. Do not use amateur radio operating abbreviations, though of course you may use recognized technical abbreviations and symbols.

(6) Remember that while it is an examination and cannot be considered too lightly, there is no reason to be over-awed by the occasion. It is not a professional examination, and every allowance will be made on that score. All that is required is for you to show that you possess sufficient knowledge about radio and amateur practice to be trusted to make proper use of a transmitting licence. There are never any "trick" questions, and you are not competing against others taking the exam.

Many of these points apply both to a multiple-choice examination and the "written" examination in use in many countries.

## The morse test

Morse tests can be taken throughout the year at the Post Office HQ in London; at any of the Post Office Coast Stations (located at Highbridge, Somerset; Whitley Bay, Northumberland; near Mablethorpe, Lincs.; near Penzance, Cornwall; near Ventnor, Isle of Wight; Broadstairs, Kent; Connel, Argyll; near Stranraer, Wigtownshire; near Amlwch, Anglesey; Stonehaven, Kincardineshire; Wick, Caithness; Ilfracombe, Devon); or at any of the Marine Radio Surveyor's Offices (located at Belfast; Cardiff; Falmouth; Glasgow; Hull; Edinburgh; Liverpool; Newcastle-on-Tyne; Southampton; Aberdeen). Tests are also held, provided there are sufficient applicants, in March and September of each year, at the Head Post Offices in Birmingham, Cambridge, Derby, Leeds and Manchester.

In the test, 36 words (average length five letters per word) must be sent and 36 words received in two periods of 3min each. Up to four errors are permitted in the copy received and up to four corrections may be made while sending; there must be no uncorrected errors in sending. 10 groups of five figures must be sent, and 10 groups copied, in two periods of 1½min each; a maximum of two receiving errors are permitted in this section, and up to two corrections made while sending.

## Learning the morse code

Newcomers who really wish to learn morse operating are few and far between. The majority view it as a necessary evil that has to be surmounted before a Class A licence can be obtained. Yet once achieved, mastery of the code opens up a new world to the short-wave enthusiast and proves a source of endless satisfaction. So much so that many amateurs spend most of their time using the morse key rather than a microphone. But it is a fascination that comes only with experience and eludes the printed page.

It was proved during the 1939–45 war that reasonably competent operators could be trained in a matter of a few months from all sections of the community, young or old, men or women. Admittedly a few exceptional individuals become proficient in a fraction of the time required by the far greater number who find the early stages rather difficult. And yet many who appear to make only slow headway at first often in the long run become the best operators. The real key to morse is perseverance and a refusal to be discouraged.

Some 70 hours of practice are usually needed to reach a morse operating speed of about 13 wpm, although some people find learning a good deal easier than others. It is usually more effective to have a large number of short practice sessions regularly, than long sessions with long gaps between.

When a newcomer starts to learn the code he gradually commits to memory the 26 letters and begins to identify these letters by thinking of them as a series of dits and dahs. When this is done he often feels that he is halfway home—he just needs a little more practice and his speed will automatically increase. In fact it seldom works out like this.

After achieving perhaps 35 letters a minute he seems to come up against a barrier and for a time makes little further progress. Some learners become discouraged at the slow progress and abandon the whole business. But if they persevere the brain gradually becomes "programmed" to the letters and responds *not* to the individual dits and dahs but to the *pattern* of sound of each *letter*, automatically. There remains, however, the problem that when an operator loses the sequence of one letter, this automatic programming seems to fail temporarily and he loses several letters. It might seem that a good operator is a person who can concentrate most intensely upon the sounds; in fact *too intense concentration is usually fatal to good copying* since it is the subconscious rather than the conscious brain that, once programmed, is the key factor. It has been estimated that to acquire the ability to transcribe a morse symbol automatically into a letter requires that the ear-brain combination should have been stimulated by hearing and reacting correctly some 40,000 times. No special way of learning materially changes this basic requirement.

Morse skill thus comes after a series of "humps" interspersed with periods of apparent rapid progress. So the first requirement, and the most important one, is a determination to learn and to devote time to regular practice.

Learning morse can be divided into three processes:

1. Memorizing the code;
2. Learning to receive at the necessary speed;
3. Learning to send at the necessary speed.

In most cases, the second and third process will be learned during the same period, although this is not essential and there are some advantages to not worrying about trying to send until good progress has been made with receiving.

## Memorizing the code

Here the question of instruction does not arise. The code can be memorized in a few days without assistance, but this does not mean it should be tackled in a haphazard manner. It is essential to think of the letter symbols in terms of *sounds* and not a visual conception of dots and dashes. Do not try to learn C as "dash-dot-dash-dot" but as "dah-di-dah-dit" which is approximately how it would sound.

Do not try memorizing the entire alphabet in one or two sessions, but take half a dozen letters at a time following an orderly sequence. One possible division is as follows:

*Lesson* 1. Letters formed of dits and dahs alone: E,I,S,H,T,M,O.
*Lesson* 2. Letters including a di-dah combination: A,U,V,W,J.
*Lesson* 3. Letters including a dah-dit combination: N,D,B,G.
*Lesson* 4. Letters including a di-dah-dit combination: R,L,F.
*Lesson* 5. Letters including a dah-di-dah combination: K,Y,C,Q.
*Lesson* 6. Letters including a central di-dit or dah-dah combination: P,X,Z.
*Lesson* 7. Figures ending in dah: 0,1,2,3,4.
*Lesson* 8. Figures ending in dit: 5,6,7,8,9.

After each lesson, once the new symbols have been thoroughly committed to memory, recapitulate all preceding lessons.

The basic time unit of the morse code is the length of one dit (for certain purposes this may be termed a baud) with the dah three times this length. The correct separation between each part of a letter symbol is one dit unit; the interval between letters is three dit units (during practice newcomers may find it beneficial to increase this length); the interval between words is five dit units.

The basic structure of the code in allocating "short codes" to the most frequently used letters speeds up transmission considerably but it must be admitted that it does make the learning process rather more difficult since the newcomer finds the uneven flow of letters disconcerting: a J, Q or Y takes more than four times as long as some of the others because of the three dahs involved in each. This can mean that if a phrase such as "it is seen" is received it will seem like a sudden speeding up of the rate of transmission and so tends to run away from the inexperienced operator.

These early memorizing lessons can be practised at any odd moment; all that is required is something to read—an advertisement hoarding, a newspaper headline, a poster. Until the full alphabet has been learned just run your eye through the text picking out each letter you should know and silently form the sound equivalent of the letters.

Only if extreme difficulty is experienced should you resort to mnemonics of words or long and short syllables (eg great grand-ma to remind you that G is dah-dah-dit; Queen Queen the Queen for Q).

Once you have memorized all the letters really well you need to start practising. This can be done in classes with instructors, or from recordings using discs or tape. Here again the importance is to practice, but not to *over-concentrate*. Never attempt to write down the actual morse symbols; if a character is not recognized just leave a space and pass on to the next letter. It is the natural tendency to continue to try puzzling out a letter too long, and this results in missing not just one but several letters. The ability not to worry about a lost letter is a difficult one for a diligent student

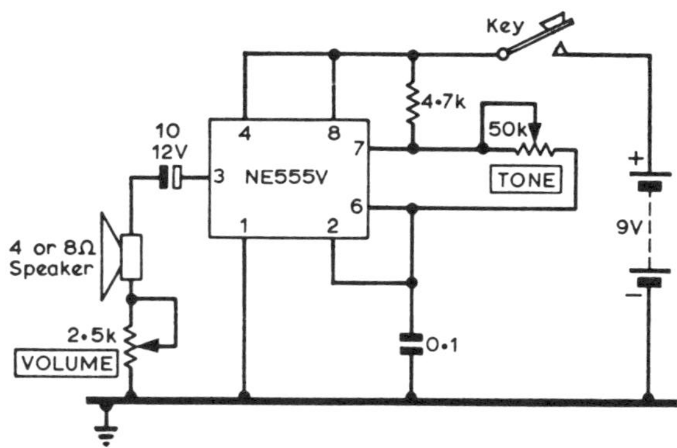


Simple code practice oscillator designed by WA5YFL, and capable of providing sufficient output for classroom instruction. The tone is adjustable from a few hundred to several thousand hertz. No on-off switch is required

to acquire, but it is essential if good progress is to be made. You are not trying to teach yourself to *think* out letters, but to *programme* the sounds into your brain so that this does the work automatically. What you want to be able to do is to let your mind wander during reception yet ending up with all the letters on the paper! Morse reception at any real speed is possible *only* if it is done automatically. One problem for the newcomer is that he becomes interested in the message coming through and starts guessing at the letters before they are sent; what you really have to do is to train your brain to jog along a little *behind* rather than in *front* of the transmission.

During training, remember that you need the ability to *copy* messages, not just to read in your mind each letter as it is transmitted. Develop a clear style of writing the messages. It is worth remembering that for experienced operators the limiting factor is the speed at which they can write out the messages (particularly when there is a succession of "short" letters) and this is one reason why for example the Navy teaches its operators to copy on a typewriter. A useful dodge when copying in block letters is to use *script* for the letter *e* instead of *E* which needs four strokes. Another useful technique is to drop all figures slightly below the line used for the letters, and this helps a reader discriminate between an I and a 1, a 2 and a Z and so on.

As gradually your receiving speed increases it is useful to begin trying to copy through interference from other signals. The ear/brain combination is a really remarkably good filter and it is possible to train yourself to be able to pick out a particular tone when this is accompanied by many other unwanted tones; the human ear can provide a filter bandwidth of about 50Hz with a large dynamic range (over 100dB), the ability to "tune" from about 200 to over 1,000Hz and all without introducing any of the "ringing" that so often mars the effects of electronic narrow-band filters! You will be surprised how quickly it is possible to ignore notes of a different audio tone to the one you want to copy; here again it is actual practice rather than concentration that is needed; if you try and concentrate on one note the brain perversely often switches to another, but is quite happy to pick out the required one if left to get on with it after plenty of practice!

## Sending morse

Learning to send good morse is the third and final stage and should be left until you have progressed sufficiently far in receiving to know how correctly formed morse should sound. Slovenly sending of morse is a habit too often acquired in the early days of training. It is a fallacy to suppose that good sending is much easier than receiving; remember that the time intervals in morse (a dit length of about 0·1s for 12wpm or only 50ms for 24wpm) are much shorter than normal human reaction times involved when the brain decides to tell the muscles to do a series of individual actions. So here again we are depending on the programming of the brain to send complete letters or even words without individual instructions about the dits and dahs.

For practice you need a good key, preferably a reasonably rugged type which can be adjusted to provide different "gaps" and "tensions". A reasonable gap (about 1mm) and medium spring tension will help to develop a steady rhythmic style of sending. It is a mistake to use some of the roughly made practice keys that are little more than "tappers". Make sure the arm is smoothly pivoted without side-play. The lightweight finger key is best put away until the danger of developing a nervous, jerky style of sending has been overcome.

There is no one correct way of holding a key. Different finger positions suit different operators; but with the majority of rounded or flattened knobs, the thumb should lie to the left of the knob or platform, the first two fingers curved slightly over the top, and the third finger resting against the right of the knob. The hold should be moderately firm but not tight and all movement should come from the wrist. The hold should be maintained throughout the period of keying and any tendency to raise the fingers between characters should be checked.

The aim of every operator should be to send morse as correctly as possible, even if he or she later intends to use one of the electronic forms of keyers that can correct for some human errors. The all-important factor is to achieve a regular pace with the correct ratio of lengths between dit and dah. The longer letters such as C or V must not be broken—ie any additional space left between parts of the letter—or it may be received as two letters. It is useful to record your sending on a tape recorder and listen to it critically.

A common problem is that of sending the right number of dits in letters such as H and S or in the figure 5. A useful practice is to use trick sentences such as *THIS IS HIS SHEET* or *HIS 5 HISSING GEESE*. Once you can get these right every time you are well on the way to overcoming any problems of dit counting.

It is worth spending those hours in practising and learning the morse code. Once the brain has been well programmed the ability will remain throughout your life-time and opens the way to communication with amateurs all over the world whether or not they speak the same language. As a communication system, morse requires less bandwidth and therefore can provide a better signal-to-noise ratio than any other systems open to the radio amateur. When used correctly it is neither slow nor old-fashioned. Many amateurs find in cw a satisfaction and enjoyment that they do not experience with phone operation. This may seem surprising and some newcomers will find it difficult to believe. It is surely worth giving up some 70 to 100 hours to acquire a skill that will last you a lifetime. It is like learning to walk and almost as natural!

## THE MORSE CODE AND SOUND EQUIVALENTS

| | | | |
|---|---|---|---|
| **A** | di-dah | **S** | di-di-dit |
| **B** | dah-di-di-dit | **T** | dah |
| **C** | dah-di-dah-dit | **U** | di-di-dah |
| **D** | dah-di-dit | **V** | di-di-di-dah |
| **E** | dit | **W** | di-dah-dah |
| **F** | di-di-dah-dit | **X** | dah-di-di-dah |
| **G** | dah-dah-dit | **Y** | dah-di-dah-dah |
| **H** | di-di-di-dit | **Z** | dah-dah-di-dit |
| **I** | di-dit | **1** | di-dah-dah-dah-dah |
| **J** | di-dah-dah-dah | **2** | di-di-dah-dah-dah |
| **K** | dah-di-dah | **3** | di-di-di-dah-dah |
| **L** | di-dah-di-dit | **4** | di-di-di-di-dah |
| **M** | dah-dah | **5** | di-di-di-di-dit |
| **N** | dah-dit | **6** | dah-di-di-di-dit |
| **O** | dah-dah-dah | **7** | dah-dah-di-di-dit |
| **P** | di-dah-dah-dit | **8** | dah-dah-dah-di-dit |
| **Q** | dah-dah-di-dah | **9** | dah-dah-dah-dah-dit |
| **R** | di-dah-dit | **0** | dah-dah-dah-dah-dah |

(0 is sometimes sent as one long dah and 9 as dah dit)

### Punctuation

| | |
|---|---|
| Question Mark | di-di-dah-dah-di-dit |
| Full Stop | di-dah-di-dah-di-dah |
| Comma* | dah-dah-di-di-dah-dah |

*Sometimes used to indicate exclamation mark.

### Procedure Signals

| | |
|---|---|
| Stroke | dah-di-di-dah-dit |
| Break sign (=) | dah-di-di-di-dah |
| End of Message (+ or $\overline{AR}$) | di-dah-di-dah-dit |
| End of Work ($\overline{VA}$) | di-di-di-dah-di-dah |
| Wait ($\overline{AS}$) | di-dah-di-di-dit |
| Preliminary call ($\overline{CT}$) | dah-di-dah-di-dah |
| Error | di-di-di-di-di-di-di-dit |
| Invitation to transmit (K) | dah-di-dah |

Note also the procedure signal used by many amateur stations inviting a named station only to transmit: $\overline{KN}$ dah-di-dah-dah-dit

• • •

One dah should be equal to three di's (dit's).
The space between parts of the same letter should be equal to one di (dit).
The space between two letters should be equal to three di's (dit's).
The space between two words should be equal to from five to seven di's (dit's).

CHAPTER 9

# Operating an amateur station

Newcomers are often puzzled by the codes and abbreviations used by radio amateurs. These codes are necessary in order to enable international communication to take place with operators who may speak little or no English, and also to save time in conveying information. When using morse telegraphy, for example, it would take far too long to spell out every word in full, but, by condensing routine phrases into code groups and by abbreviating certain commonly used words, it is possible to transmit and receive information almost as quickly as when operating on telephony.

## The value of codes

The advantage of using codes was realized by commercial operators in the early days of telegraphic communication with the result that a number of international codes—such as the Q-code—have been established for many years. Amateur operators use these codes but alter them slightly in order to meet their particular needs: in addition, they have adopted many informal abbreviations until today they possess a form of international morse language—often called "radioese"—which covers all the routine needs of amateur communication. Thus it is possible for an Englishman to enjoy radio contacts in quick succession with, for example, a German, a Spaniard, a Japanese, and a Czech: each understanding the other perfectly.

Fortunately, most of these abbreviations and codes are very easy to learn: in many cases "once seen, never forgotten." The Q-code will be found on p111 and amateur abbreviations overleaf.

## A typical telegraphy contact

The following imaginary cw contact is shown in full to give newcomers an idea of the general form of amateur operating procedure: operators, however, should guard against allowing their contacts to become too rigidly set in the "rubber-stamp" mould.

CQ CQ CQ CQ DE G4ZZZ G4ZZZ + K
G4ZZZ G4ZZZ DE GM2XXX GM2XXX + K
GM2XXX DE G4ZZZ = GM OM ES MNI TKS FER CALL = UR SIGS RST 579X = QTH LONDON = NAME HR IS JOHN NW PSE HW? + GM2XXX DE G4ZZZ K
G4ZZZ DE GM2XXX = R FB ES GM JOHN = TKS FER RPRT ES GLD TO MEET U = UR RST 558C = QTH EDINBURGH ES NAME MAC = WX COLD ES DULL = RIG IS TRANSCEIVER WID 75 WATTS INPT ES W3DZZ ANT = RX IS DOUBLE SUPER = OK? + G4ZZZ DE GM2XXX K
GM2XXX DE G4ZZZ = R OK MAC ES TKS FER ALL = SRI ABT CHIRP HR NEW VFO = RIG IS HOME BREW WID 15 WATTS ES 132 FT LONG WIRE = HR CNDX POOR FER DX = QSL VIA RSGB = NW QRU? + GM2XXX DE G4ZZZ K
G4ZZZ DE GM2XXX = MOST OK JOHN BUT NW VY HVY QRM ON FREQ = OK UR RIG ES WL SURE QSLL = NW 73 ES HPE CUAGN SN = GM JOHN ES GUD DX + G4ZZZ DE GM2XXX VA
GM2XXX DE G4ZZZ = R FB MAC ES MNI TKS FER QSO = 73 ES BCNU GB + GM2XXX DE G4ZZZ VA

Punctuation signals are seldom used, sections of a transmission being split up by means of the break (double hyphen =) sign.

Unless asked to do so, or given a readability report less than R4, words should be sent once only. Exceptions are the RST report, location and name, which are generally sent twice.

A common, but most annoying, fault is the sending of long strings of CQ calls without interspersing the station's callsign. A good rule is to send *four* (or, at the most, five) CQs followed by the station callsign *twice*, eg CQ CQ CQ CQ DE G4ZZZ G4ZZZ CQ CQ CQ CQ DE G4ZZZ G4ZZZ etc.

Amateurs often use Q-signals as nouns rather than in question and answer form. Examples are:

| | | | |
|---|---|---|---|
| **QRG** | Frequency | **QRT** | Close down |
| **QRI** | Bad note | **QRX** | Stand by |
| **QRK** | Signal strength | **QSB** | Fading |
| **QRM** | Interference from other stations | **QSD** | Bad sending |
| | | **QSL** | Verification card |
| **QRN** | Interference from atmospherics or local electrical apparatus | **QSO** | Radio contact |
| | | **QSP** | Relay message |
| | | **QSY** | Change of frequency |
| **QRO** | High power | **QTH** | Location |
| **QRP** | Low power | | |

## Telephony operation

Whereas a poor or inconsiderate cw operator is a nuisance only to his fellow enthusiasts, bad telephony operation discredits amateur radio generally. The casual listener judges our hobby by the quality of our telephony transmissions, the subjects discussed, and the procedure used.

When using telephony, operators should talk normally and avoid the excessive use of amateur abbreviations and jargon other than signal reporting codes or, when it becomes necessary, to make themselves understood by foreign amateurs. The recommended phonetic alphabet is given overleaf.

## Contests and band planning

Although amateur radio is primarily a scientific pursuit the hobby possesses a strong competitive side which is well catered for by a number of contests mostly of an annual nature. These events provide opportunities to test not only equipment but also the skill and ability of those who compete.

The types of contest vary widely, from international events extending over several weekends to concentrated three-hour hidden transmitter hunts in which physical endurance is almost as necessary as radio equipment. Almost all operating interests are catered for: dx (cw and telephony); top band; low power; portable; single operator and club and group events; vhf, uhf and microwaves.

In addition to contests, there are many other ways in which the radio amateur is able to show his prowess. Certificates and awards for various feats of operating ability and skill are issued by societies and amateur radio magazines all over the world.

In order to make the best possible use of the frequencies available to radio amateurs, the Radio Society of Great Britain suggested some years ago that a voluntary band plan should be adopted by all users of the 3·5, 7, 14, 21 and 28MHz bands. The primary purpose of the plan is to protect those who use cw telegraphy.

The band plan, which has been adopted by Region 1 of the International Amateur Radio Union, is shown on p111.

RTTY operation is recommended to take place around 3,600 ± 20kHz; 7,040 ± 5kHz; 14,090 ± 10kHz; 21,100 ± 20kHz; and 28,100 ± 50kHz. 3,500–3,510 and 3,790–3,800 reserved for intercontinental working. 3,635–3,650kHz is used by USSR stations for intercontinental working. SSTV recommended frequencies are 3,735, 7,040, 14,230, 21,340, 28,670 all ± 5kHz. For beacons 28,200–28,250kHz is recommended. For downlink of amateur space satellites 29,400–29,550 is recommended.

Band plans have also been adopted for vhf and uhf bands (see p115).

## RECOMMENDED PHONETIC ALPHABET

| | | | | | |
|---|---|---|---|---|---|
| A | Alfa | J | Juliett | S | Sierra |
| B | Bravo | K | Kilo | T | Tango |
| C | Charlie | L | Lima | U | Uniform |
| D | Delta | M | Mike | V | Victor |
| E | Echo | N | November | W | Whiskey |
| F | Foxtrot | O | Oscar | X | X-ray |
| G | Golf | P | Papa | Y | Yankee |
| H | Hotel | Q | Quebec | Z | Zulu |
| I | India | R | Romeo | | |

## AMATEUR ABBREVIATIONS

| | |
|---|---|
| **AA** | All after ... (used after a question mark to request a repetition) |
| **AB** | All before ... (see AA) |
| **BK** | Signal used to interrupt a transmission in progress |
| **BN** | All between ... and ... (see AA) |
| **C** | Yes |
| **CFM** | Confirm (or I confirm) |
| **CL** | I am closing my station |
| **CQ** | General call to all stations |
| **DE** | Used to separate the callsign of the station called from that of the calling station |
| **ER** | Here |
| **K** | Invitation to transmit |
| **NIL** | I have nothing to send you |
| **NW** | Now |
| **OK** | We agree (or It is correct) |
| **R** | Received |
| **RPT** | Repeat (or I repeat) |
| **TFC** | Traffic |
| **W** | Word(s) |
| **WA** | Word after (see AA) |
| **WB** | Word before |

**Informal amateur abbreviations**

| | |
|---|---|
| **ABT** | about |
| **ADR** | address |
| **AGN** | again |
| **ANI** | any |
| **ANT** | antenna (aerial) |
| **BA** | buffer amplifier |
| **BC** | broadcast |
| **BCI** | broadcast interference |
| **BCL** | broadcast listener |
| **BCNU** | be seeing you |
| **BD** | bad |
| **BFO** | beat frequency oscillator |
| **BK** | break-in |
| **BLV** | believe |
| **BUG** | semi-automatic key |
| **CANS** | headphones |
| **CC** | crystal-controlled |
| **CK** | check |
| **CLD** | called |
| **CNT** | cannot |
| **CO** | crystal oscillator |
| **CONDX** | conditions |
| **CPSE** | counterpoise |
| **CRD** | card |
| **CUD** | could |
| **CUAGN** | see you again |
| **CUL** | see you later |
| **CW** | continuous wave |
| **DF** | direction finding |
| **DR** | dear |
| **DX** | long distance |
| **DXCC** | DX Century Club |
| **ECO** | electron-coupled oscillator |
| **ELBUG** | electronic key |
| **ENUF** | enough |
| **ES** | and |
| **FB** | fine business |
| **FOC** | First Class Operators' Club |
| **FCC** | Federal Communications Commission |
| **FD** | frequency doubler |
| **FM** | frequency modulation |
| **FER** | for |
| **FONE** | telephone |
| **FREQ** | frequency |
| **GA** | go ahead, or good afternoon |
| **GB** | goodbye |
| **GD** | good day |
| **GE** | good evening |
| **GG** | going |
| **GLD** | glad |
| **GM** | good morning |
| **GN** | good night |
| **GND** | ground (earth) |
| **GUD** | good |
| **HAM** | amateur transmitter |
| **HI** | laughter |
| **HPE** | hope |
| **HR** | here or hear |
| **HRD** | heard |
| **HV** | have |
| **HVY** | heavy |
| **HW** | how |
| **IARU** | International Amateur Radio Union |
| **II** | repetition signal |
| **INPT** | input |
| **LID** | poor operator |
| **LSN** | listen |
| **MNI** | many |
| **MO** | master oscillator |
| **MOD** | modulation |
| **MSG** | message |
| **MTR** | meter (or metres) |
| **NBFM** | narrow band frequency modulation |
| **ND** | nothing doing |
| **NR** | number |
| **OB** | old boy |
| **OC** | old chap |
| **OM** | old man |
| **OP** | operator |
| **OT** | old timer |
| **PA** | power amplifier |
| **PP** | push-pull |
| **PSE** | please |
| **PWR** | power |
| **RAC** | rectified (raw) ac |
| **RAOTA** | Radio Amateur Old Timers' Association |
| **RCC** | Rag Chewers' Club |
| **RCVR** | receiver |
| **RPRT** | report |
| **RX** | receiver |
| **SA** | say |
| **SED** | said |
| **SIG** | signal |
| **SKED** | schedule |
| **SN** | soon |
| **SRI** | sorry |
| **SSB** | single sideband |
| **STN** | station |
| **SUM** | some |
| **SW** | short-wave |
| **SWL** | short-wave listener |
| **TFC** | traffic |
| **TKS** | thanks |
| **TMW** | tomorrow |
| **TNX** | thanks |
| **TRX** | transceiver |
| **TV** | television |
| **TVI** | television interference |
| **TX** | transmitter |
| **U** | you |
| **UR** | your |
| **VFO** | variable frequency oscillator |
| **VY** | very |
| **W** | watts |
| **WAC** | Worked all Continents |
| **WID** | with |
| **WKD** | worked |
| **WKG** | working |
| **WL** | will or well |
| **WUD** | would |
| **WX** | weather |
| **XMTR** | transmitter |
| **XYL** | wife |
| **XTAL** | crystal |
| **YF** | wife |
| **YL** | young lady |
| **73** | best regards |
| **88** | love and kisses |

## THE INTERNATIONAL Q CODE

The following Q signals taken from the official list are widely used in the Amateur Service.

| | |
|---|---|
| **QRG** | Will you tell me my exact frequency? Your exact frequency is...........kHz. |
| **QRH** | Does my frequency vary? Your frequency varies. |
| **QRI** | What is the tone of my transmission? The tone of your transmission is...........(amateur TI—T9). |
| **QRK** | What is the readability of my signals? The readability of your signals is........... (amateur RI—R5). |
| **QRL** | Are you busy? I am busy. Please do not interfere. |
| **QRM** | Are you being interfered with? I am being interfered with. |
| **QRN** | Are you troubled by static? I am troubled by static. |
| **QRO** | Shall I increase power? Increase power. |
| **QRP** | Shall I decrease power? Decrease power. |
| **QRQ** | Shall I send faster? Send faster. |
| **QRS** | Shall I send more slowly? Send more slowly. |
| **QRT** | Shall I stop sending? Stop sending. |
| **QRU** | Have you anything for me? I have nothing for you. |
| **QRV** | Are you ready? I am ready. |
| **QRX** | When will you call me again? I will call you again at...........hours. |
| **QRZ** | Who is calling me? You are being called by...........(on kHz). |
| **QSA** | What is the strength of my signals? The strength of your signals is...........(amateur SI—S9). |
| **QSB** | Are my signals fading? Your signals are fading. |
| **QSD** | Is my keying defective? Your keying is defective. |
| **QSL** | Can you give me acknowledgement of receipt? I give you acknowledgement of receipt. |
| **QSO** | Can you communicate with...........direct or by relay? I can communicate with........... direct (or by relay through...........). |
| **QSP** | Will you relay to? I will relay to........... |
| **QSV** | Shall I send a series of VVVs? Send a series of VVVs. |
| **QSY** | Shall I change to another frequency? Change to transmission on another frequency (or on...........kHz). |
| **QSZ** | Shall I send each word more than once? Send each word twice. |
| **QTH** | What is your location? My location is........... |
| **QTR** | What is the correct time? The correct time is...........hours. |

## IARU REGION 1 HF BAND PLAN

| Frequency band (kHz) | Use |
|---|---|
| 3,500–3,600 | Telegraphy only |
| 3,600–3,800 | Telegraphy and telephony |
| 7,000–7,040 | Telegraphy only |
| 7,040–7,100 | Telegraphy and telephony |
| 14,000–14,100 | Telegraphy only |
| 14,100–14,350 | Telegraphy and telephony |
| 21,000–21,150 | Telegraphy only |
| 21,150–21,450 | Telegraphy and telephony |
| 28,000–28,200 | Telegraphy only |
| 28,200–29,700 | Telegraphy and telephony |

## THE RST CODE

### Readability

| | |
|---|---|
| **R1** | Unreadable. |
| **R2** | Barely readable, occasional words distinguishable. |
| **R3** | Readable with considerable difficulty. |
| **R4** | Readable with practically no difficulty. |
| **R5** | Perfectly readable. |

### Signal strength

| | |
|---|---|
| **S1** | Faint, signals barely perceptible. |
| **S2** | Very weak signals. |
| **S3** | Weak signals. |
| **S4** | Fair signals. |
| **S5** | Fairly good signals. |
| **S6** | Good signals. |
| **S7** | Moderately strong signals. |
| **S8** | Strong signals. |
| **S9** | Extremely strong signals. |

### Tone

| | |
|---|---|
| **T1** | Extremely rough hissing note. |
| **T2** | Very rough ac note, no trace of musicality. |
| **T3** | Rough, low-pitched ac note, slightly musical. |
| **T4** | Rather rough ac note, moderately musical. |
| **T5** | Musically modulated note. |
| **T6** | Modulated note, slight trace of whistle. |
| **T7** | Near dc note, smooth ripple. |
| **T8** | Good dc note, just a trace of ripple. |
| **T9** | Purest dc note. |

If the note appears to be crystal-controlled add X after the appropriate number. Where there is chirp add C, drift add D, clicks add K.

## USA PREFIXES

Revisions made by the FCC in 1978 to the prefixes assigned to amateur stations in the USA and USA Dependencies have led to considerable confusion and ambiguity, since callsigns based on the former prefixes continue in use (p112). Similarly within the USA, "district areas" (see list of States under "W" in the main table) can no longer be positively identified since amateurs can now retain an existing callsign when moving to a different "call area".

| | |
|---|---|
| **K, N, W** | Single-letter prefixes denote stations located in the USA |
| **AA–AG, KA–KG, NA–NG, WA–WG** | Now USA but see list below |
| **AH0, KH0, NH0, WH0** | Mariana Is (KG6) |
| **AH1, KH1, NH1, WH0** | Baker Is, Howland Is,(KB6) |
| **AH2, KH2, NH2, WH2** | Guam (KG6) |
| **AH3, KH3, NH3, WH3** | Johnston Is (KJ6) |
| **AH4, KH4, NH4, WH4** | Midway Is (KM6) |
| **AH5, KH5, NH5, WH5** | Jarvis Is, Palmyra Is (KP6) |
| **AH6, KH6, NH6, WH6** | Hawaiian Is (KH6) |
| **AH7, KH7, NH7, WH7** | Kure Is |
| **AH8, KH8, NH8, WH8** | American Samoa (KS6) |
| **AH9, KH9, NH9, WH9** | Wake Is (KW6) |
| **AI–AK, KI–KK, NI–NK, WI–WK** | USA |
| **AL7, KL7, NL7, WL7** | Alaska (KL7) |
| **KM–KO, NM–NO, WM–WO** | USA |
| **KP1, NP1, WP1** | Navassa Is (KC4) |
| **KP2, NP2, WP2** | Virgin Is (KV4) |
| **KP3, NP3, WP3** | Roncador Key, Serrana Bank |
| **KP4, NP4, WP4** | Puerto Rico (KP4) |
| **KQ–KZ, NQ–NZ WQ–WZ** | USA |

## AMATEUR RADIO PREFIXES

|  |  |
|---|---|
| **A2** | Botswana |
| **A35** | Tonga |
| **A4X** | Oman |
| **A51** | Bhutan |
| **A6X** | United Arab Emirates |
| **A7X** | Qatar |
| **A9X** | Bahrein |
| **AA–AL** | See USA table |
| **AP** | Pakistan |
| **BL** | Tibet |
| **BV** | Formosa (Taiwan) |
| **BY** | China |
| **C5** | Gambia |
| **C6** | Bahamas |
| **C9** | Mozambique |
| **C21** | Nauru |
| **C31** | Andorra |
| **CE** | Chile |
| **CE9** | Antarctica (Chile) |
| **CE0A** | Easter Island |
| **CE0X** | San Felix Is |
| **CE0Z** | Juan Fernandex Is |
| **CM** | Cuba |
| **CN** | Morocco |
| **CO** | Cuba |
| **CP** | Bolivia |
| **CR9** | Macau |
| **CT1,4** | Portugal |
| **CT2** | Azores |
| **CT3** | Madeira |
| **CX** | Uruguay |
| **D2** | Angola |
| **D4** | Cape Verde Is |
| **D68** | State of the Comoros |
| **DA, DB, DC, DF, DJ, DK, DL** | West Germany |
| **DU, DX** | Philippines |
| **EA** | Spain |
| **EA6** | Balearic Is |
| **EA8** | Canary Is |
| **EA9** | Spanish North Africa (Ceuta, Melilla) |
| **EI, EJ** | Eire |
| **EL** | Liberia |
| **EP** | Iran |
| **ET3** | Ethiopia |
| **F** | France |
| **FB8W** | Crozet Is |
| **FB8X** | Kerguelen Is |
| **FB8Y** | Terre Adelie |
| **FB8Z** | Amsterdam Is, St Paul Is |
| **FC** | Corsica |
| **FG** | Guadeloupe |
| **FH8** | Mayotte Is |
| **FK** | New Caledonia |
| **FM** | Martinique |
| **FO** | French Oceania (eg Tahiti), Clipperton Is |
| **FP** | St Pierre and Miquelon Is |
| **FR** | Reunion Island, etc |
| **FS7** | St Martin |
| **FW** | Wallis and Futuna Is |
| **FY** | French Guiana, Inini |
| **G** | England |
| **GB** | United Kingdom (Exhibitions and Special Purposes) |
| **GD** | Isle of Man |
| **GI** | Northern Ireland |
| **GJ** | Jersey |
| **GM** | Scotland |
| **GU** | Guernsey, Alderney, Sark |
| **GW** | Wales |
| **H44** | Solomon Is |
| **H5** | Bophutatswana |
| **HA, HG** | Hungary |
| **HB4, 9** | Switzerland |
| **HE, HB0** | Liechtenstein |
| **HC** | Ecuador |
| **HC8** | Galapagos Is |
| **HH** | Haiti |
| **HI** | Dominican Republic |
| **HK** | Colombia |
| **HK0** | San Andres, Providencia, etc |
| **HL, HM** | Korea |
| **HP, HO** | Panama |
| **HR** | Honduras |
| **HS** | Thailand |
| **HV** | Vatican City |
| **HZ** | Saudi Arabia |
| **I, IA, IB, IC, ID, IE, IF, IG, IH, IL, IP, IZ** | Italy |
| **IS, IM** | Sardinia |
| **IT** | Sicily |
| **J28** | Djibouti |
| **J3** | Grenada |
| **J5** | Guinea-Bissau |
| **J6** | St Lucia |
| **J7** | Dominica |
| **JA, JE, JF, JG, JH, JI, JJ, JR** | Japan |
| **JD** | Bonin and Volcano Is (Minami Is, Ogasawara Is) |
| **JT** | Mongolia |
| **JW** | Svalbard |
| **JX** | Jan Mayen |
| **JY** | Jordan |
| **K** | USA (for Districts see under "W") |
| **KA–KZ** | See USA table |
| **KB6*** | Baker, Howland Is |
| **KC4*** | Navassa Is, Antarctica |
| **KC6** | Caroline Is |
| **KG4** | Guantanamo Bay |
| **KG6*** | Mariana Is, Marcus Is, Guam |
| **KH6*** | Hawaiian Is |
| **KJ6*** | Johnston Is |
| **KL7*** | Alaska |
| **KM6*** | Midway Is |
| **KP6*** | Palmyra Group, Jarvis Is |
| **KS6*** | American Samoa |
| **KV4*** | Virgin Is |
| **KW6*** | Wake Is |
| **KX6** | Marshall Is |
| **LA** | Norway |
| **LB** | Norway (Special) |
| **LU** | Argentina |
| **LX** | Luxembourg |
| **LZ** | Bulgaria |
| **M1** | San Marino |
| **N** | USA (for districts see under "W") |
| **NA–NZ** | See USA table |
| **OA** | Peru |
| **OD** | Lebanon |
| **OE** | Austria |
| **OF, OH** | Finland |
| **OH0** | Aaland Is |
| **OH0M, OJ0** | Market Reef |
| **OK, OL, OM** | Czechoslovakia |
| **ON** | Belgium |
| **OR4** | Antarctica (Belgium) |
| **OX** | Greenland |
| **OY** | Faeroes |
| **OZ** | Denmark |
| **P29** | Papua New Guinea |
| **PA, PE** | Holland |
| **PI** | Holland (Special) |
| **PJ** | Dutch West Indies |
| **PJ5** | St Eustatius |
| **PJ6** | Saba |
| **PJ7** | St Maarten |
| **PP–PY** | Brazil |
| **PY0** | Fernando de Noronha, Trinidade and Vaz Is |
| **PZ** | Surinam |
| **RA–RZ** | USSR |
| **S2** | Bangladesh |
| **S7** | Seychelles |
| **S8** | Transkei |
| **S9** | Sao Thome, Principe |
| **SL** | Sweden (Special) |
| **SM, SK** | Sweden |
| **SP, SQ** | Poland |
| **ST** | Sudan |
| **SU** | Egypt |
| **SV** | Greece |
| **SV5** | Rhodes |
| **SV9** | Crete |
| **SY** | Mt Athos |
| **T2** | Tuvalu |
| **T3K** | Kiribati |
| **T3L** | Line Is |
| **T3P** | Phoenix Is |
| **TA** | Turkey |
| **TF** | Iceland |
| **TG** | Guatemala |
| **TI** | Costa Rica |
| **TI9** | Cocos Is |
| **TJ** | Cameroons |
| **TL8** | Central African Republic |
| **TN8** | Congo Republic (formerly French Congo) |
| **TR8** | Gabon Republic |
| **TT8** | Tchad Republic |
| **TU2** | Ivory Coast |
| **TY** | Benin |
| **TZ** | Mali Republic |
| **UA, UV, UW, RA 1,3,4,6** | European Russian SFSR |
| **UA2** | Kaliningradsk |
| **UA9, 0, UW9, UZ9** | Asiatic RSFSR |
| **UB5, UT5, UY5** | Ukraine |
| **UC2** | White Russia |
| **UD6** | Azerbaijan |
| **UF6** | Georgia |
| **UG6** | Armenia |
| **UH8** | Turkoman |
| **UI8** | Uzbek |

|  |  |
|---|---|
| **UJ8** | Tadzhiz |
| **UK** | Russian club stations (see p115) |
| **UL7** | Kazakh |
| **UM8** | Kirghiz |
| **UN1** | Finno-Karelia |
| **UO5** | Moldavia |
| **UP2** | Lithuania |
| **UQ2** | Latvia |
| **UR2** | Estonia |
| **VE, VO** | Canada |
| **VE1** | Maritime Provinces |
| **VE2** | Province of Quebec |
| **VE3** | Province of Ontario |
| **VE4** | Province of Manitoba |
| **VE5** | Province of Saskatchewan |
| **VE6** | Province of Alberta |
| **VE7** | Province of British Columbia |
| **VY1** | Yukon Territories |
| **VE8M–Z** | NW Territories |
| **VO1** | Newfoundland |
| **VO2** | Labrador |
| **VK** | Australia |
| **VK0** | Heard Is, McDonald Is, Macquarie Is |
| **VK1** | Canberra |
| **VK2** | New South Wales |
| **VK3** | Victoria |
| **VK4** | Queensland |
| **VK5** | South Australia |
| **VK6** | Western Australia |
| **VK7** | Tasmania |
| **VK8** | Northern Territories |
| **VK9** | New Guinea, Norfolk Is, Papua, Cocos-Keeling and Admiralty Islands |
| **VO** | See under VE |
| **VP1** | Belize |
| **VP2** | Leeward Is, Windward Is |
| **VP2A** | Antigua and Barbuda |
| **VP2E** | Anguilla |
| **VP2H** | Anguilla |
| **VP2K** | St Kitts and Nevis |
| **VP2M** | Montserrat |
| **VP2S** | St Vincent |
| **VP2V** | British Virgin Isles |
| **VP5** | Turks and Caicos Islands |
| **VP8** | Falkland Is, S Georgia, S Orkneys, Shetland Is, Sandwich Is, Grahamland |
| **VP9** | Bermuda |
| **VQ9** | Chagos Is |
| **VR6** | Pitcairn Is |
| **VS5** | Brunei |
| **VS6** | Hong Kong |
| **VU2** | India |
| **VU5** | Andaman Is, Laccadive Is |
| **W** | United States of America (Note stations can now retain former callsigns when moving districts) (see also KH6, KL7) |
| **W1** | Connecticut, Maine, Massachusetts, New Hampshire, Rhode Island Vermont |
| **W2** | New Jersey, New York |
| **W3** | Delaware, Maryland, Pennsylvania (including District of Columbia) |
| **W4** | Alabama, Florida, Georgia, Kentucky, North Carolina, South Carolina, Tennessee, Virginia |
| **W5** | Arkansas, Louisiana, Mississippi, New Mexico, Oklahoma, Texas |
| **W6** | California |
| **W7** | Arizona, Idaho, Montana, Nevada, Oregon, Utah Washington, Wyoming |
| **W8** | Michigan, Ohio, West Virginia |
| **W9** | Illinois, Indiana, Wisconsin |
| **W0** | Colorado, Iowa, Kansas, Minnesota, Missouri, Nebraska, North Dakota, South Dakota |
| **WA–WZ** | See USA table |
| **XE, XF** | Mexico |
| **XE4** | Revilla Gigedo |
| **XT2** | Upper Volta |
| **XU** | Kampuchea (Khmer, Cambodia) |
| **XV** | Vietnam |
| **XW8** | Laos |
| **XZ** | Burma |
| **Y** | East Germany |
| **YA** | Afghanistan |
| **YB, YC, YD** | Indonesia |
| **YI** | Iraq |
| **YJ** | New Hebrides |
| **YK** | Syria |
| **YN** | Nicaragua |
| **YO** | Roumania |
| **YS** | Salvador |
| **YU** | Yugoslavia |
| **YV** | Venezuela |
| **YV0** | Aves Is |
| **ZA** | Albania |
| **ZB2** | Gibraltar |
| **ZC4** | Cyprus (British assigned) |
| **ZD7** | St Helena |
| **ZD8** | Ascension Island |
| **ZD9** | Tristan de Cunha, Gough Is |
| **ZE** | Zimbabwe (Rhodesia) |
| **ZF1** | Cayman Isles |
| **ZK1** | Cook Island |
| **ZK2** | Niue |
| **ZL, ZM** | New Zealand, Chatham Is, Kermadec Is |
| **ZL1** | Auckland District |
| **ZL2** | Wellington District |
| **ZL3** | Canterbury District |
| **ZL4** | Otago District |
| **ZL5** | New Zealand Antarctica |
| **ZM7** | Tokelau Is |
| **ZP** | Paraguay |
| **ZR, ZS** | Republic of South Africa, Prince Edward and Marion Is |
| **ZS1** | Cape District |
| **ZS2** | Cape Province (excluding ZS1) |
| **ZS3** | South West Africa |
| **ZS4** | Orange Free State |
| **ZS5** | Natal (including Zululand) |
| **ZS6** | Transvaal |
| **1S** | Spratly Is |
| **3A2** | Monaco |
| **3B6** | Chagos Is |
| **3B7** | Agalega and St Brandon |
| **3B8** | Mauritius |
| **3B9** | Rodriguez |
| **3C** | Equatorial Guinea |
| **3D2** | Fiji |
| **3D6** | Swaziland |
| **3G3** | Chile |
| **3V8** | Tunisia |
| **3W8** | Vietnam |
| **3X** | Republic of Guinea |
| **3Y** | Bouvet Is |
| **4M** | Venezuela |
| **4S7** | Sri Lanka |
| **4U1ITU** | ITU, Geneva |
| **4W** | Yemen (AR) |
| **4X, 4Z** | Israel |
| **5A** | Libya |
| **5B4** | Republic of Cyprus |
| **5H1** | Zanzibar |
| **5H3** | Tanzania (except Zanzibar) |
| **5L** | Liberia |
| **5N** | Nigeria |
| **5R8** | Malagasy Republic |
| **5T5** | Mauritania |
| **5U7** | Niger Republic |
| **5V** | Togo |
| **5W1** | Western Samoa |
| **5X5** | Uganda |
| **5Z4** | Kenya |
| **6D** | Mexico |
| **6O** | Somali Republic |
| **6W8** | Senegal Republic |
| **6Y** | Jamaica |
| **7O** | Yemen (PDR) |
| **7P8** | Lesotho |
| **7Q** | Malawi |
| **7X** | Algeria |
| **7Z** | Saudi Arabia |
| **8F** | Indonesia |
| **8J1** | Antarctica (also OR4, VK0, ZL5, VP8) |
| **8P** | Barbados |
| **8Q** | Maldives |
| **8R** | Guyana |
| **8Z4** | Iraq/Saudi Arabia Neutral Zone |
| **9A1** | San Marino |
| **9G1** | Ghana |
| **9H** | Malta |
| **9J2** | Zambia |
| **9K2** | Kuwait |
| **9L1** | Sierra Leone |
| **9M2** | Malaysia |
| **9M6** | Sabah |
| **9M8** | Sarawak |
| **9N** | Nepal |
| **9Q5** | Zaire |
| **9U5** | Burundi |
| **9V1** | Singapore |
| **9X5** | Rwanda |
| **9Y4** | Trinidad and Tobago |

*Former prefix still in use.

Note that many "special event" and commemorative prefixes are now being used for limited periods and do not conform with the above list, examples CK Canada, SQ Poland, EE Spain. See "International Allocation of Callsigns" table overleaf.

## INTERNATIONAL ALLOCATION OF CALLSIGNS

Since many "special" amateur callsigns are issued to mark special events using unfamiliar prefixes, it is useful to have available a complete list of the ITU callsign series. Any special callsign must be in accordance with this series:

**AAA–ALZ** United States of America
**AMA–AOZ** Spain
**APA–ASZ** Pakistan
**ATA–AWZ** India
**AXA–AXZ** Australia
**AYA–AZZ** Argentina
**A2A–A2Z** Botswana
**A3A–A3Z** Tonga Islands
**A4A–A4Z** Sultanate of Oman
**A5A–A5Z** Bhutan
**A6A–A6Z** United Arab Emirates
**A7A–A7Z** Qatar
**A8A–A8Z** Liberia
**A9A–A9Z** Bahrein
**BAA–BZZ** China (People's Republic)
**CAA–CEZ** Chile
**CFA–CKZ** Canada
**CLA–CMZ** Cuba
**CNA–CNZ** Morocco
**COA–COZ** Cuba
**CPA–CPZ** Bolivia
**CQA–CRZ** Portuguese overseas provinces
**CSA–CUZ** Portugal
**CVA–CXZ** Uruguay
**CYA–CZZ** Canada
**C2A–C2Z** Nauru Island
**C3A–C3Z** Andorra
**C4A–C4Z** Cyprus
**C5A–C5Z** The Gambia
**C6A–C6Z** Bahamas
**C7A–C7Z** World Meteorological Org
**C8A–C9Z** Mozambique
**DAA–DRZ** Germany (FRG)
**DSA–DTZ** Republic of Korea
**DUA–DZZ** Philippine Islands
**D2A–D3Z** Angola
**D4A–D4Z** Cape Verde
**D5A–D5Z** Liberia
**D6A–D6Z** Comoros
**D7A–D9Z** Republic of Korea
**EAA–EHZ** Spain
**EIA–EJZ** Ireland
**EKA–EKZ** USSR
**ELA–ELZ** Liberia
**EMA–EOZ** USSR
**EPA–EQZ** Iran
**ERA–ESZ** USSR
**ETA–ETZ** Ethiopia
**EUA–EWZ** Bielorussian SSR
**EXA–EZZ** USSR
**FAA–FZZ** France & overseas provinces
**GAA–GZZ** Great Britain
**HAA–HAZ** Hungary
**HBA–HBZ** Switzerland
**HCA–HDZ** Ecuador
**HEA–HEZ** Switzerland
**HFA–HFZ** Poland
**HGA–HGZ** Hungary
**HHA–HHZ** Haiti
**HIA–HIZ** Dominican Republic
**HJA–HKZ** Colombia
**HLA–HLZ** Korea
**HMA–HMZ** Korea (DPRK)
**HNA–HNZ** Iraq
**HOA–HPZ** Panama
**HQA–HRZ** Honduras
**HSA–HSZ** Thailand
**HTA–HTZ** Nicaragua
**HUA–HUZ** Salvador
**HVA–HVZ** Vatican
**HWA–HYZ** France & overseas provinces
**HZA–HZZ** Saudi Arabia
**H2A–H2Z** Cyprus
**H3A–H3Z** Panama
**H4A–H4Z** Solomon Is
**H6A–H7Z** Nicaragua
**H8A–H9Z** Panama
**IAA–IZZ** Italy
**JAA–JSZ** Japan
**JTA–JVZ** Mongolia
**JWA–JXZ** Norway
**JYA–JYZ** Jordan
**JZA–JZZ** Indonesia
**J2A–J2Z** Djibouti
**J3A–J3Z** Grenada
**J4A–J4Z** Greece
**J5A–J5Z** Guinea-Bissau
**J6A–J6Z** St Lucia
**J7A–J7Z** Dominica
**KAA–KZZ** United States of America
**LAA–LNZ** Norway
**LOA–LWZ** Argentina
**LXA–LXZ** Luxembourg
**LYA–LYZ** USSR
**LZA–LZZ** Bulgaria
**L2A–L9Z** Argentina
**MAA–MZZ** Great Britain
**NAA–NZZ** United States of America
**OAA–OCZ** Peru
**ODA–ODZ** Lebanon
**OEA–OEZ** Austria
**OFA–OJZ** Finland
**OKA–OMZ** Czechoslovakia
**ONA–OTZ** Belgium
**OUA–OZZ** Denmark
**PAA–PIZ** Netherlands
**PJA–PJZ** Neth. Antilles
**PKA–POZ** Indonesia
**PPA–PYZ** Brazil
**PZA–PZZ** Suriname
**P2A–P2Z** Papua, New Guinea
**P3A–P3Z** Cyprus
**P4A–P4Z** Neth. Antilles
**P5A–P9Z** Korea (DPRK)
**QAA–QZZ** Service abbreviations
**RAA–RZZ** USSR
**SAA–SMZ** Sweden
**SNA–SRZ** Poland
**SSA–SSM** Egypt (Arab Republic)
**SSN–STZ** Sudan
**SUA–SUZ** Egypt (Arab Republic)
**SVA–SZZ** Greece
**S2A–S3Z** Bangladesh
**S6A–S6Z** Singapore
**S7A–S7Z** Seychelles
**S9A–S9Z** Sao Thome, Principe
**TAA–TCZ** Turkey
**TDA–TDZ** Guatemala
**TEA–TEZ** Costa Rica
**TFA–TFZ** Iceland
**TGA–TGZ** Guatemala
**THA–THZ** France & overseas provinces
**TIA–TIZ** Costa Rica
**TJA–TJZ** Cameroun
**TKA–TKZ** France & overseas provinces
**TLA–TLZ** Central African Republic
**TMA–TMZ** France & overseas provinces
**TNA–TNZ** Congo Republic
**TOA–TQZ** France & overseas provinces
**TRA–TRZ** Gabon Republic
**TSA–TSZ** Tunisia
**TTA–TTZ** Chad Republic
**TUA–TUZ** Ivory Coast
**TVA–TXZ** France & overseas provinces
**TYA–TYZ** Dahomey Republic
**TZA–TZZ** Mali Republic
**T2A–T2Z** Tuvalu
**T3A–T3Z** Kiribati
**UAA–UQZ** USSR
**URA–UTZ** Ukraine SSR
**UUA–UZZ** USSR
**VAA–VGZ** Canada
**VHA–VNZ** Australia
**VOA–VOZ** Canada
**VPA–VSZ** British overseas territories
**VTA–VWZ** India
**VXA–VYZ** Canada
**VZA–VZZ** Australia
**WAA–WZZ** United States of America
**XAA–XIZ** Mexico
**XJA–XOZ** Canada
**XPA–XPZ** Denmark
**XQA–XRZ** Chile
**XSA–XSZ** China
**XTA–XTZ** Voltaic Republic
**XUA–XUZ** Kampuchea
**XVA–XVZ** Vietnam
**XWA–XWZ** Laos
**XXA–XXZ** Portuguese overseas provinces
**XYA–XZZ** Burma
**YAA–YAZ** Afghanistan
**YBA–YHZ** Indonesia
**YIA–YIZ** Iraq
**YJA–YJZ** New Hebrides
**YKA–YKZ** Syria
**YLA–YLZ** Latvia
**YMA–YMZ** Turkey
**YNA–YNZ** Nicaragua
**YOA–YRZ** Roumania
**YSA–YSZ** Salvador
**YTA–YUZ** Yugoslavia
**YVA–YYZ** Venezuela
**YZA–YZZ** Yugoslavia
**Y2A–Y9Z** Germany (GDR)
**ZAA–ZAZ** Albania
**ZBA–ZJZ** British overseas territories
**ZKA–ZMZ** New Zealand
**ZNA–ZOZ** British overseas territories
**ZPA–ZPZ** Paraguay
**ZQA–ZQZ** British overseas territories
**ZRA–ZUZ** South Africa
**ZVA–ZZZ** Brazil
**2AA–2ZZ** British overseas territories
**3AA–3AZ** Monaco
**3BA–3BZ** Mauritius
**3CA–3CZ** Equatorial Guinea
**3DA–3DM** Swaziland
**3DN–3DZ** Fiji Islands
**3EA–3FZ** Panama
**3GA–3GZ** Chile
**3HA–3UZ** China

| | | | | | |
|---|---|---|---|---|---|
| **3VA–3VZ** | Tunisia | **5RA–5SZ** | Malagasy Republic | **7ZA–7ZZ** | Saudi Arabia |
| **3WA–3WZ** | Vietnam | **5TA–5TZ** | Mauritania | **8AA–8IZ** | Indonesia |
| **3XA–3XZ** | Guinea (Republic of) | **5UA–5UZ** | Niger Republic | **8JA–8NZ** | Japan |
| **3YA–3YZ** | Norway | **5VA–5VZ** | Togo Republic | **8OA–8OZ** | Botswana |
| **3ZA–3ZZ** | Poland | **5WA–5WZ** | Western Samoa | **8PA–8PZ** | Barbados |
| **4AA–4CZ** | Mexico | **5XA–5XZ** | Uganda | **8QA–8QZ** | Maldives (Republic of) |
| **4DA–4IZ** | Philippine Islands | **5YA–5ZZ** | Kenya | **8RA–8RZ** | Guyana |
| **4JA–4LZ** | USSR | **6AA–6BZ** | Egypt (Arab Republic) | **8SA–8SZ** | Sweden |
| **4MA–4MZ** | Venezuela | **6CA–6CZ** | Syria | **8TA–8YZ** | India |
| **4NA–4OZ** | Yugoslavia | **6DA–6JZ** | Mexico | **8ZA–8ZZ** | Saudi Arabia |
| **4PA–4SZ** | Sri Lanka (Republic of) | **6KA–6NZ** | Korea (RK) | **9AA–9AZ** | San Marino |
| **4TA–4TZ** | Peru | **6OA–6OZ** | Somali Republic | **9BA–9DZ** | Iran |
| **4UA–4UZ** | United Nations | **6PA–6SZ** | Pakistan | **9EA–9FZ** | Ethiopia |
| **4VA–4VZ** | Haiti | **6TA–6UZ** | Sudan | **9GA–9GZ** | Ghana |
| **4WA–4WZ** | Yemen (Arab Republic) | **6VA–6WZ** | Senegal Republic | **9HA–9HZ** | Malta |
| **4XA–4XZ** | Israel | **6XA–6XZ** | Malagasy Republic | **9IA–9JZ** | Zambia |
| **4YA–4YZ** | ICAO | **6YA–6YZ** | Jamaica | **9KA–9KZ** | Kuwait |
| **4ZA–4ZZ** | Israel | **6ZA–6ZZ** | Liberia | **9LA–9LZ** | Sierra Leone |
| **5AA–5AZ** | Libya | **7AA–7IZ** | Indonesia | **9MA–9MZ** | Malaysia |
| **5BA–5BZ** | Cyprus | **7JA–7NZ** | Japan | **9NA–9NZ** | Nepal |
| **5CA–5GZ** | Morocco | **7OA–7OZ** | Yemen (PDR) | **9OA–9TZ** | Zaire (Republic of) |
| **5HA–5IZ** | Tanzania | **7PA–7PZ** | Lesotho | **9UA–9UZ** | Burundi |
| **5JA–5KZ** | Colombia | **7QA–7QZ** | Malawi | **9VA–9VZ** | Singapore |
| **5LA–5MZ** | Liberia | **7RA–7RZ** | Algeria | **9WA–9WZ** | Malaysia |
| **5NA–5OZ** | Nigeria | **7SA–7SZ** | Sweden | **9XA–9XZ** | Rwanda |
| **5PA–5QZ** | Denmark | **7TA–7YZ** | Algeria | **9YA–9ZZ** | Trinidad and Tobago |

## USSR CLUB STATION PREFIXES

| | | | | | |
|---|---|---|---|---|---|
| **UK1, 3, 4** | European RSFSR (equivalent UA1, 3, 4) | **UK2Q** | Latvia (UQ2) | **UK6O** | Georgia (UF6) |
| | | **UK2R** | Estonia (UR2) | **UK6Q** | Georgia (UF6) |
| **UK9, 0** | Asiatic RSFSR (equivalent UA9, 0) | **UK2S** | White Russia (UC2) | **UK6V** | Georgia (UF6) |
| | | **UK2T** | Estonia (UR2) | **UK6 (other)** | European RSFSR (UA6) |
| **UK2A** | White Russia (UC2) | **UK2W** | White Russia (UC2) | | |
| **UK2B** | Lithuania (UP2) | **UK5O** | Moldavia | **UK7** | Kazakh (UL7) |
| **UK2C** | White Russia (UC2) | **UK5 (other)** | Ukraine (UB5) | **UK8H** | Turkoman (UH8) |
| **UK2F** | Kaliningradsk (UA2) | | | **UK8J** | Tadzhiz (UJ8) |
| **UK2G** | Latvia (UQ2) | **UK6C** | Azerbaijan (UD6) | **UK8M** | Kirghiz (UM8) |
| **UK2I** | White Russia (UC2) | **UK6D** | Azerbaijan (UD6) | **UK8N** | Kirghiz (UM8) |
| **UK2L** | White Russia (UC2) | **UK6F** | Georgia (UF6) | **UK8R** | Tadzhiz (UJ8) |
| **UK2O** | White Russia (UC2) | **UK6G** | Armenia (UG6) | **UK8 (other)** | Uzbek (UI8) |
| **UK2P** | Lithuania (UP2) | **UK6K** | Azerbaijan (UD6) | | |

## UK BAND PLANS

| | |
|---|---|
| 70·025–70·100 | cw only |
| 70·100–70·700 | all modes, including ssb, nbfm, a.m., cw (70·260 national mobile calling/working frequency; 70·560 rtty) |
| 70·675–70·700 | beacon transmissions |
| 144·000–144·150 | cw only (144·000–·010 moonbounce; 144·050 cw calling) |
| 144·150–144·500 | ssb and cw (144·300 ssb calling) |
| 144·500–144·850 | all modes (144·500 sstv calling; 144·600 ± rtty; 144·650 a.m. calling; 144·700 facsimile calling; 144·750 atv talk-back; 144·800/825/850/875 Raynet) |
| 144·850–145·000 | beacons only |
| 145·000–145·175 | repeater input channels (25kHz steps from R0 on 145·000, R1 on 145·025 to R7 on 145·175) |
| 145·200–145·575 | simplex channels (25kHz steps from S8 on 145·200 to S23 on 145·575). Note mobile calling channel S20 on 145·500, 145·300 rtty and 145·250 atv talk-back. |
| 145·600–145·800 | repeater output channels (25kHz steps from R0 on 145·600 to R7 on 145·775) |
| 145·800–146·000 | satellite service |

Note that "channelization" begins at 145MHz

| | |
|---|---|
| 432·000–432·150 | cw only (432·000–·010 moonbounce; 432·050 cw calling) |
| 432·150–432·500 | ssb and cw (432·200 and 432·300 ssb calling) |
| 432·500–432·800 | all modes (432·600 ± rtty-fm; 432·700 facsimile calling) |
| 432·800–433·000 | beacons only |
| 433·000–433·350 | repeater output channels (25kHz steps from RB0 on 433·000 to RB14 on 433·350 but note SU12 on 433·300 for simplex rtty local working with fm or ask) |
| 433·375–433·500 | simplex channels (25kHz steps from SU15 on 433·375 to SU20 on 433·500). Note that SU20 is fixed/mobile calling channel |
| 434·600–434·950 | repeater input channels (25kHz steps from RB0 on 434·600 to RB14 on 434·950) |
| 435·000–438·000 | satellite service |

No channelization below 433MHz. Amateur television (434–440MHz) stations should endeavour to use frequencies not causing interference to other users, particularly the satellite service.

CHAPTER 10

# The RSGB and the radio amateur

The Radio Society of Great Britain (RSGB) is the national society of radio amateurs in the United Kingdom. The majority of its 25,000 members hold UK amateur transmitting licences; the others either hope to do so later or are interested primarily in the receiving side of amateur radio. The Society has several thousand overseas members in over 150 countries.

HRH The Prince Philip, Duke of Edinburgh, KG, is the Society's patron. Many well-known people are members of the Society, including King Hussein I of Jordan (JY1), Lord Wallace of Coslany and actor-impresario Brian Rix (G2DQU), who is a Honorary Vice-President.

The Society acts as the spokesman for the radio amateur in the UK and is one of the founder members of the International Amateur Radio Union, the world-wide association of over 100 national radio societies.

The Society was founded (as The London Wireless Club) in 1913 but soon attracted members throughout the country. The name "Radio Society of Great Britain" was formally adopted in 1922.

The Society helps its members in many ways. Of particular importance is the provision of information on technical matters and on the various activities and events of concern to amateurs. Since 1925 the RSGB has published a monthly journal—*Radio Communication* (formerly *RSGB Bulletin*)—the longest-established and largest magazine devoted to amateur radio in the UK. All members of the Society are sent its journal monthly.

The Society is directed by its Council who are composed primarily of elected Corporate members. Any such member of the Society may be nominated to serve on the Council provided he/she has been a member for over five years and is proposed by not less than 10 Corporate members. This Council comprises of 18 persons, of whom seven are elected on a geographical basis and thus provide representation from all parts of the UK. Additionally there are 20 regional representatives elected by members in their own region who are responsible to the geographically elected members of the Council. A large number of area representatives, nominated by the membership, operate in close liaison with the regional representatives and these provide Society support at local level. The RSGB also has several hundred affiliated societies, groups and clubs within the UK. These clubs form the nucleus of local amateur radio activity and the Society provides a considerable level of support for these clubs in many ways.

The Society maintains about 20 full-time staff who are involved with day-to-day operations at all levels. All the normal company tasks have to be performed: accounting, sales promotion, public media relations, journal and book production, stock control and a host of other functions.

The Society is proud of its 60-year-plus role in the development of amateur radio and of the many eminent scientists who have been connected with it—including Marconi, Lodge and Fleming. It was the organized radio amateurs who originally demonstrated that wavelengths below 100m could provide world-wide communication on low power. But the Society looks to the future, rather than the past, and concentrates its efforts on the effective organization of amateur radio activities both nationally and internationally to provide the greatest opportunities for useful experimental work, and to encourage general interest in and enjoyment of the hobby.

## Why you should join

All those who are actively engaged in any form of radio research, experimentation or communication are eligible for election as Corporate Members. This includes anyone who builds or operates amateur radio equipment. You do not have to be engaged professionally in radio—but equally this does not debar you from joining. Many members do in fact work in the electronics field—often as the result of their interest in amateur radio—but for very many others radio is purely a spare-time hobby. If you are under 18 years of age, a genuine interest in experimental radio makes you eligible for Associate Membership. Associate Members have all of the privileges of full membership except they do not vote in the annual Council election or at meetings.

The following are just a few of the many reasons why, if you are really interested in amateur radio, you should join the RSGB immediately.

You will receive every month a copy of *Radio Communication*, recognized as providing a complete and accurate survey of every phase of amateur radio activity. Containing at least 64 pages each month, the journal is noted internationally for the high standard of its technical and constructional articles, written by Britain's leading radio amateurs, and the wide scope of its news coverage. The needs of newcomers are not overlooked in the selection of technical articles.

**HRH Prince Philip operates the Townsville Amateur Radio Club station VK4TC and talks to the crew of the Las Balsas raft expedition. Prince Philip is the Patron of the Radio Society of Great Britain**

You will be able to use the world's largest and most comprehensive free QSL Bureau operated by the Society. Use of this efficient bureau will save the active amateur and listener a great deal of trouble and expense. QSL cards are sent and received from the bureau in batches. This eliminates the need for stamping, addressing and posting of individual cards. The bureau distributes the cards via other national societies' bureaux to amateurs throughout the world. Full details of how the RSGB QSL Bureau operates are sent to every member on election.

You will receive a certificate of membership and a lapel badge which identifies you as part of the amateur radio movement. Members who do not hold amateur transmitting licences are given special identification numbers for use in connection with the QSL Bureau, beginning BRS (British Receiving Station) or ORS (Overseas Receiving Station), followed by a number.

You will be encouraged to contribute, according to your interests, to the advancement of amateur radio. Many members serve as local representatives or on local or national committees, or pass on to other members through *Radio Communication* or by lectures the results of their experiments and observations. The members in fact *are* the Society.

The Society helps organize many special scientific studies and tests, and has set up a Radio Amateur Emergency Network in collaboration with the British Red Cross Society, the St John Ambulance Brigade, the County Emergency Planning Officers and the Police.

The Society maintains special committees dealing with such matters as television interference, licences and questions concerned with the Town and Country Planning Act, which are often able to help with related problems.

The Society is recognized as the representative of the amateur radio movement in all negotiations with the Home Office (Radio Regulatory Division) on matters affecting the issue of amateur transmitting licences.

## How to join the RSGB

Joining the RSGB is simple, but there are of course certain formalities to be observed. As explained earlier, anyone with an active and genuine interest in amateur radio is warmly welcome to apply for membership.

All applicants, for both Corporate and Associate membership, should be proposed by a Corporate Member of the Society to whom they are personally known. The member simply completes the proposal on the application form, and you will find that he will be glad to do this.

The Society however fully recognizes that many newcomers to amateur radio—who are most welcome as members—may not know or be in touch with other members. In such cases a brief reference in writing should be submitted from a suitable person who can vouch for your interest in amateur radio.

In order to apply for Associate membership, you must be under 18 years of age. There is also a reduced-rate student Corporate membership.

An application form for membership can be obtained on request from RSGB headquarters.

All applications are placed before the Council at its regular meetings.

The subscription rates are given on the application form, and the first year's subscription should be sent with the completed form. After having been a Corporate Member for five consecutive years, subject to the approval of Council, a member can commute all future annual subscriptions by a payment of a fee determined by the Council.

All correspondence should be sent to the Radio Society of Great Britain, 35 Doughty Street, London WC1N 2AE. The telephone number is 01-837 8688.

## What the Society does

Some of the important activities of the Society have already been described, but there are many other ways in which the Society helps radio amateurs and all those interested in amateur radio. A few are outlined below.

### Publications

One main publishing activity of the Society is in producing the monthly *Radio Communication*. But the Society also produces many books and other publications to help the amateur. Because the Society is anxious to disseminate sound technical information as widely as possible, many of these publications are issued at prices below what a commercial publishing organization would have to charge. An example is the RSGB *Radio Communication Handbook*, a large comprehensive book in two volumes covering principles, design and construction of all types of amateur equipment. Some publications of the Society are specially written for newcomers to help them obtain transmitting licences; these include this book, *A Guide to Amateur Radio*, and the *Radio Amateurs' Examination Manual*; morse code instruction cassette tapes have also been produced. The Society also provides facilities for obtaining a selection of the many amateur radio publications issued in the USA where there are over 300,000 radio amateurs.

Technical guidance and a large number of practical hints and circuits will be found in *Amateur Radio Techniques* published by the RSGB. For vhf and Class B enthusiasts there is the *VHF-UHF Manual*. Other publications include *Amateur Radio Operating Manual*, *Radio Data Reference Book*, *Test Equipment for the Radio Amateur*, *TVI Manual* etc. Members receive a discount when purchasing Society publications.

### Meetings

Official Society meetings are held throughout the British Isles. Local meetings are held by RSGB Groups and by affiliated societies. Mobile rallies and specialized conventions are also held regularly. A list of affiliated societies and clubs is given in the *RSGB Amateur Radio Call Book*.

### Frequencies

The Society maintains close liaison with the Home Office on all matters affecting licence facilities and the frequencies assigned to amateur radio, and regularly sends official representatives to the important World Radio Conferences of the International Telecommunication Union and other conferences where decisions vital to the future of amateur radio are taken.

### Contests and field days

Many interesting tests, contests and field days—many of which are open to listeners as well as transmitting members—are held each year. Trophies or certificates are awarded to leading entrants.

### Achievement certificates

A number of certificates representing graded degrees of achievement in amateur radio operating (receiving as well as transmitting) are issued by the Society. These include the DX Listeners' Century Award, The Worked the British Commonwealth Award, 4-2-70 Squares Award and the Commonwealth DX Certificate. The rules governing all awards are available from RSGB headquarters.

### Slow morse transmissions

The Society sponsors the transmission by amateurs throughout the country of morse practice lessons intended for beginners. Details appear periodically in *Radio Communication*.

**RSGB NEWS BULLETIN SERVICE**

The RSGB news bulletin, callsign GB2RS, is broadcast every Sunday morning on hf and vhf, giving almost complete coverage of the British Isles. Its main purpose is to provide an outlet for amateur radio news items and announcements.

Bulletins are transmitted at 30min intervals (starting at 0900 local time) from many sites in the UK and Channel Islands on 3,650kHz ssb (alternatively 3,640 and 3,660kHz in some Scottish regions); 7,047·5kHz a.m. (at 1100 hours only); 144·250MHz ssb (horizontal polarization); 145·525MHz (S21) fm (vertical polarization). Time/location schedules appear regularly in *Radio Communication*, or are available by sending an sae to the Membership Services Officer at RSGB HQ.

An rtty news bulletin, callsign GB2ATG, is also transmitted every Sunday at 1200 and 1900 on 3·590MHz and at 1230 and 1245 on 144·6MHz. This bulletin carries items of interest to rtty enthusiasts.

**Beacon stations**
A number of beacon stations on hf, vhf, uhf and microwave bands are operated by or in conjunction with the RSGB and details are published regularly in *Radio Communication*.

* * *

Society members have a committee to help members overcome any problems caused by breakthrough to television reception; collate information relating to applications to local authorities for antenna installations; maintain an "intruder watch" to detect breaches of international regulations in the use of amateur frequencies by commercial or broadcasting stations; run a recorded lecture library; and help provide many other facilities.

* * *

In brief, the Society supports and encourages all activities "For the advancement of amateur radio". It welcomes within its ranks all those who share this aim.

# THE ROLE OF THE AMATEUR

No introduction to amateur radio and the RSGB can be considered complete without at least a few very brief notes on how this remarkable international hobby has developed through the years; and on how it has already contributed a vast amount of original work to the practice of radio communication and to the self-training of countless thousands of radio operators and engineers. This work, it should not be forgotten, has been financed entirely by the amateurs themselves, without any form of government subsidy, and arises solely from the very real interest to be found in such work as a hobby.

## How amateur radio began

Although, from the very earliest days, amateur experimenters followed in the wake of Hertz, Loomis, Lodge, Marconi and Popov in developing telegraphy without wires, amateur radio may be said to have been officially recognized in the UK in 1904, when the first "Wireless Telegraphy Act" made necessary the registration of all radio receiving and transmitting apparatus, but made specific provision for the use of such equipment for experimental purposes.

By 1914 some 1,000 amateur experimental permits with three-letter callsigns (MXA, etc) had been issued in Britain. Meanwhile, in 1912, after a period of growing congestion of the radio frequencies, the first specifically amateur radio licences were issued in the USA allowing unrestricted operation on "200m and down"—wavelengths then believed by most professional communication engineers to be valueless for other than short-distance working.

British enthusiasts soon felt the need for some organization to cater specifically for their interests and 1913 saw the formation of the Wireless Society of London with members throughout the country. It was this Society which in 1922 changed its name to the Radio Society of Great Britain.

Almost all of these early amateur stations used spark transmitters and crystal receivers but even before 1914 primitive attempts were being made at telephony as well as telegraphy operation, and some of the larger stations were already covering distances to be measured in tens of miles.

On the outbreak of war in 1914 all British experimental stations were closed down and there was a total ban on amateur receiving as well as transmitting. Indeed it was only after a tremendous struggle that the first British amateur licences to use the present style of callsigns (apart from the lack of an international prefix) were issued late in 1920. At first many of these stations used wavelengths of 1,000m and later 440m, but these permits, like the American ones, made provision for operation below 200m. Soon spark transmitters were replaced by valves, and a good deal of operation was on phone.

## The dawn of international dx

The early 'twenties saw the famous series of Trans-Atlantic and Trans-Ocean Tests in which the RSGB co-operated closely with the American and Commonwealth societies and which culminated in the opening of the shortwaves (below 200m) for long-distance, low-power, two-way working. The first American amateur stations were heard in Britain in November 1921 and next year amateur stations in London (operated by the RSGB) and Manchester were heard in the USA. A year later, in November 1923, the first amateur two-way trans-Atlantic contact took place on about 110m between France and America—and within a few days British amateurs reduced wavelength and began to work across the Atlantic.

With the valves and components then available it was no simple matter to achieve operation on wavelengths of 100m and below. But by the following autumn, two-way contacts between Britain and New Zealand on about 80m represented the longest span possible on the globe.

Soon operation around 40 and 20m was producing long-distance contacts at all times of the day and night, and thereby began a flood of commercial stations opening up to exploit these discoveries which had stemmed directly from amateur radio. By 1928 the Atlantic was spanned on 10m, and even 5m was being used by the amateurs before the end of the 'twenties.

But the continuing necessity for an effective organization to look after the interests of the amateurs was again strikingly shown in the mid-'twenties when an attempt was made by the authorities—alarmed at the ease with which the amateurs were working one another throughout the world—to introduce a ban on international working by British amateurs. Fortunately this ban was soon circumscribed—though its effect was to remain for many years in the banning of the signal "CQ" by British stations until 1946.

## Amateurs and broadcasting

Meanwhile British amateurs had played a decisive role in the institution of regular broadcasting in the UK. Although much interest had been aroused by some experimental broadcasts made by the Marconi Company in 1920, the British government hesitated and forbade further broadcasts. This led in December

1921 to the presentation by the Wireless Society of London of a strongly worded petition urging regular transmissions. This petition was almost certainly the decisive factor which led to the agreement that broadcasts should be started from Writtle (2MT) and that a British Broadcasting Company should be formed. As one journal put it: "Regular broadcasting in this country was initiated not only at the request of, but through the insistence of the experimental amateur"—though one of the first results of broadcasting was that the British amateur soon lost the use of 440m (1,000m had already been lost to aircraft communication). Later, a British amateur, the late Mr Gerald Marcuse, 2NM, was to be instrumental in starting British broadcasting on hf to the Commonwealth.

The year 1925 saw the formation of the amateur's international organization (IARU) which forms a link between the various national societies which exist in almost all countries of the world, and the start of the *T and R Bulletin* (now *Radio Communication*) as the journal of the "Transmitter and Relay" Section of the RSGB—the section which soon afterwards guided the main body to concentrating almost exclusively on amateur radio activities.

During the 'thirties membership of the RSGB increased from about 1,000 to almost 4,000 by 1939, and a full-time secretariat was set up. In 1933 the first edition of *A Guide to Amateur Radio* was published, to be followed in 1938 by the first edition of the *RSGB Amateur Radio Handbook* (now *Radio Communication Handbook*).

Once again, on the approach of war, amateur licences were withdrawn in Britain (31 August 1939) and remained in suspension until January 1946. But throughout the war years the spirit of amateur radio was kept very much alive by the many amateurs who used their radio operating and technical skills on behalf of their country, and indeed by the end of the war membership of the RSGB had more than doubled to some 9,000 members!

And so this time, thanks to the high regard in which amateur radio was now held and the effective organization kept in existence by the RSGB, activities were resumed within three months of the end of the war, with licences noticeably more liberal than the pre-war experimental permits. Soon there were more amateurs than ever before, not only in the UK but throughout the world.

But amateur radio again faced an extremely difficult period in the late 'forties and 'fifties when, with the spread of tv, it seemed for a time that many amateur activities might be seriously curtailed by the formidable problem of operating stations alongside domestic television receivers without causing interference. Active measures by the amateurs, encouraged by the RSGB and the USA society ARRL, succeeded to a considerable extent in reducing this problem.

## Amateurs today

Today the number of licences in the UK stands at an all-time high of over 25,000, and amateur radio enjoys facilities which should ensure its future growth. But the enormous growth of other forms of radio communication in the hf and vhf spectrums—in many cases using techniques which have come from the amateurs—means that the demand for rf allocations grows constantly more pressing—and this underlines the necessity for radio amateurs to maintain strong national and international organizations to keep careful watch on the position: the RSGB is recognized throughout the world as one of the leading organizations among those which have striven effectively over many years to improve or retain the vital operating privileges.

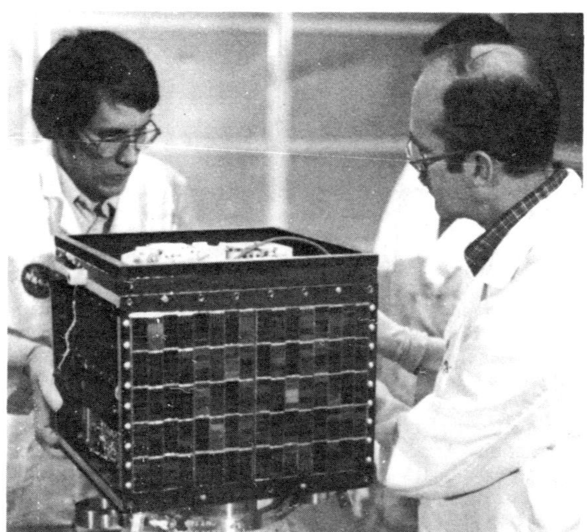

The amateur radio satellite Oscar 8 pictured during final preparation. Launched in 1978, this satellite is providing thousands of amateurs all over the world with intercontinental contacts on vhf

Amateurs continue to play a significant role in the development of practical hf and vhf radio communication, and in such allied fields as radio astronomy and the scientific study of radio propagation. Equally important is that many of the scientists, engineers and technicians in this sphere owe their initial interest to the hobby; while those amateurs without professional connection with electronics form a most useful body, within the community, of knowledgeable opinion on radio communication.

One example of the current work of the amateurs in keeping in the forefront of communication technology is illustrated by the fact that since December 1961 a series of successful communication satellites—called Oscars from "orbital satellites carrying amateur radio"—have been orbited. Although the launchings have been made by the USA services (an indication of the regard in which amateur radio is held there) the actual satellites have been designed, constructed and used operationally by amateur enthusiasts. Oscar 7 was successfully launched in November 1974 and Oscar 8 in March 1978.

One sometimes hears it suggested that as amateur radio is now a relatively "ancient" hobby, already well past its first half-century of exciting years, there can be little scope left for fresh experimental work. Nothing could be farther from the truth. Today, perhaps more than at any other time for many years, we stand at the brink of breath-taking new developments in radio communication. Particularly is this true in the tremendous number of new applications of transistors and other semiconductor devices; in improved methods of modulation; and in the microwave regions. Already it is possible to build a complete two-way phone station for world-wide operation in less space than is occupied by the average domestic hi-fi audio amplifer. Tomorrow . . . who knows?

CHAPTER 11

# International amateur radio organizations

For more than 50 years in most countries of the world, amateur radio has been run on a well-organized and effective basis through the medium of the local licensing authorities and national societies devoted to the hobby. Most of these societies maintain very close liaison with the local licensing authorities and also with each other through an organization called the International Amateur Radio Union whose membership is made up of the national societies of all the major countries (with the exception of China). The national societies publish journals or newsletters devoted to local amateur radio operation, operate QSL bureaux, organize conventions and meetings and organize and run hf and vhf contests. Notes on some of the English-language societies are given later, but it should be appreciated that such societies exist in almost every country where amateur radio operation is permitted or encouraged. The use of the radio spectrum is administered by individual countries in accordance with the *Radio Regulations* of the International Telecommunications Union, and the world is divided into three regions:

*Region* 1 comprises Europe, Africa, and the USSR and Turkey.
*Region* 2 is the Western Hemisphere, comprising the Americas, Greenland and all countries under the control of the Federal Communications Commission.
*Region* 3 is Australasia, Oceania and Asia, except those territories located in Regions 1 and 2.

It will be noted that the frequency allocations are appreciably more generous in Regions 2 and 3 compared with Region 1, particularly in respect of the limits of the 3·5MHz band, 7MHz in Region 2, the 50MHz band in Regions 2 and 3, the availability of 146 to 148MHz in Regions 2 and 3, the 220MHz band in Region 2 only, and the much wider 420MHz band in Regions 2 and 3. Local licensing authorities sometimes impose additional restrictions on the frequencies available in their country, or occasionally make special arrangements for additional frequencies: for example some countries permit amateur operation in the industrial allocation on 27MHz, similarly the UK has a special allocation on 70MHz.

**BASIC ITU AMATEUR FREQUENCY ALLOCATIONS**

| Region 1 | Region 2 | Region 3 |
|---|---|---|
| (1,800–2,000kHz)† | 1,800–2,000kHz† | 1,800–2,000kHz† |
| 3,500–3,800kHz | 3,500–4,000kHz | 3,500–3,900kHz |
| 7,000–7,100kHz | 7,000–7,300kHz | 7,000–7,100kHz |
| 14,000–14,350kHz | 14,000–14,350kHz | 14,000–14,350kHz |
| 21,000–21,450kHz | 21,000–21,450kHz | 21,000–21,450kHz |
| 28,000–29,700kHz | 28,000–29,700kHz | 28,000–29,700kHz |
| 144–146MHz | 50–54MHz | 50–54MHz |
| 430–440MHz | 144–148MHz | 144–148MHz |
| Microwave bands | 220–225MHz | 420–450MHz |
| | 420–450MHz | Microwave bands |
| | Microwave bands | |

†Only available in some countries and often under special restrictions.

## International Amateur Radio Union

The International Amateur Radio Union (IARU) was founded in Paris in 1925 with the American Radio Relay League (ARRL) providing headquarters administrative facilities. Although originally intended to have a world-wide membership made up of individual amateurs, a revised constitution in 1929 restructured the Union so that the members became the national amateur radio society of each member country.

In 1950 the IARU Region 1 Division was formed to promote the special interests of member societies in the International Telecommunications Region 1 (Europe, Africa and parts of Asia) and to represent their interests at the ITU radio conferences; subsequently similar regional divisions were established for Region 2 (Western Hemisphere) and Region 3 (Australasia, Oceania and parts of Asia). Over 40 countries are represented in Region 1 and a regular publication, *Region 1 News*, is issued by the secretary, Roy F. Stevens, G2BVN. There is close contact between member societies in each region and between the regional divisions and IARU headquarters in the USA.

## Australia

The IARU member-society for Australia is the Wireless Institute of Australia which, having been founded in 1910, is the longest-established of all member societies. The address of the Institute is PO Box 150, Toorak, Victoria 3142 (registered office Above 474 Toorak Road, Toorak, Victoria 3142), and it publishes a well-established and informative monthly journal, *Amateur Radio* (editorial address PO Box 2611W, Melbourne, 3001).

Licences are issued in Australia by the Postmaster General (Australian Post Office), generally in accordance with Region 3 ITU frequency allocations. A novice licence was introduced in 1975. Licensing authority: Postal & Telecommunications Dept, RFMD, GPO Box 5412CC, Melbourne 3001. VHF phone-only licences issued.

## Canada

The IARU member-society for Canada is the Canadian Division of the American Radio Relay League (see USA) although several specifically Canadian amateur radio societies and groups exist and publish journals.

Amateur licences are issued by the Department of Communications and all amateurs are required to operate on cw and hf during the first six months (stations being initially licensed for telephony only above 50MHz). The minimum morse code test speed is 10wpm and the advanced amateur radio operator's certificate requires a 15wpm morse test and more difficult oral and written tests. Prospective amateurs living in remote areas may obtain a provisional licence by certifying that they can meet technical and operating requirements.

Bands available to Canadian amateurs include 1,800–2,000kHz (but with geographical restrictions), 3,500–4,000kHz, 7,000–7,300kHz, 14,000–14,350kHz, 21,000–21,450kHz, 28,000–29,700kHz, 50–54MHz, 144–148MHz, 200–225MHz,

420–450MHz plus microwave bands. There is compulsory band-planning with rather broader telephony sub-bands than in the USA. Licensing authority: Department of Communications, 300 Slater Street, Ottawa, Canada. No technician or novice licences issued.

## Irish Republic

The IARU member-society for the Irish Republic is the Irish Radio Transmitters Society. The honorary secretary is Mr J. Ryan, 23 Dollymount Grove, Clontarf, Dublin 3. A monthly newsheet is published. Licences are issued by the Department of Posts and Telegraphs. Frequency allocations are very similar to those of the UK. Licensing authority: General Branch (Radio), GPO, Dublin 1. VHF phone-only licences issued.

## New Zealand

The IARU member-society is the New Zealand Association of Radio Transmitters, Box 1459, Christchurch. It was founded in 1926 and publishes an excellent monthly journal, *Break-in* (editorial address Box 1733, Christchurch). Amateur licences are issued by the Post Office in various grades (Grade 1 provides full operating privileges whereas Grade III is restricted to 144MHz and above for telephony, television and telecontrol) and 3·5MHz Novice licences. Citizen's Band licences are issued for spot frequencies in 26 and 465MHz bands for type-approved equipment only.

Amateur frequencies (Grade 1 licences) include: 1,875–1,900kHz, 3,500–3,900kHz, 7,000–7,100kHz, 14,000–14,350kHz, 21,000–21,450kHz, 26,960–27,230kHz, 28,000–29,700kHz, 51–53MHz, 144–148MHz, 420–449MHz and microwave bands. Licensing authority: Chief Radio Inspector, PO District Engineer's Office, Cambridge Terrace, Wellington.

## South Africa

The IARU member-society is the South African Radio League (Suid-Afrikaanse Radioliga) founded in 1925, whose headquarters address is PO Box 3911, Cape Town 8000 (Room 430 4th Floor, CTC Building, Plain Street, Cape Town). The League publishes a monthly bilingual journal, *Radio-ZS*. Licences are issued by the Postmaster General, Ministry of Posts and Telegraphs. Frequency allocations are basically those of Region 1. Licensing authority: Director of Telecommunications, Private Bag X74, 0001 Pretoria. VHF phone-only licences (ZR prefix) issued.

## United States of America

The IARU member-society is the American Radio Relay League (ARRL) founded in 1914, with administrative headquarters at ARRL, Newington, Connecticut, USA 06111. The League publishes a monthly journal, *QST*, the doyen of the regular amateur journals, an annual handbook, *The Radio Amateur's Handbook* (over 50 editions), and many other publications.

Licences are issued by the Federal Communications Commission in five classes: Novice, Technician, General ("Conditional" where examination is taken by mail), Advanced and Amateur Extra Class. Over 300,000 amateur licences of which about 8·5 per cent Novice, 19·1 per cent Technician, 42·5 per cent General or Conditional, 23 per cent Advanced and 5 per cent Extra class are currently issued. Licensing authority: FCC, Box 1020, Gettysburg, PA 17325. Note that compulsory band-planning includes following cw-only sub-bands: 3,500–3,775kHz; 7,000–7,150kHz; 14,000–14,200kHz; 21,000–21,250kHz; and 28,000–28,500kHz with some band segments available only to certain licence categories.

Commercial publications devoted entirely to amateur radio are published monthly and include: *CQ*, *Ham Radio* and *73 Magazine*. Many clubs, societies and groups exist devoted to specialized aspects of the hobby.

## United Kingdom

The IARU member-society is the Radio Society of Great Britain, founded in 1913, whose address is 35 Doughty Street, London, WC1N 2AE. The Society publishes a monthly journal *Radio Communication*, and many other publications.

Amateur licences issued by Home Office (Radio Regulatory Division).

A commercial monthly publication devoted to amateur radio is *Short Wave Magazine*. Many specialized clubs and societies issue newsletters and bulletins, including societies representing the Royal Signals, Royal Navy and Royal Air Force.

The British Amateur Television Club publishes *CQ-TV* devoted to amateur tv (membership secretary Brian Summers, 13 Church Street, Gainsborough, Lincs, UK).

The British Amateur Radio Teleprinter Group issues the *BARTG Newsletter* (editor J. B. Hodgson, 234 Gillingham Road, Gillingham, Kent ME7 4QT).

CHAPTER 12

# Fundamentals of electronics

Many of those who wish to understand amateur radio have already been interested in other aspects of radio and electronics, and come to the hobby with a working grasp of the principles and terminology of electricity and electronics. Such knowledge has been assumed in these pages since to provide a detailed introduction to these subjects requires a book in itself, and indeed many such books are readily available. The *Guide* therefore concentrates on the practical aspects of amateur radio equipment, operating practices and codes, licences and regulations.

However more and more people are being attracted to the hobby who, initially, would not claim to have more than a hazy idea of electrical and electronic principles, although eager to learn, or to refresh their memory of what they may have learned at school or elsewhere. In this chapter a brief outline is therefore given, written largely in the form of an *aide mémoire*, of fundamental principles, such as those required for radio amateurs' examinations in the UK or needed to understand the more practical guidance given elsewhere in this book.

## Electricity

The principles of electricity are based on aspects of atomic physics and the movement of tiny electric *charges* from atom to atom. Any sustained movement of electric charges, which may represent an excess or deficiency of *electrons*, results in an electric *current*. Materials in which a current can be readily induced to flow (usually metals but also some liquids) are called *conductors*; those in which such movement occurs only with extreme difficulty are *insulators*; and there are also some rather unusual substances between the two, called *semiconductors*, which may be changed from insulators into conductors by quite a small change in temperature. Of the metals, silver, copper, gold and aluminium are the best conductors.

An *electron* is a tiny particle that is virtually weightless and carries a *negative* charge, and a flow of electrons can be induced in a conductor by a *potential difference* (electromotive force) applied along it. A potential difference is measured in *volts* (V) or millivolts (mV, thousandths of a volt). Electron flow is from negative to positive polarity. The electrons form the *charge carriers*, but alternatively these may be in the form of positive charges (*holes*) in solid-state devices.

The current flow is measured in terms of the number of electrons flowing per second, and the practical unit is the ampere (A) although this represents the flow during one second of an extremely large number of electrons ($10^{18}$) and in electronics currents are more usually measured in *milliamps* (mA where 1,000mA = 1A) or even microamps ($\mu$A, where 1,000$\mu$A = 1mA).

There are several practical methods of generating electricity: *chemical* (as in batteries); *electromagnetic* (by moving conductors in a magnetic field); *piezo-electric* (mechanical pressure on crystalline substances including quartz and certain ceramics); *photoelectric* including photovoltaic cells to form solar arrays to generate electricity from sunlight; and *thermoelectric* using the Peltier effect to generate electricity from heat.

## Resistance

All conductors, even thick copper wire, offer some impediment to current flow by absorbing energy in the form of heat. This characteristic is called *resistance* and is measured in ohms (Ω) or often in kilohms (kΩ) or even megohms (MΩ) where 1MΩ = 1,000kΩ = 1,000,000Ω. The resistance of metallic conductors normally increases with temperature but that of a semiconductor material (eg silicon, germanium, gallium arsenide) tends to fall rapidly with rising temperature.

A fundamental relationship exists in any electrical circuit between voltage, current and resistance, a relationship known as Ohm's law. This states that the ratio of the voltage applied across a resistance to the current flowing through that resistance is a constant. That is to say:

$$\frac{\text{Voltage } (V)}{\text{Current } (I)} = \text{a constant} = \text{Resistance } (R).$$

This can be expressed in several ways: $R = V/I$; $V = I \times R$; or $I = V/R$. Thus if any two of these are known, the third can be found by calculation.

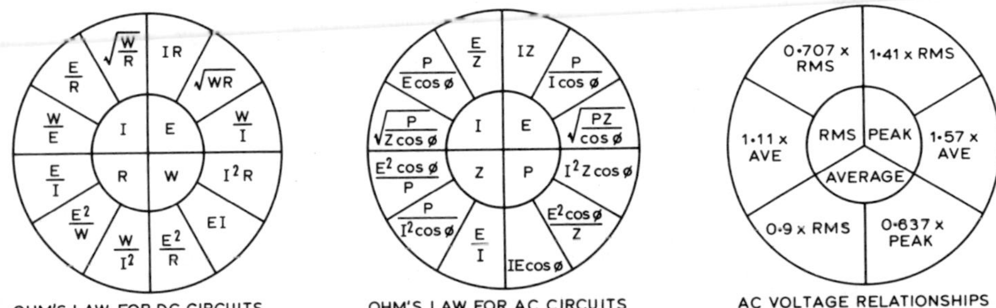


**Fig 1. Diagrammatic representation of Ohm's law and ac voltage relationships. W, watts; E, volts; I, amps; R, ohms; Z, impedance (ohms); $\phi$, phase angle**

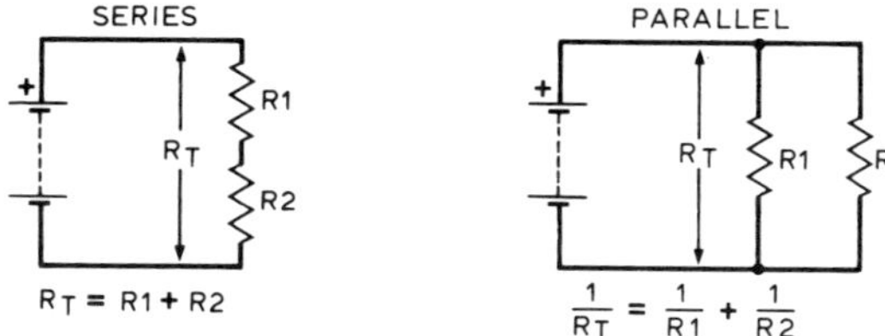


Fig 2. Resistors in series and parallel

The amount of energy absorbed in the resistance of an electrical circuit represents *power*, measured in watts (W), where 1W represents 1A flowing in a 1Ω resistor. This by Ohm's law will happen when a potential difference of 1V is applied across such a resistor. We can thus show that:

$$\text{Power } (W) = V \times I, \quad \text{or} \quad I^2R \quad \text{or} \quad V^2/R$$

**Series and parallel**

When a number of resistors are connected in a chain (that is to say in *series*) and a potential difference applied across the two outer ends, the same current must flow through them all, and the total resistance amounts to the sum of them all. That is to say $R_{(\text{total})} = R_1 + R_2 + R_3 \ldots$. In such an arrangement the total voltage will be divided across each resistor in the proportion of that resistance to the total resistance.

However if the resistors are each connected across the full supply voltage, in *parallel*, each resistor now offers an alternative path for electron flow. The total current will be the sum of the individual currents, but the total resistance of such an arrangement must be less than that of any of the individual resistors. This is given mathematically by the expression:

$$\frac{1}{R_{(\text{total})}} = \frac{1}{R_1} + \frac{1}{R_2} + \frac{1}{R_3} \cdots$$

Where only two resistors are concerned this may be written as:

$$R_{(\text{total})} = \frac{R_1 \times R_2}{R_1 + R_2}$$

We have noted that a circuit consisting of two or more resistors in series will exhibit a voltage across each particular resistor governed by the proportion of the total resistance it represents and the total voltage across the chain. By connecting suitable resistors across a power supply we can obtain any desired distribution of voltage across each, providing a potential divider, but it must be noted that if in turn this new voltage is used with a further load resistor etc, this further load will affect the voltage distribution so that such a *potential divider* does not provide a constant (*regulated*) voltage under conditions of fluctuating load.

## SI units

Units of measurement used in the Radio Amateurs' Examination are those of the International System of Units (SI units) and comprise essentially a small number of internationally-accepted base units from which all other units can be derived. The following are the base units:

| | |
|---|---|
| length | metre (m) |
| mass | kilogram (kg) |
| time | second (s) |
| electric current | ampere (A) |
| temperature | kelvin (K) |
| luminous intensity | candela (cd) |

Among the important derived units found in radio communication are:

| | |
|---|---|
| frequency | hertz (Hz) |
| power | watt (W) |
| electric charge (quantity of electricity) | coulomb (C) |
| electric potential (electromotive force) | volt (V) |
| electric capacitance | farad (F) |
| electric resistance | ohm (Ω) |
| inductance | henry (H) |
| magnetic flux | weber (Wb) |
| magnetic flux density | tesla (T) |

These units are often expressed as multiples or submultiples using a series of standard prefixes:

| | | | |
|---|---|---|---|
| giga (G) | $10^9$ | milli (m) | $10^{-3}$ |
| mega (M) | $10^6$ | micro (μ) | $10^{-6}$ |
| kilo (k) | $10^3$ | nano (n) | $10^{-9}$ |
| | | pico (p) | $10^{-12}$ |

It should be noted that although kilo represents a multiple of 1,000, the abbreviation k is correctly shown by a lower case and not a capital letter: we thus have GHz, MHz but kHz.

## Direct and alternating current

In radio and electronics we are concerned not only with circuits in which there is a steady flow of electrons from the negative to positive poles of the supply, but also with currents and voltages that are fluctuating in magnitude or where the current regularly reverses direction due to changing polarity of the source. Such currents may be:

(1) Alternating currents (ac) where the direction and magnitude changes in a regular, periodic manner. A well-known example is that of equipment fed from mains supplies which reverse direction 100 or 120 times per second. As noted below, this represents 50 or 60Hz.

(2) Fluctuating direct currents in which the current is continuously varying but does not reverse in direction (ie the supply source does not drop below zero volts).

(3) Pulsating dc in which the current varies regularly between zero and maximum in a series of pulses but again does not change direction of flow.

Since an alternating current or voltage is constantly changing (often it follows a *sine* waveform) it is necessary to define the conditions under which voltage, or current or power is measured. A sine wave current rises regularly from zero to maximum, then decreases similarly back to zero, changes direction and rises to a maximum in the reverse direction, before falling once again to zero, following a mathematical law. The number in each second of such complete *cycles* is termed the *frequency* of the supply and is measured in hertz (1Hz = 1 cycle per second). With such sine wave ac it is normal to specify voltage not as the peak values at the crest of the waveform but as a *root-mean-square* (rms) value. This is approximately 0·7 of the peak voltage and represents the same total energy that would be delivered to the same load by an equivalent dc supply. For example the domestic ac mains supply has an instantaneous voltage that is continuously varying between 0 and ±338V but the heating effect of the supply when connected to a resistive load would be the same as that of 240V dc supply; thus 240V is the rms value; 338V is the *peak* value; while the peak-to-peak value represents the difference between the maximum positive and the maximum negative instantaneous voltage, in this case 2 × 338 or 676V peak-to-peak. Mathematically the peak value represents the rms value multiplied by the square root of 2: that is to say $E_{\text{peak}} = \sqrt{2} \times E_{\text{rms}} = 1{\cdot}414E_{\text{rms}}$ and $E_{\text{peak-to-peak}} = 2\sqrt{2}E_{\text{rms}} = 2{\cdot}828E_{\text{rms}}$.

The same relationships apply also to current in a resistive load. With ac there is a further relationship which may sometimes be required, the *average* value, although this is of relatively little practical importance. For a 240V rms supply, the average value would be 0·9 × $E_{rms}$ or 0·637 × $E_{peak}$, that is to say 216V average.

If we connect an ac supply across a resistor or resistive load, then we can use Ohm's law. In practice, however, we may wish to include capacitance and inductance in the circuit. When a capacitor is connected across a dc supply, there is an initial surge of energy into the component but then no further current; however this is not the case with ac as the reversals of polarity mean that energy first flows in and then out of the capacitor so that current is continuously flowing in the external circuit. With an inductor, energy is stored in the form of the magnetic field in such a way as to oppose any change in current, so that such a component presents more impediment to an ac supply than to dc. Such components thus offer a resistance or *impedance* which differ from those offered to a dc supply; they are said to be *reactive* and although this reactance is measured in ohms, the value depends upon the frequency of the ac supply.

**Phase**

With ac, reactive components disturb the *phase* or timing of the relationship which normally exists between current and voltage, due to their providing storage of energy which means that voltage is available at those times when the voltage of the ac supply is nominally zero. The reactance offered by a capacitor falls as the frequency increases, offering infinite resistance to dc which may in such cases be thought of as ac with a frequency of 0Hz. On the other hand a coil (inductor) offers more opposition to current flow as the frequency of the supply increases (at dc opposing steady current flow only to the extent of the ohmic resistance of the winding).

Energy is stored in a capacitor as an *electrostatic field*, representing stress in the dielectric between the plates of the capacitor. If a dc potential is applied to a capacitor current will flow temporarily; when the potential is removed current will continue to flow until the capacitor is discharged.

When an alternating voltage is applied across a capacitor, current flows into and out of the component with every change of potential, so that current flows continuously in an external circuit. However when an alternating voltage is applied to a circuit containing either a capacitor or an inductor, the timing relationship (*phase*) between voltage and current is disturbed (in other words current will not rise and fall exactly in step with voltage rise and fall), and this results in what is termed a *phase shift*. An inductive reactance causes the current flow to *lag* behind voltage changes; a capacitive reactance causes current flow to *lead* voltage changes. A circuit containing only pure capacitance for example will cause a phase shift of exactly 90° so that current flow is maximum when voltage is minimum.

## Capacitance

Electrical energy in quantity of charge is measured in *coulombs*, a coulomb representing a very large number of electrons (1A for one second so that coulombs = As). A capacitor stores an electric charge in the form of electrons; the measure of the number of electrons it can store under a given electrical pressure is its *capacitance* and this is measured in *farads*. In practice the farad (a farad represents 1 coulomb stored under a potential difference of 1 volt) is too large for practical purposes and submultiples such as microfarads, nanofarads and picofarads are used. Since a capacitor forms a reservoir for storing electrons, capacitance can be defined as the property of a circuit element to oppose any change in voltage.

In simplest form a capacitor consists of two conductive plates separated by a non-conducting *dielectric* material. The capacitance is governed by: area of one plate; spacing between the plates; and the *dielectric constant* of the material separating the plates. Dielectric constant is expressed as relative to that of air (ie air is said to have a dielectric constant of 1, whereas for example mica has a dielectric constant of 5–7). Capacitors may comprise a number of interleaved plates to increase the total area between the plates; a four-plate capacitor would have three times the capacitance of a two-plate unit since an inner plate will have twice the effect of an outer plate.

If a number of capacitors are connected in parallel, the total capacitance will be the sum of the individual components: $C_{(total)} = C_1 + C_2 + C_3 \ldots$ etc.

If a number of capacitors are connected in series, the total capacitance will always be less than the lowest individual value:

$$\frac{1}{C_{(total)}} = \frac{1}{C_1} + \frac{1}{C_2} + \frac{1}{C_3} \cdots$$

## Inductance

When a current flows in a wire a magnetic field begins to expand outwards from the centre of the wire. The magnetic lines of force induce a *back emf* in the wire itself. This represents a voltage opposing any increase in current in the circuit. A similar effect, in the reverse sense, occurs when the current is interrupted: in this case the induced voltage tends to maintain the current temporarily. *Inductance* is defined as this property of a circuit to oppose any change of current.

A straight wire carrying current results in only a weak magnetic field and would normally possess very little inductive reactance except at very high radio frequencies (at 100MHz a 1in length of 23swg wire would have a reactance of about 16Ω). The inductance of a wire is much increased by winding it in the form of a coil or solenoid, termed an *inductor*, and can also be increased by inserting a core of suitable material. Most materials when used as cores show increasing losses as frequency increases and, whereas iron laminations may be used at low frequencies, iron-dust or special ceramic ferrites may be necessary at high radio frequencies to reduce losses and provide a high-*Q* component. A particular form of core of increasing importance in radio applications is the *toroid* and similar related closed loop cores which can provide high values of inductance with very little wire and virtually no leakage of external magnetic fields.

The value of inductance of the solenoid form of inductor depends on a number of factors including radius of the winding, number of turns and length of winding.

The unit of measurement of inductance is the *henry* defined as the amount of inductance required to produce a back-emf of one volt when the current is changing at an average of one ampere per second. This represents a large unit (often hundreds of turns of wire) but inductors of several henrys are found in some applications; in radio-frequency filters and resonant circuits millihenrys and submultiples of millihenrys are used.

If a number of inductors are connected in series, the total inductance will be the sum of them all: ie $L_{(total)} = L_1 + L_2 + L_3$ etc.

If a number of inductors are connected in parallel, the total inductance will be less than the lowest individual value.

## Time constant

The combination of capacitance-resistance or inductance-resistance gives rise to a delay in circuit conditions reaching a steady state. For example if a capacitor is charged from a supply of *V* volts, through a resistor *R* ohms, the potential across the capacitor will take a finite time to build up and indeed will never theoretically quite reach *V*, due to the voltage drop that occurs across *R*. In the time *R* × *C* (where time is in seconds, *R* in ohms

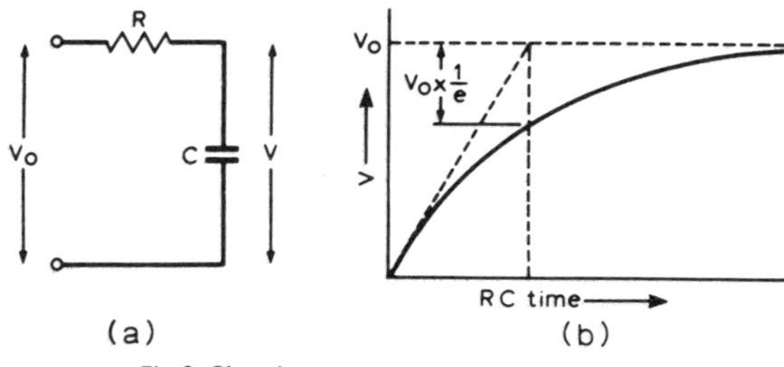


Fig 3. Charging a capacitor through a resistor

and $C$ in farads) the voltage across the capacitor reaches two-thirds (0·667) of $V$, following what is termed an exponential curve. The product $R \times C$ or $R \times L$ is termed the *time-constant* of the circuit and is expressed in seconds (or submultiples such as milliseconds or microseconds).

## Impedance

The opposition to current flow offered by reactive components is governed by frequency and can be calculated as follows: inductive reactance (ie reactance of an inductor) is termed $X_L$ and is equal to $2\pi fL$ where $f$ is in hertz and $L$ in henrys: such a reactance is considered as having a positive value. The reactance of a capacitor ($X_C$) is called capacitive reactance and is equal to $1/(2\pi fC)$ where $C$ is in farads, and this is considered to be a negative value.

When reactances are combined with resistances, the rules governing the current flow (ie Ohm's law) need some modification, although if we think of the total resistance plus reactance as *impedance* then Ohm's law remains valid *at the particular frequency concerned*. This is considered below in connection with resonance of circuits containing both inductive and capacitive reactance.

To denote the "goodness" of a reactive component we can use the term $Q$ which is the ratio of reactance to resistance. With inductors a $Q$ of 100 (ie a component which offers 100Ω inductive reactance to every 1Ω dc resistance) would represent a high-quality component, but for capacitors $Q$s of many thousands are normal.

When a number of inductances or capacitances are connected together in series, the total impedance is equal to the sum of the individual impedances provided that all reactances are of the same type (ie all capacitive or all inductive). However if positive and negative reactances (ie a mixture of capacitive and inductive reactances) are connected in series, the "sign" of the reactance must be carefully observed. For example if an inductive reactance of 100Ω is connected in series with a capacitive reactance of 50Ω, then the result is not 150Ω but 50Ω.

### Resonant frequency

The reactance obtained in this way is frequency conscious in an interesting way: as the frequency increases a positive reactance will increase and a negative reactance will decrease. There will thus be one frequency (resonant frequency) at which such a series combination of capacitance and inductance will have zero reactance, the two reactances exactly balancing out. It is also important to note that if an ac voltage at this resonant frequency is applied across the circuit, the alternating voltage at the junction between the two components will be considerably higher than the applied voltage, due to the *magnification factor*, and this in turn will be dependent upon the $Q$ of the circuit.

If instead of connecting the components in series we connect them in parallel, current rather than reactance reduces to zero and such a circuit represents to an external supply of resonant frequency a very *high impedance*, although oscillatory currents will flow between capacitor and inductor.

For inductors and capacitors of known value, the resonant frequency can be calculated:

$$f_{(\text{resonant})} = \frac{1}{2\pi f\sqrt{(LC)}}$$ with units of henrys, farads and hertz.

It should be noted that since $f$ depends inversely on the square root of $L \times C$, this product needs to decrease by *four* times to *double* the frequency. If the value of one component is fixed, the other needs to be varied over a span of four times to tune the circuit over an octave.

A tuned circuit presents a very high or very low impedance not only precisely at its resonant frequency but also at closely adjacent frequencies, changing impedance as the circuit is detuned at a rate determined by the $Q$ of the circuit. The higher the $Q$ the greater will be the *selectivity* of the resonant circuit and the narrower the width of the band of frequencies (*bandwidth*) passed or rejected by the circuit. A combination of two or more resonant circuits can be coupled together to modify the *response curve* (ie the selectivity characteristics plotted graphically).

## Filters

Inductors and capacitors may be used to form *filters* designed to pass or to greatly *attenuate* signals at specific frequencies or bands of frequencies. Such filters include *high-pass filters* which readily pass all frequencies above a specific frequency but attenuate signals below this critical frequency; *low-pass filters* which pass frequencies lower than the critical frequency but attenuate those above it; *band-pass filters* which pass frequencies between two specified frequencies but reject frequencies outside these limits; and *band-reject filters* which reject frequencies between two frequencies while passing those outside this range. In all such filters, more sharply defined characteristics can be achieved by increasing the number of $LC$ sections, $L$ being a common abbreviation for an inductance, $C$ for capacitance.

## AC circuits

Inductors and capacitors both provide *energy storage*. When a current flows in an inductor it generates a *magnetic field* and any sudden cessation of current results in an *induced* voltage (due to an effect termed *mutual inductance*) which tends momentarily to keep current flowing, or in other words to oppose a change of current to the extent of the energy stored as a magnetic field; note that induced current flow of this type occurs only when current in the inductor is changing, due either to the application or removal of a dc potential, or continuously while an ac potential is applied.

It has been noted that in an ac circuit containing positive and negative reactances, the impedance must take into account the "sign" of the reactance (positive for inductive reactance, negative for capacitive reactance). For example a series circuit comprising 100Ω inductive reactance and 50Ω capacitive reactance would have an impedance the same as that of a single inductor of 50Ω reactance. Such calculations apply only at one specific frequency and assume that the impedance is comprised

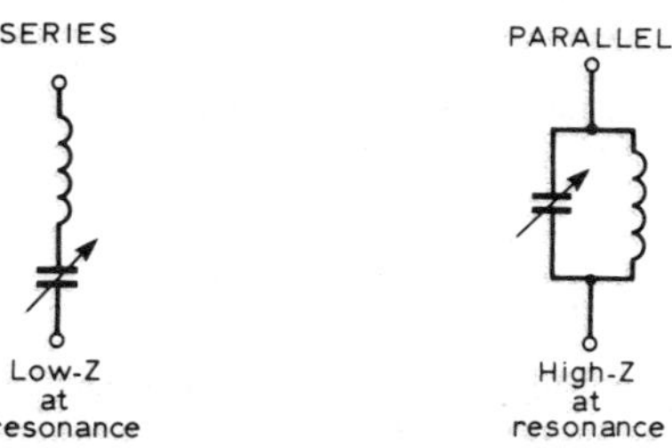


Fig 4. Series and parallel resonance circuits

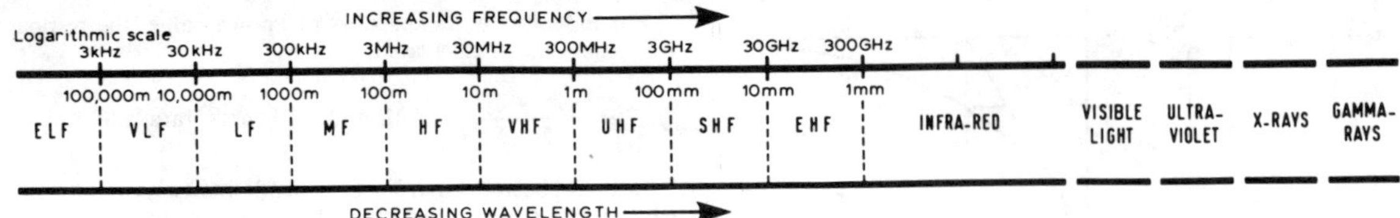


Fig 5. The electromagnetic spectrum

entirely of reactance. In practice, the circuit will inevitably also contain resistance and although this, like reactance, is measured in ohms the two quantities cannot simply be added together, as we must take into account the question of phase. Reactance acts as opposition to current flow but at what may be considered in a direction at right angles (90° phase difference) to the opposition of pure resistance, involving the mathematical concepts of vectors or complex numbers.

The impedance ($Z$) of a circuit containing resistance ($R$) and inductance ($L$) may be calculated from the expression $Z = \sqrt{(R^2 + L^2)}$, or with capacitance ($C$) $Z = \sqrt{(R^2 + C^2)}$. For a circuit containing $R$, $C$ and $L$ the impedance $Z = \sqrt{\{R^2 + (L + C)^2\}}$ in this case, remembering that $L$ will be a positive reactance, $C$ a negative reactance so that $L + C$ may be thought of as $L - C$ if both reactances are given a positive value.

A convenient method of distinguishing between resistive and reactive elements of an ac or rf circuit is to use the symbol j (which stands for $\sqrt{-1}$) in front of a reactance. For example 36 + j45Ω implies a circuit containing 36Ω resistance and 45Ω reactance. While its impedance could be obtained from the above expression as $\sqrt{\{(36)^2 + (45)^2\}}$ or approximately 58Ω, it is often more useful to keep the terms separate and then, if required, to balance out the reactive component by connecting into the circuit a compensating reactance of similar value but opposite sign. Such techniques are very important in the matching of antennas to transmitters but calculations involving complex numbers are not required in the UK Radio Amateurs' Examination.

## The transformer

When two inductors are mounted close together so that their electromagnetic fields interact, we have a *transformer*. In practice a transformer usually consists of two or more windings on a single core designed so that the magnetic fields are concentrated to maximize magnetic coupling between windings, although for some applications only loose coupling between windings may be used.

The winding used to generate the magnetic flux is called the *primary*; that in which current is induced is the *secondary* or secondaries (there may be several secondary windings in a single transformer). At low frequencies, cores may be in the form of thin plates (*laminations*) designed to reduce *eddy losses* and forming closed magnetic loops. At higher frequencies cores may be dust-iron, brass, ceramic ferrites or windings may be air-cored.

The voltage induced in a secondary winding is proportional to the ratio of turns on the secondary to the turns on the primary: alternating voltages can thus be *stepped-up* or *stepped-down* as determined by the windings. Since the power output from all secondary windings cannot exceed that fed to the primary winding (in fact there will be some losses in the transformer), a step-up transformer is designed to provide more volts but at corresponding lower current.

Since all sources of electric power (batteries, power supplies, transformer windings, etc) possess internal resistance or impedance, it is not possible to obtain unlimited power from them, although for example the impedance of domestic mains supplies is very low. Maximum power is delivered to an external load when the value of the internal resistance or impedance of the source exactly *matches* (ie equals) that of the load. A transformer is often used to match the impedance of an ac load to that of the ac power source (for example a loudspeaker winding to the output impedance of the audio amplifer). When a load of impedance $Z_s$ is connected across the secondary winding of a transformer, this impedance will be *reflected back* across the primary winding by the square of the turns ratio (ie $Z_p = n^2 Z_s$ where $n$ is the ratio of turns of primary to secondary windings).

In radio-frequency transformers one or both windings may be resonant tuned circuits and such an arrangement functions effectively only over a limited range of frequencies, forming a band-pass filter as well as an impedance matching device. Tuned circuits may as in a transformer be inductively coupled but can also conveniently be coupled together by means of capacitors, forming either *top* or *bottom* coupled resonant band-pass filters. Reactive components in series can also provide impedance matching circuits by making use of the magnification factor and by ac potential dividers. Examples include the pi ($\pi$) coupler, L-networks etc used to match transmitters to antennas.

## Frequencies and radio waves

The 240V ac mains supplies in the UK and 220V or 240V supplies in Europe have a frequency of 50Hz, whereas in North America mains supplies are usually 110V, 60Hz.

Electrical signals at frequencies between about 30Hz and 15,000Hz (15kHz) when fed to a transducer such as a loudspeaker or headphones result in audible sound waves; similarly sound waves of such frequencies can be converted into electrical signals by means of a microphone. Electrical signals ranging from kilohertz right up to gigahertz can be converted into radio waves (and vice versa) by means of an antenna. It should be noted that sound waves (pressure waves in gases and solids) are inherently different from radio waves and that an electrical signal of say 10kHz could be converted into either a high-pitched audio signal or a very low frequency radio wave by using appropriate transducers (ie loudspeaker or antenna).

Radio waves are electromagnetic waves at frequencies up to roughly 500GHz; at even higher frequencies electromagnetic waves take the form of infra-red waves; visible light; ultra-violet waves; X-rays; and gamma waves.

All electromagnetic waves, including radio waves, travel at a constant velocity of 299,790,000m/s. For practical purposes this is taken as 300,000,000m/s (186,000 miles per second). A radio wave thus takes roughly one seventh of a second to travel right round the world, or about a second to reach the moon.

Since radio waves travel at fixed velocity, the time taken to supply one complete cycle of energy can be thought of in terms of the distance travelled during this period by the initial wave, ie by its *wavelength* (often abbreviated $\lambda$, lambda). In other words, the frequency (Hz) multiplied by wavelength in metres is equivalent to velocity of radiation in metres per second: ie

$f \times \lambda = 300{,}000{,}000$ where $f$ is in hertz (Hz)
$f \times \lambda = 300{,}000$ where $f$ is in kilohertz (kHz)
$f \times \lambda = 300$ where $f$ is in megahertz (MHz)

This can be expressed as $f$(kHz) = 300,000/$\lambda$ or $\lambda$ (metres) = 300,000/$f$(kHz). This means that if we know the frequency of a radio signal we can always calculate its wavelength and vice versa, and that as frequency goes up, the wavelength shortens. For example, 3,000kHz is equivalent to 100m; 30,000kHz (30MHz) is equivalent to 10m; and 300MHz is equal to 1m.

A radio transmitting station is like a lighthouse but sending out invisible rather than visible electromagnetic waves in all, or specific, directions. But just as a steadily shining light tells us nothing other than that it is there, to provide information the radio wave must be varied in some predetermined way. For instance it may be switched on or off according to a simple code (eg morse code), or its "brightness" (amplitude) or frequency varied in accordance with the information, a process termed *modulation*.

At the receiving end, this information must be recovered from the incoming radio waves, first by causing them to provide electrical signals at the radio frequency (by means of an antenna) and then by *demodulation* to separate the information from the signals at radio frequency. In practice it is also essential to provide the receiver with a means of distinguishing between electrical signals of different frequency, and this may take the form of variable *tuning* of resonant frequency circuits arranged to provide the required degree of *selectivity*.

Theoretically, in free space, the intensity of all electromagnetic waves diminishes with distance from the transmitter according to the *inverse square law* (double the distance, and the intensity reduces to one quarter; three times the distance and the signal diminishes to one ninth etc). However to consider whether a useful signal can be received over a given distance, it is necessary to take into account not only the inverse square law but also the *propagation characteristics* at the radio frequency involved. For example propagation characteristics of mf signals (300–3,000kHz) differ significantly from those at hf (3–30MHz), vhf (30–300MHz) and uhf (300MHz–3GHz) etc.

## Active devices

The component parts of electronic equipment are *passive* (resistors, inductances etc) or *active* (thermionic valves or tubes, transistors, diodes) or may act as a *transducer* for converting sound waves into electrical signals (microphones) or vice versa (loudspeakers, headphones) or electrical signals into radio waves and vice versa (antennas or aerials). Passive components or transducers however do not amplify an electrical signal, and (crystal sets apart) it is generally required to amplify, *demodulate* and often to *frequency-change* the weak electrical signals provided by an antenna. From 1910 to 1950 (roughly), *valves* were the only practical means of amplifying signals; the development since 1947 of *transistors* has provided an alternative (solid-state) approach, and very large numbers of such devices, together with associated passive components, can now be manufactured in the form of an *integrated circuit* or *microcircuit*. Valves, however, continue to find application primarily in the field of *rf power amplification*.

### Thermionic valves

When a material is heated in a vacuum or inert gas some free electrons escape from the confines of the material (*cathode*) and form a surrounding electron cloud. If now a nearby metallic *plate* (*anode*) is held at positive potential to the cathode some free electrons will be attracted towards it (*thermionic emission*) and a current flows in the external circuit through the source of positive potential. Since such a device is unidirectional, it can be used for rectification or demodulation (*detection*). It is called a *diode* (two-electrode device).

If an open wire mesh or *grid* is imposed between cathode and anode (plate), the flow of electrons will be controlled by relatively small changes of voltage (*bias*) applied to the grid. If the grid is sufficiently *negative* with respect to the cathode, all flow of electrons to the anode ceases, and the device is said to have reached *cut-off*. Typically a change of 1V in grid potential may result in a change of the anode current of 3mA, ie 3mA/V and this is termed the *mutual conductance* or *slope* of the valve. Since there are now three electrodes, it is called a *triode* valve. It would normally require a change of much more than 1V applied to the anode to produce a change of 3mA. By passing the anode current through a load resistor, it becomes possible for a small change of voltage applied to the grid to result in a much greater change of potential across the load, so that we can achieve voltage and/or power amplification.

Since a valve amplifier provides an output signal of greater voltage than the input voltage, it is possible to feed back a proportion of the output to the input in such a way that the input signal is reinforced in phase (*positive feedback*). Such a system can be used to reinforce the natural oscillatory energy in a tuned circuit so that it will *oscillate* continuously, generating a continuous signal at a required frequency: such an arrangement is termed an *oscillator*. It is also possible to feed back energy so that amplification is reduced (*negative feedback*) and this technique may be used to reduce distortion of signals passing through an amplifying stage.

To overcome limitations (*Miller effect*) of the triode valve, a further grid electrode, forming an electrostatic shield, may be interposed between grid and anode: the device is then a four-electrode or *tetrode* valve (also known as a *screen-grid valve*). Similarly a further additional grid electrode may be used (to overcome *secondary emission*) as in the *pentode*, or beam-forming electrodes used (*beam tetrode*). For some applications more than one electrode structure may be used in a valve (eg in valves used for frequency conversion).

When two signals at different frequencies are *mixed* together in a non-linear device, then a whole new series of output signals are produced. Such a process of mixing or heterodyning signals is analogous to the *modulating* process whereby information at audio frequency is imposed on a radio frequency *carrier*.

By selecting in the output circuit a specific beat signal, it is possible to change the frequency of an incoming signal to a more convenient frequency without losing the information carried on the incoming signal, provided that the beat signal is itself unmodulated (ie in the form of a frequency generated in a stable local oscillator).

The anode *current* of a valve is, as we have seen, controlled by the *voltage* on the grid; the valve is thus a *voltage-controlled* device and, if no grid current flows, represents a high impedance; drawing little power from the input signal when biased conventionally.

The input signal is applied between grid and cathode; the output taken from between the anode/anode load connection and cathode. The cathode is effectively at the common earth (ground) potential to the signal. However other configurations can be used including *grounded-grid* and *cathode-follower* arrangements; a special configuration is the *cascode*, comprising a combination of grounded-cathode and grounded-grid amplifiers.

### Semiconductors

A valve functions by voltages acting on electric currents produced in a vacuum or inert gas; the transistor takes advantage of the basic crystalline structure of *semiconductors* with accurately defined degrees of impurity. Current flow is within the crystalline structure. The transistor requires neither electrode heating nor high electrode potentials to attract electrons, and is thus inherently more suitable for operation from batteries. On the other hand the characteristics are more affected by change of temperature.

Semiconductor materials in a pure state have few free charge carriers and are poor conductors. By *doping* the material with suitable impurities it is possible to produce many charge carriers, which may be electrons or *holes* (gaps in atoms from which electrons are missing). A semiconductor material with free electrons is termed *n-type*; that with many holes is termed *p-type*. In practice such materials are formed side by side to form *pn junctions*. In n-type material the *majority carriers* are electrons but there are also some holes forming *minority carriers*; in p-type materials the position is reversed. At a pn junction there will be a *potential hill* or *barrier* that prevents further movement of charge carriers across the junction unless this is connected across a battery; then if this biasing is of appropriate polarity (*forward biased*) the potential barrier is reduced and current flows freely; however with the alternative polarities (*reversed biased*) the barrier is increased and no current flows, until a point is reached (*zener breakdown*) where the junction, in effect, breaks down and current again flows. Provided this zener breakdown voltage is not reached a pn junction provides a unidirectional device comparable to a diode valve. Silicon and germanium diodes are widely used for power rectification, signal demodulators etc.

**Bipolar transistors**

A *bipolar transistor* comprises a sandwich of three such materials (forming two pn junctions) of either pnp or npn form, with the outer sections forming the *emitter* and *collector*, and the inner the *base*. When correct potentials are applied charge carriers flow from emitter to base and also from emitter to collector although in this case it will be much affected by the emitter/base current. In fact the base region forms a connecting link and controls the emitter/collector current which depends on the biasing *current* flowing in this region, a small change in base current resulting in an almost equal change in collector current. Since the emitter-base junction is forward biased it will represent a low resistance path; however the base/collector pn junction is reverse-biased and represents a relatively high-resistance path. If then we can use a low-resistance current to increase similarly current flow in a higher resistance circuit working into a load resistor, we can obtain voltage amplification. The base/emitter impedance is low and the device is *current* operated; this makes it important to provide *impedance matching* in the input circuits to bipolar transistors. Most modern silicon transistors are of npn form and the collector is positive to the emitter, with the base at an intermediate positive level. If no current flows in the base/emitter junction the device is virtually *cut-off*.

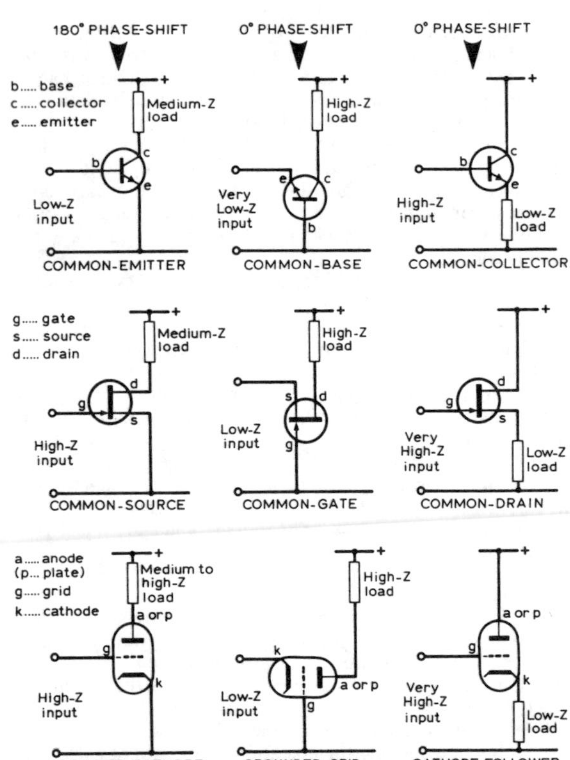


**Fig 6. Amplifier circuit connections**

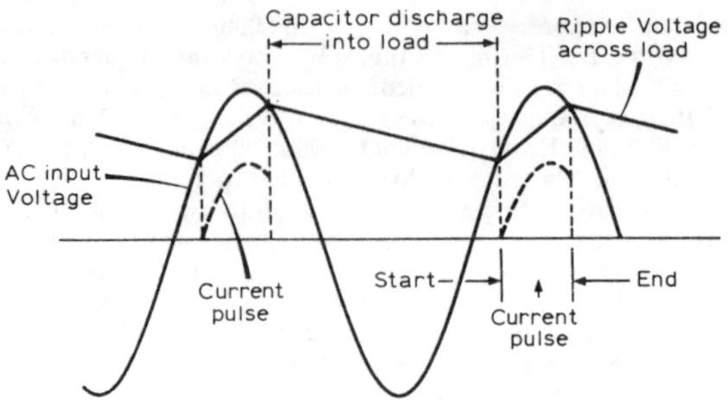


**Fig 7. Approximate waveforms of a half-wave capacitor input circuit for a power supply**

**Field-effect transistors**

An alternative form of semiconductor device is the *field-effect transistor*; this is a *unipolar* device in which the action depends solely on majority carriers (although like the bipolar transistor these may be electrons or holes). The action is more closely akin to thermionic valves, with the charge carriers flowing from *sources* to *drain* through a narrow channel with the flow governed by the field (voltage) applied to the *gate*. The device is thus voltage controlled and has an extremely high input impedance, and is used in circuit configurations very similar to those developed for valves. Although early field-effect transistors were suitable only for *small-signal* applications, more recently power devices using a "V-groove" structure have been developed.

Integrated circuits (ICS) may be based on bipolar transistors, the field-effect approach (*cmos*) or occasionally a combination of the two.

## Rectification and power supplies

Alternating currents can be converted into pulsating dc by means of *rectifiers* which permit current to flow freely in one direction only (*forward direction*) by presenting a much higher resistance to current flow in the opposite direction (*reverse direction*). Such components include thermionic and semiconductor *diodes* and the process is termed *rectification*. The pulsating dc output may then be *smoothed* by means of *ripple filters*, which represent a form of low-pass filter, and comprising large-value reservoir capacitors and smoothing choke or resistor. A smoothing filter with a capacitor as the first component is termed a *capacitor-input* filter; if a choke or *swinging choke* is used as the first component, the arrangement is termed a *choke-input filter*. Various configurations of *half-wave*, *full-wave*, *full-wave bridge* and *voltage-multiplying* arrangements are widely used for rectification, each arrangement offering advantages and disadvantages.

Both the rectifying circuit and the ripple filter characteristics affect the degree of *ripple* remaining on the dc output. Additionally a power supply may incorporate *voltage regulation* to enable a fairly constant output voltage to be achieved over a wide variation of current (*load*) taken from the unit.

If the output of a power supply is to be fully isolated from the ac mains supply a *mains transformer* with separate primary and secondary winding (ie not an *autotransformer*) must be incorporated in the unit: this transformer also provides a convenient means of obtaining any required dc output from the supply.

Where a higher output voltage is required from a dc supply (for example a 12V car battery) a transformer cannot be used directly on dc and it is necessary first to convert (*chop*) the dc into a repetitive series of pulses or sine-wave oscillating form which can then be stepped up by transformer action, and then subsequently rectified and smoothed. Such a unit may be called a *dc–dc converter* or *inverter*. If the repetitive series of pulses or oscillatory waveform is a relatively high frequency, eg 400Hz, it becomes possible to use much smaller cores in the transformer and smaller value components in the ripple filter. Such techniques are therefore also now found applied to ac mains supplies in what are termed *switched-mode* power units.

## Digital electronics

Spreading out from the field of computers and data processing, digital electronics are beginning to play an important part in amateur radio equipment. A *digital* system is one in which all the waveforms are selected from a restricted number, as opposed to *analogue* systems in which the various waveforms may have an infinite number of shapes and amplitudes. A *binary* system is a digital system having only two different states which can be represented by "1" and "0" levels, as in an "on" and "off" electric circuit. The morse code could be defined as a form of digital non-return-to-zero binary code, and illustrates that for successful digital transmission a receiver has only to distinguish between the presence or absence of a signal. It can thus tolerate a great deal of distortion of the pulse waveforms before it becomes unreadable, whereas any change of the relative levels of an analogue waveform results in degradation or distortion of a greater or lesser degree. Digital techniques form the basis of manual or automatic telegraphy and high-speed data systems, but also increasingly for many forms of control and signal-processing applications. By a process of sampling and coding (quantization) it is possible to convert any analogue waveform, including speech, into a coded digital system of pulses (pulse-code modulation), although this requires high pulse rates and as a result requires more *bandwidth* than an analogue system; digital speech transmission is still seldom used by amateurs.

Although digital forms of telegraphy have a long history, it was the development of electronic computers and data-processing equipment that underlined the value of using digital techniques for processing and calculations, and many of the terms used stem directly from the world of data processing, including the particular form of mathematics (Boolean algebra).

The binary "0" and "1" two-level systems utilize the advantages of simple on-off *switching* and *gating* processes in which an input signal(s) can control or initiate complex operations. A *gate* is a form of "open" and "shut" switching based on the characteristics of diodes and transistors, and gates are increasingly implemented in the form of integrated circuits based on a number of *logic families*. Although the basic circuits are very simple, they are often used in very large numbers in a block-like or modular approach. For this reason logic diagrams are often used rather than conventional circuit diagrams to explain the system.

A logic circuit is one that can recognize and react to a change in its input conditions. For example an AND gate provides an output "1" only when two or more "1" signals are applied simultaneously

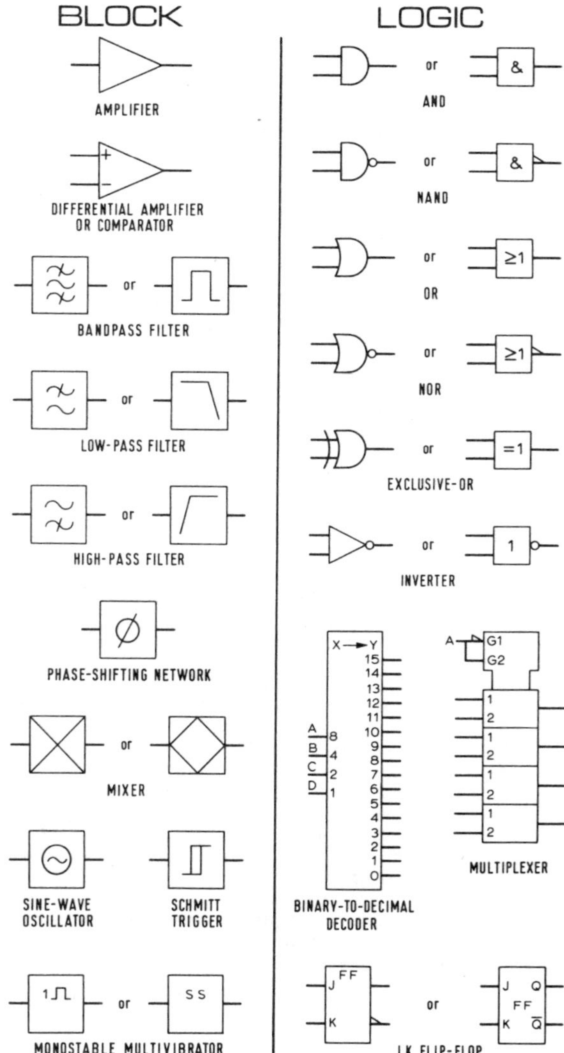


**Fig 8. Block function graphical symbols and "box" logic**

to its input circuits; an OR gate provides an output "1" when an input "1" is applied to any of its input circuits. If the circuit inverts the input/output pulses (ie provides a "0" when a "1" input is applied and vice versa) it is known as an *inverter*; a NAND gate is thus a combination of an AND gate followed by an inverter, while similarly a NOR gate is an OR gate combined with an inverter.

It is often helpful to list all possible combinations of input and output states of a logic gate in a *truth table*. A truth table representing simple two-input AND, OR, NAND and NOR gates would be:

| Input A | Input B | AND output | OR output | NAND output | NOR output |
|---|---|---|---|---|---|
| 0 | 0 | 0 | 0 | 1 | 1 |
| 0 | 1 | 0 | 1 | 1 | 0 |
| 1 | 0 | 0 | 1 | 1 | 0 |
| 1 | 1 | 1 | 1 | 0 | 0 |

A single binary digit (ie a "0" or a "1") is termed a *bit*, and the

## TECHNICAL ABBREVIATIONS AND SYMBOLS

The following list includes technical abbreviations and symbols commonly found in amateur radio journals.

|  |  |
|---|---|
| A | ampere (unit of current) |
| A1, A2, etc | types of emission, see p105 |
| ac | alternating current |
| a/d | analogue-to-digital (converter) |
| af | audio frequency |
| afc | automatic frequency control |
| afsk | audio frequency shift keying |
| agc | automatic gain control |
| alc | automatic level control |
| a.m. | amplitude modulation |
| anl | automatic noise limiter |
| ant | antenna (aerial) |
| ASCII | 8-bit data code (American Standard Code for Information Interchange) |
| atu | antenna tuning unit |
| avc | automatic volume control |
| baud | measure of rtty/data/telegraphy speed (1 baud equals 1 bit/s and 50 bauds approximately 66wpm) |
| bcd | binary coded decimal |
| bci | interference to broadcast reception |
| bfo | beat frequency oscillator |
| C | capacitor or capacitance |
| ccd | charge-coupled device |
| cmos | (or cos/mos or cosmos) complementary-symmetry metal oxide semiconductor (logic family) |
| coax | coaxial cable |
| cro | cathode-ray oscilloscope |
| crt | cathode-ray tube |
| c/s | cycles per second (hertz) |
| ct | centre tap |
| cw | continuous wave emission (A1) often used to denote telegraphy |
| d/a | digital-to-analogue (converter) |
| dB | decibel |
| dBd | antenna gain, reference dipole |
| dBi | antenna gain, reference isotropic source |
| dBm | decibel gain, reference 1mW |
| dBW | decibel gain, reference 1W |
| dc | direct current |
| dcc | double cotton covered (insulation of wire) |
| dil | (or dip) dual-in-line ic package |
| dpdt | double pole double throw (switch) |
| dpst | double pole single throw (switch) |
| dsb | double sideband emission with suppressed carrier |
| dtl | diode-transistor logic |
| ecl | emitter-coupled logic |
| eht | extra high tension |
| emc | electromagnetic compatibility |
| eme | earth-moon-earth ("moonbounce") |
| emf | electromotive force |
| emi | electromagnetic interference |
| enam | enamelled (insulation of wire) |
| eprom | erasable programmable read-only memory |
| erp | effective radiated power |
| F | farad (unit of capacitance) |
| f | frequency |
| fax | facsimile |
| fet | field effect transistor |
| fifo | first-in, first-out memory |
| fm | frequency modulation |
| fot | optimum working frequency |
| fsd | full scale deflection (of meter) |
| fsk | frequency-shift keying |
| gc | conversion conductance |
| GHz | gigahertz (MHz × 1,000) |
| gm | mutual conductance |
| gmt | Greenwich mean time |
| H | henry (unit of inductance) |
| Hz | hertz (c/s) |
| hex | device with six similar circuits |
| hf | high frequency (10-100m, 3-30MHz) |
| hfo | heterodyne frequency oscillator |
| hp | high pass (filter) |
| hpf | high-pass filter *or* highest possible frequency |
| I | symbol for current |
| Ia | anode current (of a valve) |
| ic | integrated circuit |
| id | inner diameter |
| Ig | grid current (of a valve) |
| Ig2 | screen-grid current (of a valve) |
| Ih | heater current (of a valve) |
| i.f. | intermediate frequency |
| ift | intermediate-frequency transformer |
| imd | intermodulation distortion |
| irt | incremental receiver tuning (also rit) |
| jfet | junction field-effect transistor |
| k | kilo (1,000 times) |
| kHz | kilohertz (Hz × 1,000) |
| kV | kilovolts (volts × 1,000) |
| kW | kilowatts (watts × 1,000) |
| kΩ | kilohms (ohms × 1,000) |
| lcd | liquid crystal display |
| led | light-emitting diode |
| lp | low pass (filter) |
| lsb | lower sideband |
| lsi | large-scale integration |
| lt | low tension |
| luf (or luhf) | lowest usable (high) frequency |
| mcw | modulated continuous wave emission |
| mf | medium frequency (0·3-3MHz) |
| mic | microphone |
| mosfet | metal-oxide-semiconductor field effect transistor |
| ms | meteor scatter |
| msi | medium-scale integration |
| muf | maximum usable frequency |
| mult | (frequency) multiplier |
| MΩ | megohms (ohms × 1,000,000) |
| MHz | megahertz (Hz × 1,000,000) |
| mA | milliampere (ampere ÷ 1,000) |
| mA/V | milliamperes per volt |
| mH | millihenry (henry ÷ 1,000) |
| mV | millivolt (volt ÷ 1,000) |
| nbfm | narrow band frequency modulation |
| nf | noise figure |
| ns | nanosecond |
| od | outside diameter |
| op-amp | operational amplifier |
| osc | oscillator |
| owf | optimum working frequency |
| pa | power amplifier |
| pcb | printed circuit board |
| pd | potential difference (or power doubler) |
| p.e.p. | peak envelope power |
| pF | picofarad ($\mu$F ÷ 1,000,000) |
| piv | peak-inverse voltage |
| pll | phase-locked loop |
| pm | permanent magnet |
| pm | phase modulation |
| pot | potentiometer |
| pp | push-pull |
| p-p | peak-to-peak |
| pri | primary winding (of transformer) |
| prom | programmable read-only memory |
| psk | phase-shift keying |
| ptt | push-to-talk |
| pvc | polyvinyl chloride |
| quad | device containing four similar circuits |
| R | resistor or resistance |
| ram | random-access memory |
| rect | rectifier |
| rit | receiver incremental tuning |
| rf | radio frequency |
| rfc | radio-frequency choke |
| rfi | radio-frequency interference |
| rms | root mean square |
| rom | read-only memory |
| rt | radiotelephony |
| rtl | resistor-transistor logic |
| rtty | radioteletype |
| s | second |
| scc | single cotton covered (insulation) |
| scfm | subcarrier frequency modulation (for sstv) |
| scr | silicon-controlled rectifier |
| sec | secondary winding (of transformer) |
| sg | screen-grid |
| shf | super high frequency (3-30GHz) |
| sm | silver mica (capacitor) |
| snr | signal-to-noise ratio |
| spdt | single pole double throw (switch) |

| | |
|---|---|
| spst | single pole single throw (switch) |
| ssb (or sssc) | single sideband, suppressed carrier, emission |
| sstv | slow-scan television |
| swg | standard wire gauge |
| swr | standing-wave ratio |
| te (or tep) | transequatorial propagation |
| t-r | transmit-receive |
| tpi | turns per inch |
| trf | tuned radio-frequency (usually indicating "straight" receiver) |
| ttl | transistor-transistor logic |
| tu | terminal unit (for rtty) |
| tvi | interference to television reception |
| uart | universal asynchronous receiver and transmitter |
| uhf | ultra high frequency (300-3,000MHz) |
| ujt | unijunction transistor |
| usb | upper sideband |
| utc | co-ordinated universal time (approx gmt) |
| V | volt (unit of potential difference) |
| vco | voltage-controlled oscillator |
| vcxo | voltage-controlled crystal oscillator |
| vdu | visual display unit |
| vfo | variable frequency oscillator |
| vhf | very high frequency (30-300MHz) |
| vox | voice-operated switching device |
| vr | voltage regulator |
| vxo | variable-frequency crystal oscillator |
| W | watt (unit of power) |
| wpm | words per minute |
| w/t | wireless-telegraphy |
| ww | wire-wound (resistor) |
| Z | symbol for impedance |
| $\lambda$ | (lambda) wavelength |
| $\mu$ | (mu) micro- (prefix meaning 1/1,000,000th) |
| $\mu$A | microampere (ampere ÷ 1,000,000) |
| $\mu$H | microhenry (henry ÷ 1,000,000) |
| $\mu$F | microfarad (farad ÷ 1,000,000) |
| $\mu$V | microvolt (volt ÷ 1,000,000) |
| $\Omega$ | (omega) ohms |

waveforms are essentially a series of pulses (pulse-trains) at a specific *bit-rate* (number of bits per second).

The basic circuit blocks of digital electronics include gates, inverters, clocks, counters, shift registers, etc. A clock is a source of timing signals and is comparable to an oscillator in an analogue circuit except that the output is arranged in the form of square-wave on-off pulses; digital clocks may be variable or crystal controlled. A shift-register is a number of bistable "flip flops" (circuits that change from one state to the other when triggered by an input pulse) connected so that whenever a pulse is applied to the first element, then the existing state of that element is immediately transferred to the next circuit element and so on. A shift register can thus store any particular sequence of "0"s and "1"s, and the sequence can be recovered from the output of the shift register; it is thus a form of *memory* (and can for example be used to store a sequence of morse characters to provide automatic callsign transmission) just as a piece of magnetic tape can provide a memory. Various combinations of flip-flops can form counters and dividers; for instance a *decade divider* is a component that is arranged so that when 10 pulses are connected to the input, only one pulse is delivered from the output, a facility that can be very useful for the construction of frequency calibrators since a 100kHz "clock" will provide an output rich in harmonics at 10kHz intervals. Similarly we can arrange to "count" the number of pulses and then display them as decimal numbers by various digital display techniques. Many modern amateur receivers and transmitters use such systems to provide the operator with a "frequency readout"; or again there are many measuring instruments which now use displays rather than an analogue meter pointer.

Integrated circuits may contain a considerable number of gates or other component elements designated by "quad" for four elements, "hex" for six elements etc. Certain "families" of logic circuits have come into widespread use designated, for example, as "ttl" (transistor-transistor logic), "ecl" (emitter-coupled logic), and "cmos" (logic based on field-effect rather than bipolar transistors), each offering advantages and disadvantages. Since components based on different logic families cannot readily be connected together without special interface circuits (since the potentials of the supply line, or the "1" levels may differ), often an entire system will be based as far as is practicable on a single logic family. For example, many amateurs use low-cost ttl devices and these require a supply voltage of 5V; or alternatively cmos devices which are more tolerant of supply voltage and which consume virtually no power when remaining in the same state, but which cannot be switched or toggled at such a high rate as ttl. In addition there are a number of variations within a similar logic family, and in particular various forms of ttl exist which offer either higher speeds or lower power consumption than the standard ttl.

In amateur radio equipment such digital techniques have been widely adopted for various forms of frequency synthesizers (in which any required output frequency is generated and stabilized against a reference frequency, partly by digital techniques and partly by such techniques as phase-locked loops whereby a voltage-controlled oscillator is kept locked to an output derived from a stable crystal-controlled oscillator), frequency displays and readouts (generally using semiconductor displays made from matrices of light-emitting diodes (LEDs) under the control of binary-to-decimal ic converter devices), electronic keyers and automatic keyers, automatic band-scanning systems and so on.

## METRIC CONVERSION

1 **milli-inch** ("one thou") is equal to 25·4 micrometres ($\mu$m)
1 **inch** (in) is equal to 25·4mm, or 2·54cm
1 **foot** (ft) is 12in and equal to 30·48cm or 0·3048m
1 **yard** (yd) is 3ft and equal to 0·91m
1 **mile** (1760yd or 5280ft) is equal to 1·609km
1 **nautical mile** is equal to 1·853km
1 **square inch** (sq in or in$^2$) is equal to 645mm$^2$ or 6·45cm$^2$
1 **square foot** (sq ft or ft$^2$) is equal to 929cm$^2$ or 0·0929m$^2$
1 **square yard** (sq yd or yd$^2$) is equal to 0·836m$^2$
1 **cubic inch** (cu in or in$^3$) is equal to 16·39cm$^3$
1 **cubic foot** (cu ft or ft$^3$) is equal to 28·32dm$^3$
1 **inch per second** (in/s) is equal to 2·54cm/s
1 **pound** (lb) weight is 16oz and equal to 0·454kg
1 **ounce** (oz) weight is equal to 28·35g
1 **hundredweight** (cwt) is equal to 50·8kg
1 **UK ton** is equal to 1·016 metric tonnes (ie 1000kg is equal to 1 metric tonne)
1 **gallon** is equal to 4·546 litres
1 **centimetre** is equal to 0·394in
1 **metre** is equal to 39·37in or 3·281ft or 1·094yd
1 **kilometre** is equal to 0·621 miles
1 **litre** is equal to 0·22 gallons
1 **pound per cubic inch** is equal to 27·68g/cm$^3$
1 **British thermal unit** is equal to 1·055 kilojoules (kJ)
1 **kilowatt-hour** is equal to 3·6 megajoules (MJ)

**COMPARISON OF CENTIGRADE AND FAHRENHEIT THERMOMETER SCALES**

| Centigrade | Fahrenheit | Centigrade | Fahrenheit |
|---|---|---|---|
| −50 | −58 | +80 | +176 |
| −45 | −49 | +85 | +185 |
| −40 | −40 | +90 | +194 |
| −35 | −31 | +95 | +203 |
| −30 | −22 | +100 | +212 |
| −25 | −13 | +105 | +221 |
| −20 | −4 | +110 | +230 |
| −15 | +5 | +115 | +239 |
| −10 | +14 | +120 | +248 |
| −5 | +23 | +125 | +257 |
| 0 | +32 | +130 | +266 |
| +5 | +41 | +135 | +275 |
| +10 | +50 | +140 | +284 |
| +15 | +59 | +145 | +293 |
| +20 | +68 | +150 | +302 |
| +25 | +77 | +155 | +311 |
| +30 | +86 | +160 | +320 |
| +35 | +95 | +165 | +329 |
| +40 | +104 | +170 | +338 |
| +45 | +113 | +175 | +347 |
| +50 | +122 | +180 | +356 |
| +55 | +131 | +185 | +365 |
| +60 | +140 | +190 | +374 |
| +65 | +149 | +195 | +383 |
| +70 | +158 | +200 | +392 |
| +75 | +167 | | |

Digital circuits which perform a single function are termed *wired logic*, unlike a true electronic computer which can perform a variety of different functions by means of stored programs.

The development of large-scale integration (lsi) in which very many circuit elements can be formed together on a single chip of silicon has made possible a dramatic reduction in the size and cost of small multi-function computers. A microcomputer can be considered as consisting of a *central processing unit* (cpu), which performs all the data processing and which can now be manufactured as a single *microprocessor* ic, plus a *memory* and here again it is now possible to obtain single ICs capable of storing many thousands of bits of information. It is thus now practical to design systems which can carry out complex coding and decoding automatically; for example some amateurs use systems in which incoming rtty, cw or data signals are decoded automatically and displayed on a television screen as alphanumeric characters (visual display units); or which can convert one code into another so that the same vdu can display messages received by morse, five-unit rtty code or eight-bit ASCII code. Alternatively a microprocessor can be used within a receiver to provide variable coarse or fine tuning with the same knob; or to allow a receiver to be tuned to any required frequency by providing instructions through a keypad, or to return to a desired channel which has been memorized.

In fact computer systems can be devised which could virtually take over or greatly simplify the operation of an amateur station (even entering up the log and filling in QSL cards!). Nevertheless it should be recognized that, for example, automatic decoding of incoming cw loses the much greater flexibility of the human brain which has the ability to recognize symbols badly mutilated by interference and noise. Furthermore there is a stage beyond which "automation" of an amateur station would be pointless, since much of the enjoyment of amateur radio operating is found in the development and use of human skills. For such reasons, designers should seek to provide amateurs with *computer-assisted* rather than *computer-controlled* systems, and recognize that many of the digital techniques, although fascinating to watch in action, may not in the long-term contribute greatly to the effectiveness of the station in working over long distances and in passing information at rates which suit human operators.

APPENDIX 1

# Sample RAE questions

The following list of 40 sample items, prepared by the City and Guilds of London Institute, illustrate the kind of questions found in the UK Radio Amateurs' Examination, although not necessarily representative of the entire scope in either content or difficulty. A key to the correct answers is provided at the end of the section.

The RAE comprises two papers: 765-1-01 *Licensing Conditions and Interference* lasting 1 hour and containing 35 items (about 66 per cent on licensing conditions, 34 per cent on transmitter interference); and paper 765-1-02 *Operating Practices, Procedures and Theory* lasting 1 hour 45 minutes and containing 60 items (8 per cent on operating practices and procedures; 18 per cent on electrical theory; 12 per cent on semiconductors; 15 per cent on radio receivers; 13 per cent on transmitters; 24 per cent on propagation and aerials; and 10 per cent on measurement).

In the following sample items 1–12 cover syllabus sections for paper 765-1-01 and items 13–40 are applicable to paper 765-1-02.

1. The callsign prefix of the Isle of Man is

(a) GM
(b) GD
(c) GI
(d) GW.

2. The holder of an Amateur Licence A may only use signals which

(a) are listed in the Radio Regulations
(b) form part of, or relate to, the transmission of messages to and from amateur radio stations
(c) are in codes and cyphers, which have been notified to the Secretary of State, Home Office or his agents, at least four weeks prior to their use
(d) are recognized by the Secretary of State, Home Office as being necessary to the Amateur Service.

3. The maximum dc power input permitted on the 1·8–2MHz band is

(a) 10W
(b) 50W
(c) 150W
(d) 400W.

4. The Amateur Licence states that the Licensee shall test his transmissions for radiation of harmonics and other spurious emissions and record such tests in the log

(a) once a month
(b) once every three months
(c) at the request of a Home Office official
(d) from time to time.

5. Entries in the Station log book need not include

(a) date
(b) time of commencement of operation
(c) signature of the Licensee
(d) time of closing down the Station.

6. In the case of a prolonged contact amateurs in the United Kingdom should make a station identification at intervals not exceeding

(a) 10 minutes
(b) 15 minutes
(c) 20 minutes
(d) 30 minutes.

7. Poor frequency stability of an amateur transmitter can result in

(a) the generation of parasitic oscillations
(b) operation outside the amateur bands
(c) a reduction in the power output
(d) difficult adjustment of the power amplifier stage.

8. The radiation of harmonics from an amateur transmitter may be caused by

(a) the power amplifier stage being over-driven
(b) keying a high current circuit in the transmitter
(c) the power supply to the driver stage being unregulated
(d) rf being induced in the mains supply to the transmitter.

9. The circuit diagram shown in Fig 1 may be used to

(a) detect the presence of harmonic radiation
(b) ensure that an amateur transmission is free from frequency instability
(c) check that the transmission is not unnecessarily broad
(d) measure accurately the percentage depths of modulation.

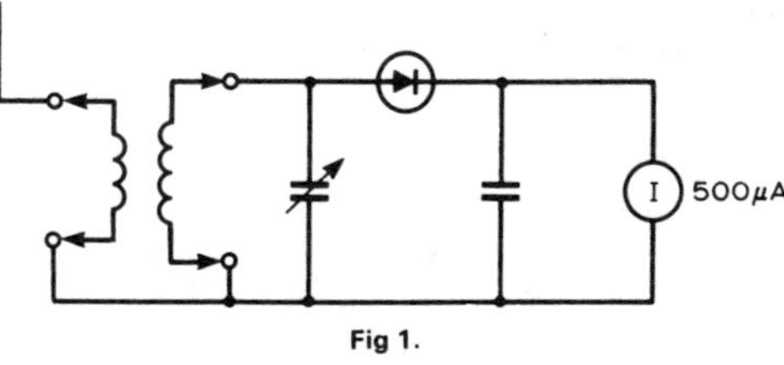


Fig 1.

10. A transmitter would be likely to radiate harmonics if

(a) the supply voltage was unregulated
(b) it was overmodulated
(c) the output stage was overdriven
(d) there were short-circuited turns in the aerial tuning unit.

11. The circuit configuration shown in Fig 2 would normally be used
(a) as a precaution against overmodulation
(b) to limit the modulating frequency range
(c) to match a crystal microphone to the amplifier input circuit
(d) to increase the high frequency response of the modulator.

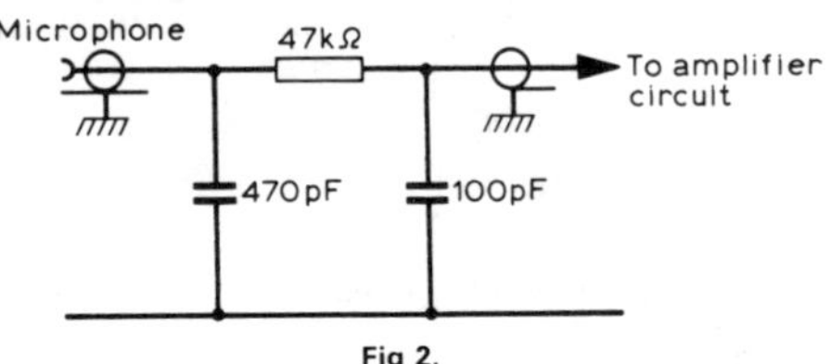


Fig 2.

12. A transmitter operating in the band 21MHz to 21·450MHz has a frequency tolerance of 100 parts in one million and a radiated bandwidth of 6kHz when using emissions of Type A3. If the frequency checking equipment at the station has a frequency tolerance of 10 parts in one million what is the lowest frequency a Licensee can use that ensures no emission below 21MHz?
(a) 21005·1kHz
(b) 21053·1kHz
(c) 21008·31kHz
(d) 21005·31kHz.

13. It is good operating practice when making calls by radio telephony in difficult reception conditions to
(a) pronounce each letter and figure slowly and deliberately using a phonetic alphabet for all letters
(b) use the phonetic alphabet for any letters which might be difficult to receive
(c) make long calls of at least 10 callsigns in order to take advantage of breaks in interference
(d) make the callsign of the called station continuously with sufficient spaces between each call to receive.

14. The information missing from the station log shown in Fig 3 is the
(a) class of emission
(b) power
(c) signature of the Licensee
(d) precise frequency (not frequency band).

| Date | GMT Time Start | GMT Time End | Band | Station Called/Worked or Heard | Calling Station |
|---|---|---|---|---|---|
| 8.2.80 | 1417 | | 144 | COMMENCEMENT OF OPERATING | |
| | 1420 | 1427 | 144 | G2ZZZ | G5XXX |
| | | 1430 | 144 | STATION CLOSED DOWN | |

Fig 3.

15. Fig 4 shows the relationship between $V$ and $I$ in a dc circuit. The resistance of the circuit is
(a) 8Ω
(b) 2Ω
(c) 0·5Ω
(d) 0·2Ω.

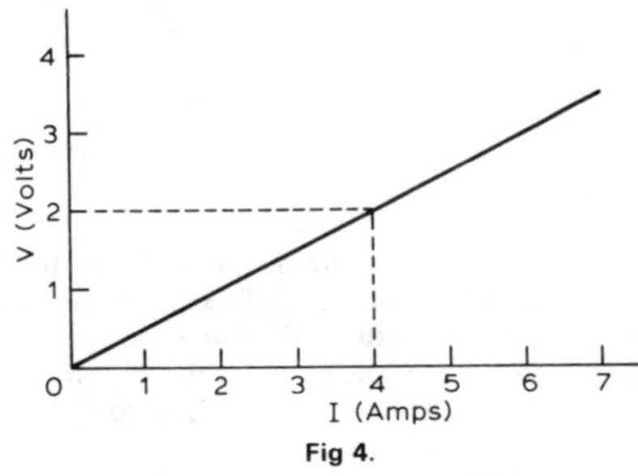


Fig 4.

16. The effect of a capacitor in an ac circuit containing capacitance only is that the
(a) current in the circuit varies inversely to the applied frequency
(b) current through the capacitor lags the applied voltage by 90°
(c) current through the capacitor leads the applied voltage by 180°
(d) current through the capacitor leads the applied voltage by 90°

17. The international Q-code symbol used on telegraphy to indicate interference caused by other stations is
(a) QRA
(b) QRM
(c) QRN
(d) QSB.

18. The spelling of the word LIBYAN using the recommended phonetic alphabet is
(a) Love India Baker Yankee Alpha November
(b) Lima India Bravo Yankee Alpha November
(c) Love India Bravo Yankee Able November
(d) Lima India Bravo Yankee Alpha Nan.

19. The potential difference between points A and B shown in Fig 5 is
(a) 27V
(b) 13·5V
(c) 6V
(d) 0V.

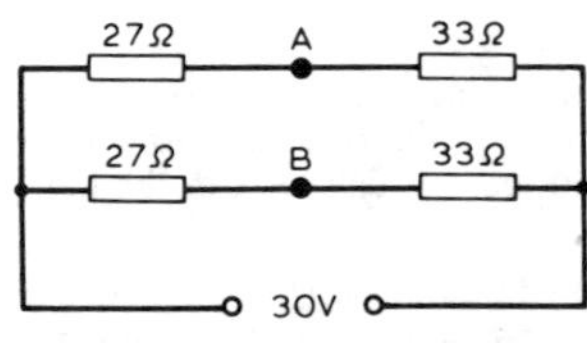


Fig 5.

20. A charge of 300$\mu$C is stored in a capacitor. If the voltage across it is 200V its capacitance is
   (a) 1·5$\mu$F
   (b) 5$\mu$F
   (c) 6$\mu$F
   (d) 15$\mu$F.

21. The value of C in Fig 6 can be found from
   (a) $\frac{fI}{V}$
   (b) $fVI$
   (c) $\frac{2\pi fV}{I}$
   (d) $\frac{I}{2\pi fV}$

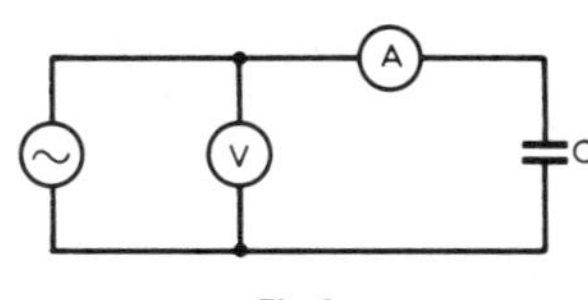

Fig 6.

22. An alternating voltage of sinusoidal waveform has a peak-to-peak value of 500V. The rms value would be
   (a) 707V
   (b) 353·5V
   (c) 177V
   (d) 141·4V.

23. In a series-resonant circuit, to halve the resonant frequency the LC product must be
   (a) halved
   (b) doubled
   (c) quadrupled
   (d) tripled.

24. A transistor rf amplifier is operating in Class C initially with no drive. If the drive power is then increased, the collector current will
   (a) slightly decrease
   (b) considerably decrease
   (c) stay the same
   (d) considerably increase.

25. The purpose of the coupling capacitor in a two-stage transistor amplifier is to
   (a) provide a phase shift of signal
   (b) improve the frequency response of the amplifier
   (c) provide a constant voltage bias to the next stage
   (d) isolate the dc voltages in the two stages.

26. The output voltage ripple frequency of a 50Hz full-wave mains bridge rectifier is
   (a) 25Hz
   (b) 50Hz
   (c) 100Hz
   (d) 150Hz.

27. The main use of agc in a radio receiver is to
   (a) stabilize the oscillator frequency
   (b) ensure accurate tuning
   (c) provide a more even output level
   (d) reduce interference.

28. Refer to Fig 7. The circuit connected in parallel to the rf input stage of a superhet receiver would be effective in
   (a) decreasing the gain
   (b) reducing the bandwidth
   (c) bypassing an interfering signal
   (d) increasing the gain of a receiver.

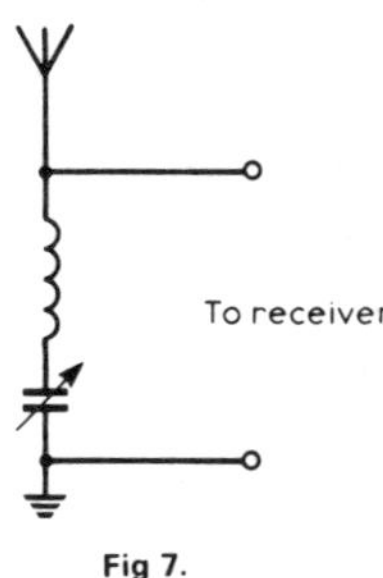

Fig 7.

29. The purpose of a discriminator in a fm receiver is to
   (a) limit the peaks of incoming signals
   (b) attenuate signals on adjacent channels
   (c) demodulate the incoming signal
   (d) reduce the bandwidth.

30. Class A3J amplitude modulation uses
   (a) only the upper sideband with a bandwidth of about 4kHz
   (b) either one of the sidebands with the carrier suppressed
   (c) morse telegraphy using either sideband
   (d) one telephony sideband with reduced carrier.

31. The frequency developed in a crystal oscillator circuit depends principally on
   (a) a combination of the various components in the circuit
   (b) the physical dimensions of the quartz crystal
   (c) the values of the external circuit
   (d) the crystal holder.

32. A push-pull stage in a transmitter is used as a frequency multiplier. The strongest output harmonic will be the
   (a) third
   (b) second
   (c) fourth
   (d) fifth.

33. A double sideband telephony transmitter is 100 per cent modulated by a sine wave. If the carrier and one sideband is suppressed, by what ratio can the power in the remaining sideband be increased?
   (a) 25:1
   (b) 12:1
   (c) 4:1
   (d) 2:1.

34. In a transverse electromagnetic wave (where E represents the electric component, H the magnetic component and $v$ the direction of propagation)
   (a) E is always vertical and at right angles to $v$
   (b) E and H are at right angles to each other and at right angles to $v$
   (c) H is always vertical and at right angles to E
   (d) H and E are 180° out of phase.

35. The major mode of propagation of radio waves over long distances at frequencies above 50MHz is by
   (a) the ground wave
   (b) reflection by the ionosphere
   (c) refraction by the ionosphere
   (d) refraction by the troposphere.

36. The type of propagation used for radio contacts between the British Isles and Canada on the 14MHz band is
   (a) ground wave
   (b) direct wave
   (c) ionospheric
   (d) tropospheric.

37. A 14MHz signal will fade to a minimum at a point 4,000km away when
   (a) the sky wave and ground wave arrive in phase
   (b) the sky wave and ground wave arrive out of phase
   (c) two or more sky waves arrive in phase
   (d) two or more sky waves arrive out of phase.

38. The type of meter used for measurement of current at 3MHz is
   (a) polarized moving iron
   (b) electrodynamic type having two coils
   (c) moving coil with parallel resistor
   (d) moving coil with thermocouple.

39. Fig 8 shows an aerial commonly used for mobile operation on the 144MHz band. If the physical length, $L$, is approximately $\frac{5}{8}$ wavelength, it is likely that the electrical length will be
   (a) a full wavelength
   (b) $\frac{3}{4}$ wavelength
   (c) $\frac{1}{2}$ wavelength
   (d) $\frac{1}{4}$ wavelength.

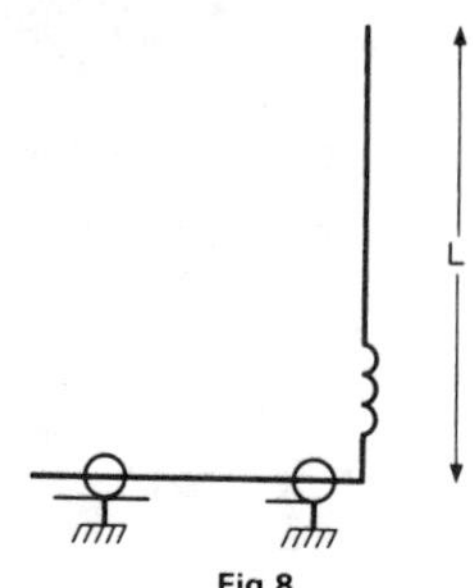

Fig 8.

40. A transmitter is operating in the 145MHz band and its frequency is measured using a digital frequency meter. If the accuracy of the frequency meter is 3 parts in 100,000, which one of the following frequencies is the most accurate?
   (a) 145·8946MHz
   (b) 145·894MHz
   (c) 145·89MHz
   (d) 145·9MHz.

ANSWER KEY

| | | | | |
|---|---|---|---|---|
| 1—b | 9—a | 17—b | 25—d | 33—c |
| 2—b | 10—c | 18—b | 26—c | 34—b |
| 3—a | 11—b | 19—d | 27—c | 35—d |
| 4—d | 12—d | 20—a | 28—c | 36—c |
| 5—c | 13—a | 21—d | 29—c | 37—d |
| 6—b | 14—a | 22—c | 30—b | 38—d |
| 7—b | 15—c | 23—c | 31—b | 39—b |
| 8—a | 16—d | 24—d | 32—a | 40—c |

If you have an enquiry about these sample items, please write to OTS, City and Guilds of London Institute, 46 Britannia Street, London WC1X 9RG.

APPENDIX 2

# Safety pointers

Construction and operation of radio equipment poses few real hazards other than those stemming from an injudicious attitude towards high voltages, such as those involved in the high-tension supplies for thermionic valves (particularly in transmitter power amplifier stages) and in any connection to ac mains supplies. It should also be appreciated that even low-voltage 12V supplies, if capable of delivering very high peak currents, can present potential hazards: for example from the high energy stored in very-large-capacitance electrolytic capacitors (eg up to 100,000$\mu$F) that may be used in such supplies or from the high peak currents that can be delivered from car batteries. Such hazards can involve, for example, the risk of hot or molten metal when a high-energy supply is accidentally short-circuited by a thin screwdriver blade or by a wedding-ring or other personal jewellery.

A number of people, if exposed to solder-flux fumes over a considerable period of time, tend to develop asthmatic symptoms of coughing and wheezing; constructors who do much soldering are advised to heed medical advice to ensure that this is done where effective exhaust ventilation is available.

Similarly a number of the chemicals used in construction and/or servicing can have toxic effects unless handled carefully in accordance with any instructions given by the manufacturers; in particular carbon tetrachloride, formerly frequently used for cleaning quartz crystals, switch contacts etc must be regarded as a dangerous substance.

Some silicon planar transistors contain a very toxic substance —beryllia—and discarded transistors should never be cracked open; this substance is also sometimes used in the mounting washers that form part of the heat sinking. Then again, the white powder (cadmium salts of organic acids) that may appear on cadmium-plated items of electronic equipment, particularly when exposed to humid atmospheres, can be dangerous if ingested: cadmium plating may include screwheads and switches.

Care should always be taken when installing and using mobile transmitters of significant power output that these are never used when the vehicle, or a vehicle in the immediate locality, is being refuelled (ie never use a transmitter in the confines of a service station); similarly transmitters must not be used near to blasting operations, explosives, oil refineries etc because of the possibility of the transmitter inducing small sparks in resonant metal structures etc.

Care should also be taken that a transmitter does not affect the increasing amount of electronic equipment now being fitted in cars and commercial vehicles; the possibility of rf-induced accidents, although remote, could exist where electronic devices are used as part of fuel-injection, anti-skidding and similar systems.

Mains-operated domestic equipment in the UK should normally conform to the specifications of British Standard BS415; although this does not appear to apply directly to amateur radio equipment. Many of the equipments imported into the UK do not conform to BS415 and may, for example, be found to have single-pole switches in the "neutral" mains lead (a potentially dangerous practice).

## Electromagnetic radiation

For many years there has existed a safety limit for continuous exposure to non-ionizing radiation, such as radio waves: this is equivalent to a power flux of 10mW/cm$^2$. This power limit is most unlikely to be exceeded in amateur radio stations where the antenna is mounted externally at reasonable height above ground; nevertheless it could be exceeded, for example, in the line of fire of a vhf or uhf beam antenna if this is in or very close to the operating position as may occur in portable operation; or if a moderately high-power rig was used hand-held etc. The primary purpose of the 10mW/cm$^2$ limit is to prevent localized overheating of sensitive organs, and especially the eyes.

Non-ionizing radiation differs from the much more dangerous ionizing radiation (X-rays, nuclear radiation, gamma rays) in that there is insufficient energy to displace electrons from atoms; it does however, particularly when absorbed in tissue with high water content, produce a localized heating effect, a fact used in microwave ovens.

In recent years there has been some public concern, particularly in the USA, that other biological effects of a less-readily defined nature may result from protracted absorption of radio waves at power levels rather below 10mW/cm$^2$, based on some experiments carried out with small animals. There is no convincing evidence from any scientific studies carried out in Western countries of athermal effects on humans, although the matter is again a subject of research studies.

Pending the results of these studies, radio amateurs may wish to observe the following precautions:

1. Avoid high-power transmitting equipment with antennas in the shack within 3m (10ft) of the operating position or a family living area.
2. Avoid direct radiation to the eye from a microwave transmitter (in particular avoid looking down a waveguide, horn or parabolic dish antenna at close range).
3. Avoid prolonged close contact with any antenna radiating more than minimal amounts of energy.
4. Women in the early months of pregnancy, or those who may become pregnant, should avoid prolonged contact with very strong hf, vhf and uhf fields.

While these recommendations may prove to be over-cautious, to follow them will cause few problems and will allow confident and enjoyable operation of all amateur radio equipment. The recommended standard of 10mW/cm$^2$ should, of course, always be observed.

## SAFETY RECOMMENDATIONS FOR THE AMATEUR RADIO STATION

1. All equipment should be controlled by one master switch, the position of which should be well known to others in the house or club.
2. All equipment should be properly connected to a good and permanent earth. *(Note A.)*
3. Wiring should be adequately insulated, especially where voltages greater than 500V are used. Terminals should be suitably protected.
4. Transformers operating at more than 100V rms should be fitted with an earthed screen between the primary and secondary windings.
5. Capacitors of more than 0·01 $\mu$F capacitance operating in power packs, modulators, etc (other than for rf bypass or coupling) should have a bleeder resistor connected across their terminals. The value of the bleeder resistor should be low enough to ensure rapid discharge. A value of 1/C megohms (where C is in microfarads) is recommended. The use of earthed probe leads for discharge capacitors in case the bleeder resistor is defective is also recommended. *(Note B)*. Low leakage capacitors, such as paper and oil filled types, should be stored with their terminals short-circuited to prevent static charging.
6. Indicator lamps should be installed showing that the equipment is live. These should be clearly visible at the operating and test position. Faulty indicator lamps should be replaced immediately. Gas filled (neon) lamps are more reliable than filament types.
7. Double-pole switches should be used for breaking mains circuits on equipment. Fuses of correct rating should be connected to the equipment side of each switch. *(Note C.)* Always switch off before changing a fuse. The use of ac/dc equipment should be avoided.
8. In metal enclosed equipment install primary circuit breakers, such as micro-switches, which operate when the door or lid is opened. Check their operation frequently.
9. Test prods and test lamps should be of the insulated pattern.
10. A rubber mat should be used when the equipment is installed on a floor that is likely to become damp.
11. Switch off before making any adjustments. If adjustments must be made while the equipment is live, use one hand only and keep the other in your pocket. Never attempt two-handed work without switching off first. Use good quality insulated tools for adjustments.
12. Do not wear headphones while making internal adjustments on live equipment.
13. Ensure that the metal cases of microphones, morse keys, etc, are properly connected to the chassis.
14. Do not use meters with metal zero adjusting screws in high voltage circuits. Beware of live shafts projecting through panels particularly when grub screws are used in control knobs.
15. Antennas should not, under any circumstances, be connected to the mains or other ht source. Where feeders are connected through a capacitor, which may have ht on the other side, a low resistance dc path to earth should be provided (rf choke).

*Note A.*—Owing to the common use of plastic water main and sections of plastic pipe in effecting repairs, it is no longer safe to assume that a mains water pipe is effectively connected to earth. Steps must be taken, therefore, to ensure that the earth connection is of sufficiently low resistance to provide safety in the event of a fault. Checks should be made whenever repairs are made to the mains water system in the building.

*Note B.*—A "wandering earth lead" or an "insulated earthed probe lead" is an insulated lead permanently connected at one end to the chassis of the equipment; at the other end a suitable length of bare wire is provided for touch contacting the high potential terminals to be discharged.

*Note C.*—Where necessary, surge-proof fuses can be used.

# Index

L



M



N



O



P



Q



R



S



T



V



W